*A Red Bird in a Brown Bag*

# A RED BIRD IN A BROWN BAG

## The Function and Evolution of Colorful Plumage in the House Finch

Geoffrey E. Hill

OXFORD
UNIVERSITY PRESS
2002

# OXFORD
UNIVERSITY PRESS

Oxford  New York
Auckland  Bangkok  Buenos Aires  Cape Town  Chennai
Dar es Salaam  Delhi  Hong Kong  Istanbul  Karachi  Kolkata
Kuala Lumpur  Madrid  Melbourne  Mexico City  Mumbai  Nairobi
São Paulo  Shanghai  Singapore  Taipei  Tokyo  Toronto

and an associated company in Berlin

Copyright © 2002 by Oxford University Press, Inc.

Published by Oxford University Press, Inc.
198 Madison Avenue, New York, New York 10016

www.oup.com

Oxford is a registered trademark of Oxford University Press

Library of Congress Cataloging-in-Publication Data
Hill, Geoffrey E. (Geoffrey Edward)
   A red bird in a brown bag : the function and evolution of colorful plumage in the House
   Finch / Geoffrey E. Hill.
      p.   cm.
   Includes bibliographic references (p. )
   ISBN 0-19-514848-7
   1. House finch—Color.   2. Sexual selection in animals.   I. Title.
   QL696.P246H56   2001
   598.8'83—dc21                                                        2001051022

9 8 7 6 5 4 3 2 1
Printed in the United States of America
on acid-free paper

To my mother,

Sally Anne Hill,

whose sacrifices, encouragement, and support
made this book possible

# PREFACE

It was a Saturday afternoon in 1987. I don't know the specific date although I think it was in January or February. I was accompanying my wife, Carolyn, on a shopping trip to the little mall near our Northwood apartment in Ann Arbor, Michigan. Carolyn, then as now, put up with all sorts of nonsense from her ornithologist husband—decades of education for a pauper's wage, alarm clocks in the pre-dawn hours, long trips to obscure locations, dead birds or parts of birds in the freezer, and so on—so, even though I hated shopping, as a sort of payback and a means to reduce my guilt, I escorted her on various trips to malls and shops. On this particular day my mind was searching, as it had been for more than a year, for the ideal question and organism to serve as the focus of my doctoral dissertation. I felt as if I was getting nowhere and that time was running short.

For my master's research at the University of New Mexico, I had spent three years testing a plethora of hypotheses for why some male passerine birds wait until their second breeding season to fully develop their nuptial plumage, spending their first season in a drab and often female-like plumage. I found it to be a frustrating endeavor. At first I thought the frustration was a natural consequence of the difficulties of field work with Black-headed Grosbeaks and the challenge of trying to test hypotheses and conduct experiments with an organism that was hard to catch, hard to watch, and hard to manipulate in ways pertinent to testing hypotheses for the function of plumage coloration. But the more I became immersed in the plumage-color and sexual-selection literature, the more I came to realize that the problem was not in the process; the problem was in the approach. The handful of biologists who were interested in the topic of ornamental coloration were debating fine nuances of selection on first-year plumage, when as a scientific community we knew virtually nothing about the general function and control of ornamental plumage coloration. We were trying to construct a scaffold of explanation not merely on an unstable foundation but with no foundation at all.

But how to set things straight? How to pull back to the beginning and test the basic function and proximate control of plumage coloration? Over the weeks preceding my shopping trip, as a first-year doctoral student at the University of Michigan, I had been fixating on and then discarding a long string of potential plumage-related projects on an array of species—Malurid wrens, Vermilion Flycatchers, Indigo Buntings, Red-winged Blackbirds, and so on. Try as I might, I couldn't seem to extricate myself from the topic of delayed plumage maturation. I was looking for the ideal bird population where I could conduct definitive experiments on the role of female choice in the evolution and maintenance of plumage color. I needed a common, easily catchable bird. It had to do well in captivity. And, perhaps most importantly, it had to have substantial variation in ornamental plumage coloration that was not completely determined by age. I was beginning to think that I was demanding too much and as I was ambling about the pet store, I was getting close to settling for a less than ideal project (back to delayed plumage maturation).

My wife had just set up a large fish tank in our apartment and she was now stocking the tank with various cichlids. Each purchase demanded long periods of pacing back and forth in front of the cichlid tanks at the pet shop deciding which fellow would join the others in the tank at home. As a biologist, you might think I would have been standing beside her, equally intent on getting just the right ecological balance for the tank. I wasn't. I was over in the bird section pondering how "Song Food" could improve Canary song. I ended up staring at the half-dozen Canaries on display in the store as I thought about a passage in an ornithology text by Welty that stuck in my mind. The passage said: "Canaries will, in successive molts, gradually change from yellow to intense orange if fed red peppers" (Welty 1982:55). Suddenly, things fell into place. I remembered a lecture, given by Randy Thornhill when I was a master's student at the University of New Mexico, in which he talked about guppy color being determined by a scarce red pigment found in algae. I remembered him telling us that males compete for the algae and only the best males are able to eat a lot of algae and become bright red. (This of course was an account of John Endler's classic work. I didn't know much about Endler at the time, but he would soon become one of my science heroes.) What if the same thing was going on in bird plumage? What if plumage redness was determined by diet? That would be really exciting because then we could explain variation in male coloration and provide a rationale for why females should choose brightly colored males.

I knew right away I had my dissertation project, but I didn't yet have my study organism. My first idea was to study Canaries themselves, but there were many drawbacks to Canaries as a study organism. Pet-store birds were hundreds of generations removed from the wild and had been under intense artificial selection for plumage coloration and/or song for centuries. It would be a stretch to assume that the behaviors of captive Canaries were the same as the behaviors of wild Canaries. Fieldwork in the Canary Islands would be exciting but not very practical. I would spend most of my time and large amounts of money just getting to a study site. The adventure of studying wild Canaries was appealing, but I was looking for a much more accessible study bird. Then, within a day or two of my epiphany in the pet store, I found a paper by Alan Brush and Dennis Power that sealed my fate: "House Finch pigmentation: carotenoid metabolism and the effect of diet," published in *The Auk* in 1976. This was the bird. House Finches had extremely variable plumage coloration, and the variation in male coloration was not specifically tied to age. Moreover, the biochemical basis for variation in color was presented by Brush and Power. All accounts indicated that they could be kept in cages for experiments, and they were literally right outside my office window—my advisor, Bob Payne, was catching them in traps on the windowsill of the lab.

So began my study of the House Finch, which has lasted fifteen years to date and is giving every indication of being a life-long pursuit. One might imagine that I would have long ago exhausted the insight that House Finches have to offer to theories of sexual selection and the evolution of plumage coloration. To the contrary, however, what appeared to be a simple signaling system has proven to be more complex than I ever imagined when I scored the plumage coloration of my first male House Finch in 1987. We (meaning myself, my students, my professional colleagues with an academic interest in ornamental plumage, and anyone else who wonders why some birds are colorful) still know very little about the physiology of red pigmentation in vertebrates in general, in birds more specifically, or in House Finches in particular. The role of diet in determining natural variation in plumage coloration remains an open debate with few studies and amazingly little data available to back up any point of view. We have little understanding of the role of hormones in plumage pigmentation. Only very recently have we discovered that plumage redness in House Finches is inversely related to dominance—brightly colored males are routinely defeated by drab males in contests for food. We are still not sure why. The role of disease in determining expression of plumage coloration is relatively unstudied, the role of nutrition even more so. House Finches exist not as a single population but as many populations that can be divided into about fifteen subspecies. There may actually be two species of House Finches (a northern and a southern House Finch), but no systematist has considered the problem for sixty years. Males in the various subspecies display fascinating variation in expression of ornamental plumage, and it is this variation in expression that holds the key to understanding the evolution of ornamental plumage in House Finches and perhaps birds in general.

This book is an account of studies of the function and evolution of colorful plumage in the House Finch. It is meant for anyone interested in birds beyond the tick on a checklist. I have tried to make the text readable and accessible to serious amateurs as well as professional ornithologists, behavioral ecologists, and evolutionary biologists. To help readers who may have little background in ornithology, evolutionary biology, or statistics, I have included a glossary that defines some key terms related to these topics. This book is about House Finches, but it is not about everything ever written or discovered about House Finches. To attempt to be comprehensive in all things finchy would mean superficial coverage of topics and a book that would be very thick and unread. Rather, I've focused exclusively on topics that relate either directly or indirectly to carotenoid-based plumage coloration. I hope that I will stimulate thinking not just about how plumage coloration functions in the communication system of House Finches, but also about broader issues of the proximate control, function, and evolution of ornamental traits in general.

Throughout this book the output from statistical analyses is given only in the legends of the figures. Readers with no training in statistical analysis of data can ignore the statistical summaries and simply accept my verbal summaries of the outcome of comparisons, or the reader can consult the statistics section of my glossary for help with some of the terms and concepts. For most of the studies that I present in this book, I am reproducing results that have been published in journal articles. Each of these articles was written at a different time, often with different collaborators, and so the statistics used, even for very similar comparisons, are sometimes different.

This book is divided into three parts. Part 1, which contains three chapters, provides some necessary background. First, the stage is set for modern studies of the function of plumage coloration with a review of the extensive work that was devoted to the topic at the end of the nineteenth and beginning of the twentieth centuries. This is followed by an introduction to the House Finch, including a description of its breeding biology and an overview of the study sites and study techniques that were the bases for much of the data presented in the book. This first part of the book concludes with a description of how color is quantified and an outline of the approaches to color quantification that I've used in studies of the House Finch.

Part 2 has six chapters and focuses on the proximate control and present function of plumage coloration. It begins with a review of carotenoid physiology in general and the pigmentary basis of plumage coloration in the House Finch in particular. This is the most technically detailed and, in some ways, demanding chapter of the book, and readers who are not interested in details of carotenoid physiology may want to read the summary paragraph to chapter 4 and move on. For biologists focused on a fundamental understanding of how carotenoid-based coloration functions as a signal of quality, however, I think that it will be well worth wading through the details of chapter 4. This review of carotenoid physiology is followed by a chapter on the environmental factors that control individual expression of plumage coloration. I build from this understanding of proximate control to a look at plumage redness in relation to female mate choice and the advantages to males of being red and to females for choosing red mates. I review how plumage color affects dominance interactions, and the role of hormones in mediating aggressive behavior, parental behavior, immune response, and plumage coloration. This part concludes with an account of the costs and benefits of ornamental color display in female House Finches.

Part 3 takes an explicitly evolutionary approach to the study of plumage coloration using biogeography and phylogeny to test hypotheses for why specific forms of plumage color display have evolved. This section begins with a description of the numerous subspecies of House Finches found in North America, including the interesting ways in which subspecies differ in expression of ornamental traits. In chapter 11, I then present my efforts to use the geographic variation in expression of plumage coloration among populations of House Finches to conduct comparative tests of hypotheses for the evolution of ornamental display. Part 3 and the book end with the Epilogue, in which I pull together the ideas of the book and discuss what I think are the greatest gaps in our understanding of carotenoid-based ornamental displays with the hope that this book might serve as a starting point for future studies of colorful plumage.

## Acknowledgments

I think that what I have enjoyed most about my study of plumage coloration in the House Finch is that it has taken me to the most unexpected places, from atop a volcano in Hawaii to the crowded markets of central Mexico, from the great trays of specimens at the National Museum of Natural History to an ultracold freezer packed with frozen tubes of cultured disease. It has also led to an association with a most extraordinary group of colleagues and collaborators. It goes without saying that a decade of work on a topic is not accomplished in isolation, and I am indebted to the many people who helped make this book possible. First and foremost are the outstanding graduate students who have worked with me on finches over the years: Cathy Stockton-Shields, Marjie Tobias, Blue Brawner, Paul Nolan, Andrew Stoehr, Kevin McGraw, Caron Inouye, Anne Dervan, and Renee Duckworth. The theses and projects of these students constitute the bulk of the analyses presented in this book. Alex Badyaev, my postdoc at the time this book was written, has been an invaluable colleague, inviting me to participate in the exciting studies of a House Finch population that he has initiated in Montana, and providing diverse original insight into how natural and sexual selection can shape the morphology and behavior of finches. The ideas presented in this book benefited greatly from discussions with Alex. Bob Montgomerie, my postdoc mentor, both worked with me on studies of the condition dependency of carotenoid coloration and read this entire book in manuscript form, providing many useful comments. Anne Dervan, Lynn Siefferman, and Barb Ballentine also proofed the entire book. The original finch research team at Auburn—Kevin McGraw, Andrew Stoehr, and Paul Nolan—deserve special thanks. They read and proofed every sentence of the book, checking numbers and facts, challenging unsupported or tenuous ideas, and supplying unpublished data and new analyses on demand, even when the requests became excessive. Finally, my mother, who gave up a career as a fiction writer so she could work to support her children, read the entire manuscript. She not only improved the prose of the text substantially, but she also provided an invaluable layperson's perspective, ruthlessly weeding out jargon and unclear presentations of biological concepts.

I am indebted to Chris Burney, a student from my ornithology class at Auburn, who drew the illustrations in this book. Chris's wonderful drawings help bring the text to life.

From the day I banded my first House Finch to the day this book was mailed to the publisher, my wife Carolyn has been tremendously supportive of my work. Carolyn and my children, Trevor and Savannah, not only put up with my long hours spent on this project, they were a constant source of inspiration, particularly when my energy to complete the project began to wane.

During my graduate studies, the Museum of Zoology at the University of Michigan provided outstanding facilities and financial support for my research. Likewise, Auburn University has been tremendously supportive of my work. Not only has my department and college provided monetary support and a huge aviary where I could conduct my work with finches, but the grounds and facility crews have given my students and me unrestricted access to the campus buildings, including the football stadium, for our studies. Anyone who has worked at a large southern university knows that keys to the football stadium are like keys to the cathedral. Without the support of the grounds and facility crews we could not have conducted the work presented in this book.

Over the past fifteen years, financial support for my studies of House Finches has come from many sources. As a graduate student, I was supported by grants from the Frank M. Chapman Memorial Fund from the American Museum of Natural History, the Animal Behavior Society, Sigma Xi, the Alexander Wetmore and Josselyn Van Tyne funds from the American Ornithologists' Union, the Hawaii Audubon Society, and the Hinsdale-Walker fund from the Museum of Zoology. At Auburn, I have been supported by the now defunct Department of Zoology and Wildlife Science and by the new Department of Biological Sciences as well as by the College of Science and Mathematics and the Alabama Agricultural Experiment Station. During the latter years of my House Finch research and during the preparation for this book, support of my research came primarily from grants from the National Science Foundation (IBN-9722171 and DEB-0077804).

Geoff Hill
*Auburn, Alabama*

# CONTENTS

# PART 1

*Prelude*

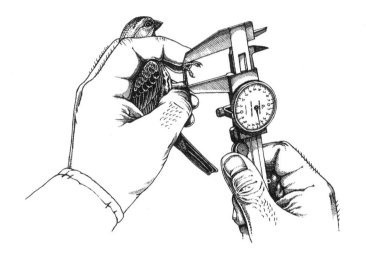

# 1 Darwinism and Wallacism

## A Brief Account of the Long History of the Study of Plumage Coloration

I can see no good reason to doubt that female birds, by selecting, during thousands of generations, the most melodious or beautiful males, according to their own standards of beauty, might produce a marked effect.

—C. Darwin (1859:193), in *The Origin of Species*

Mr. Darwin has devoted four chapters in his Descent of Man to the colours of birds, their decorative plumage, and its display at the pairing season; . . . Any one who reads these most interesting chapters will admit, that the fact of display is demonstrated. . . But it by no means follows that slight differences in the shape, pattern, or colours of ornamental plumes are what lead a female to give the preference to one male over another; still less that all females of a species, or the great majority of them, over a wide area of country, and for many successive generations, prefer exactly the same modification of the colour or ornament.

—A. R. Wallace (1889:285), in *Darwinism*

*When I was a boy, maybe seven or eight, I saw a Scarlet Tanager (a species with brilliant red plumage) in the woodlot beside my house. I ran home and with my mother consulted the greatest reference on birds available to my family—an ancient set of Collier's Encyclopedias. Miraculously, among the fifteen or twenty birds pictured was a scarlet tanager. I was astounded that we had found my mystery bird pictured in the encyclopedia.*

*"But why is it so red?" I asked.*

*"God made it red. It is part of God's plan." was my mother's answer, the same answer she had gotten to such questions when she was a little girl. It was a comforting and satisfying answer to me as a child as it had been for her as a child.*

*As I grew older and stumbled back to the question of why birds have bright colors, I remembered my mother's answer, but I could no longer find any comfort in it. As a matter of fact, the more I came to think about it over the years, the more I became sure that my mother's answer was really no answer at all, because the same response could have been given to virtually any question.*

*Later, in high school, I was introduced to the idea of evolution by natural selection. The phenotypes of organisms are shaped by survival of the fittest, I was told. Organisms have a particular size, shape, and color because that is the optimal size, shape, and color for survival. But why red birds? My high school teachers had no satisfactory answers.*

*As an undergraduate at Indiana University, I was first introduced to the idea of sexual selection. Showy traits and weaponry result from a struggle for mates. So, males are red because females like red males, I was told. But why would females make such a choice? I was, by this point, getting answers of all sorts, but not the answer I was looking for.*

*In graduate school, I finally realized that the simple and naive question of an eight-year-old was actually a very good question indeed. I decided that answering my own question would be a worthwhile endeavor. I'm still working on it.*

## Plumage Coloration Then and Now

My studies of colorful plumage of male House Finches follow a long history of human fascination with and admiration of brilliant and elaborate plumage. Birds were the first creatures to be cultivated not because they provided any useful product or service, but simply because they were enjoyable to look at or listen to. In the eighteenth and nineteenth centuries, as academic focus on natural history increased, attempts to explain colorful plumage centered on chronicling the beauty, complexity, and diversity of avian ornaments as a means of proving the existence of a Creator. Paley's *Natural Theology*, published in 1802, and the *Bridgewater Treatise*, commissioned by the Earl of Bridgewater and published as eight volumes in the 1830s, marked the culmination of this type of natural theology. The colorful plumage and song of birds played a central role in this effort to demonstrate "the power, wisdom, and goodness of God" (Kirby 1833). Paley, Kirby and others argued the feathers of birds were so beautiful, so intricate in design, and so diverse in pattern and color that they could not have come to be by chance; they had to be the product of intelligent design. The existence of these ornamental traits, as carefully documented by these naturalists, along with other natural phenomena including especially the human mind, was the ultimate proof of the existence of God.

It was in this era of natural theology in the middle of the nineteenth century that both Charles Darwin and Alfred Russell Wallace developed the radical concept

Charles Darwin

of evolution by natural selection and forever changed the way biologists viewed the natural world (Darwin 1859, Darwin and Wallace 1858, Wallace 1864, 1870). Darwin and Wallace loom as the greatest figures in the long history of the study of plumage coloration both because they established the primary explanations for colorful plumage—the ideas of evolution by natural and sexual selection—and because they wrote voluminously on the topic, documenting for the first time in detail the diversity of color display and important aspects of plumage coloration like the variable differences among the sexes. Only recently, with the publication of the correspondence of Darwin and Wallace and two excellent syntheses on the contributions of these men to the resolution of the question of why some birds have colorful plumage (Blaisdell 1992, Cronin 1991), are evolutionary biologists and behavioral ecologists rediscovering the details and sophistication of the discussions carried on by Darwin and Wallace regarding the function and evolution of colorful plumage.

Much has been made of the debate between Wallace and Darwin, particularly related to sexual selection (Blaisdell 1992, Cronin 1991), but they agreed about much more than they disputed. The most fundamental point of agreement between Darwin and Wallace was that natural selection is the primary mechanism for the evolution of organisms. Natural selection is the process that promotes the maintenance and spread in a population of traits that enhance survival or reproduction. Natural selection is simply the perpetuation, through the struggle for existence, of the genes of those individuals who survive and reproduce well and the elimination from the population of the genes of those individuals who survive or reproduce poorly. The result is change in gene frequency of the population through time, which by definition is evolution. Genes code for phenotype (all observable features of organisms), so change in gene frequency generally means change in observable characteristics of organisms over time. Extended over many generations, this process molds organisms to their environments, and, when populations become isolated, it leads to speciation. It is a simple process. It is a mindless process. And yet, it is the process that gives us the vertebrate eye, the human mind, and millions of species of organisms. Without a creator or purposeful design, we can get the evolution of sophisticated, complicated, and well-adapted organisms. And we get the incredible diversity of life on earth.

The resolution of natural selection as the primary mechanism of evolution was the crowning achievement of both Darwin and Wallace. The theory of evolution through natural selection provided a unifying explanation for an enormous array of natural observations. Patterns of biogeography, the hierarchical nature of species similarities, sequential changes in taxon morphology over geologic time, and the seemingly perfect fit between many organisms and their environment all were explained by natural selection. The list of problems solved and concepts united was vast compared to the small number of observations that seemed to contradict the theory. And yet, it was the apparent contradiction to natural selection theory posed by ornamental traits, particularly the bright coloration of insects and birds, that became the focus of discussion and debate for Darwin, Wallace, and many biologists of the late nineteenth century (see Blaisdell 1992, Cronin 1991 for detailed accounts). Thus, discussion of the evolution of ornamental plumage coloration began at the very inception of evolutionary thinking, and the study of colorful plumage played a central role in the refinement of an understanding of how evolution works.

To provide an explanation for the existence of traits that seemed to defy explanation by natural selection—that is, traits that seemed to serve no function in promoting the survival or fecundity of organisms and indeed that might incur a survival cost—Darwin proposed the idea of sexual selection. He first presented the idea in his *Origin of Species* (Darwin 1859:136), but he laid out his ideas in full in his book *The Descent of Man and Selection in Relation to Sex* (Darwin 1871). Sexual selection, as outlined by Darwin and as still defined today, is subtly different from natural selection. Rather than promoting traits that enhance survival or fecundity, sexual selection is a process that promotes the maintenance and spread of traits that aid in the attraction or defense of mates. Thus, sexual selection can work in two distinct manners: through direct competition among individuals for access to mates or through mate-choice. For most animal species, females will be the limiting sex, so contests for mates will be among males and choice of mates will be made primarily by females. Sexual selection, then, is the process by which traits that enhance fighting ability such as large body size, horns, or spurs and traits that enhance attractiveness as a mate such as long plumes, complex song, or bright coloration spread, are maintained, and are elaborated in populations.

Male-male competition is easy to observe directly in nature and its importance in shaping the evolution of large body size and weaponry in males has never been doubted, although many biologists, including Wallace, saw it simply as a form of natural selection. Sexual selection driven by female mate choice, on the other hand, requires that members of the choosing sex, usually females, are capable of and inclined to make subtle discriminations among males based on the size or quality of their ornamental display. It was this perceived need for an "aesthetic sense" in non-human animals that caused Wallace (1878) and eventually all leading evolutionary biologists to reject completely the idea that female mate choice could drive the evolution of ornamental traits. Of course, it was pointed out by Darwin and a few contemporaries that an aesthetic sense was not necessary for female to respond to certain male displays. All that was required was a simple stimulus/response mechanism, analogous to the stimulus/response mechanism that caused animals to choose one food item over another or one habitat type over another (Morgan 1900). Nevertheless, of all Darwin's ideas, his insistence on a primary role for female choice in the evolution of ornamental traits evoked the most controversy and earned him the most ridicule. As Huxley stated in his review of sexual selection fifty years after the publication of *Descent of Man*, "None of Darwin's theories have been so heavily attacked as that of sexual selection" (Huxley 1938).

From Darwin's publication of *Origin of Species* until Wallace's publication of his autobiography in 1905, Wallace was perhaps the most influential critic of the idea that the bright coloration of animals could be the outcome of female mate choice. Wallace saw no reason to invoke what to him was an unsubstantiated assumption that females of non-human animals were capable of and inclined to discriminate among males based on the quality of their ornaments (Wallace 1878, 1889). Instead, Wallace searched for explanations of colorful plumage that would allow such traits to be understood as utilitarian, not ornamental and extravagant. In his studies of bird coloration, Wallace did not focus exclusively or even primarily on gaudy plumages. Rather Wallace focused much of his research on the subtle differences among species and individuals in explicitly non-ornamental traits like the buff, brown, gray, and green plumage of birds (Wallace 1878, 1889). In these subtle plumages of birds, Wallace found widespread evidence that natural selection had shaped the appearance of birds to increase their survival. Whenever it could be done with reasonable credibility, Wallace also explained colorful displays in animals as the product of natural selection. He explained the bright coloration and bold patterns of most animals as functioning in crypsis (red sunbirds feeding on red flowers), threat displays (both within and between species), or mimicry, but mimicry was used as an explanation for the coloration of insects much more than for the colors of birds.

For a time after the publication of *Descent of Man*, Wallace accepted that sexual selection was a necessary explanation for some bright plumages (e.g. Wallace 1870:154). In his most influential writing, however, he proposed that female mate choice played little role in the evolution of colorful plumage (Wallace 1878, 1889), and eventually he rejected the idea completely as an explanation for any plumage coloration (Wallace 1905:17). To explain those colors that did not seem to function in crypsis, in mimicry, or as threat displays, Wallace suggested that they served as markers for species recognition, both to help social species maintain cohesiveness of groups and also to help individuals mate with individuals of the same species (although not to help with mate

Alfred Russell Wallace

choice within a species). Wallace thought that species recognition was a very important and widespread function of colorful plumage: "I am inclined to believe that its necessity [recognition function] has had a more widespread influence in determining the diversities of animal coloration than any other cause" (Wallace 1889:217).

For those traits that could not be explained through crypsis, mimicry, threat displays, or as recognition markers, Wallace still rejected female mate choice as an explanation. Instead, he proposed a non-adaptive explanation. He pointed out that colors abound in nature and many of the most brilliant colors of living things—the brightly colored tissues within the bodies of organisms—certainly are not the result of selection. Thus, Wallace hypothesized that all living things have a tendency to be brightly colored "as a necessary result of the highly complex chemical constitution of animal tissues and fluids" (Wallace 1889: 297). According to Wallace, without the constraining hand of natural selection, all animals would be brilliantly colored. To explain the distribution of coloration in animals, therefore, we have only to explain the lack of coloration in females of dimorphic species and of males of cryptic species. Brilliant coloration was the normal state of nature and required no explanation. With this combination of natural selection and no selection, Wallace concluded that there was no need to invoke female mate choice to explain any plumage traits (Wallace 1889: 297–298). He promoted this view effectively for forty years.

Until his death in 1882, Darwin served as the leading advocate for the idea that female mate choice maintained and drove the evolution of the ornamental plumage of males of many species of birds. In the years immediately following Darwin's death, biologists writing on animal coloration devoted less attention to sexual selection than to topics like aggression display (which was not considered sexual selection), mimicry, crypsis, and other manifestations of natural selection. Still, the idea of female mate choice driving the evolution of ornamental traits was discussed in some detail. For example, reviews of coloration in animals by Poulton (1890) and Beddard (1892: 263) provided a discussion of sexual selection, and both authors accepted the idea that sexual selection is a likely explanation for the bright coloration of at least some species of birds. Although Romanes concluded that Wallace was correct in rejecting Darwin's ideas regarding female mate choice, he devoted chapters to a discussion of sexual selection by female choice in both the first (Romanes 1892) and second (Romanes 1895) volumes of his series *Darwin, and after Darwin*. A few biologists continued to defend the idea into the twentieth century (Finn 1907, Winterbottom 1929).

By the time that Cott published his monograph on animal coloration in 1940, however, a work so formidable that in the introduction Julian Huxley called it "the

E. B. Poulton

final word on the subject," sexual selection as an explanation for ornamental plumage was gone. The terms "sexual selection" and "mate choice" do not appear in the extensive index of Cott's book. In 438 pages of text, Cott cites Charles Darwin only four times, each for a specific natural history example, not with regard to any conceptual issue (Cott 1940). Hingston (1933), in a book on animal coloration, at least acknowledges Darwin's theory of sexual selection in the preface, but he rejects the hypothesis outright. There is no mention of sexual selection in the main text of the book. This was the state of the study of plumage coloration for the last quarter of the nineteenth century and the first three quarters of the twentieth century. Real interest in the topic would not be rekindled until the 1970s.

## To the Victor Go the Spoils

Darwin and Wallace debated the idea of female mate choice and sexual selection for over twenty years, occasionally in public forums but mostly in private correspondence (see Blaisdell 1992, Cronin 1991 for detailed accounts). Each tried, but never succeeded, to bring the other to his point of view. In hindsight and from a world view in which sexual selection dominates thinking related to ornamental traits, it is easy to see Wallace as shortsighted, working against a better understanding of the natural world, and just plain wrong. This is how I was introduced to Wallace—as the other guy who almost scooped Darwin and then turned to a metaphysical explanation of the human mind, an ultimate betrayal of the theory of natural selection. When I mention Wallace to my colleagues, rarely is their perception of him positive.

But Wallace's contribution to our understanding of the evolution of plumage coloration in birds was enormous. Far more convincingly than Darwin, Wallace showed how most plumage coloration supported the theory of evolution by natural selection. Most species, most of the time, are colored in ways that appear to enhance their survival and fecundity. Wallace provided an explanation for sexual dichromatism and drab female plumage that stands today as a triumph of the power of the comparative method in addressing evolutionary questions. Through his knowledge of the nesting biology of birds, Wallace showed that species with exposed nests in which the female alone incubates almost invariably have drab female plumage whether the male is colorful or not. Retesting and confirmation of this idea have only lately occurred (Johnson 1991, Martin and Badyaev 1996). Wallace also was the first to set forth the idea that colorful plumage functions as a signal of species recognition. This became the most widespread explanation for colorful plumage for over seventy years, and it remains a too-often-ignored hypothesis in modern treatments of plumage coloration. Wallace also foreshadowed the now popular and well-supported idea that ornamental plumage could serve as a reliable signal of condition in his discussions of vital energy (Wallace 1878, 1889), an idea that will be the central theme in this book in my discussion of plumage coloration in the House Finch.

With regard to the central role of mate choice in the evolution of ornamental plumage, Darwin was, as seems always to be the case, correct. Looking back at Darwin's work with 130 years and literally thousands of studies on selection with which to judge his insight, one has to be staggered by the range of correct explana-tions of the natural world that he put forth. In the case of female mate choice, however, Darwin's answer, while correct, was grossly incomplete. Despite his insistence on the importance of female mate choice in the evolution of ornamental traits in males, Darwin never provided a good explanation for why females should choose males based on ornament expression. Darwin simply took the existence of female choice as a given and built his argument from there. Because there was only vague anecdotal evidence for non-human mate choice, it is understandable that Wallace, Huxley, Cott, Mayr, Lack and all other evolutionary biologists who con-templated colorful plumage in the first half of the twentieth century proposed explanations firmly rooted in natural selection: threat display, mimicry, crypsis, and especially species recognition. Until recently, female mate choice seemed unmeasureable and hence untestable; invoking female choice as an explanation for traits seemed antithetical to the process of science.

## Intersexual Selection Vindicated

Following Darwin's death, only Ronald Fisher quietly carried forth and further developed the idea that female mate choice could drive the evolution of ornamental traits. In 1915, Fisher first published a brief sketch of how female mate preferences for male display traits could evolve and lead to the elaboration of such traits. In this paper, Fisher sketched out ideas for both an indicator mechanism of sexual selec-tion and a runaway process (these ideas are discussed in detail in chapter 11), but the paper generated no published discussion among evolutionary biologists. Fisher expanded his ideas for how female choice could lead to the evolution of ornamental traits in 1930 in his book *The Genetical Theory Of Natural Selection* (Fisher 1930). Although Fisher was one of the most influential evolutionary biologists through the first half of the twentieth century, sexual selection always remained a secondary topic of study for him, and his great insight into sexual selection went unnoticed until publication of the second edition of *The Genetical Theory Of Natural Selection* in 1958. Still, although some theoretical interest in the topic of sexual selection began to appear immediately after publication of the second edition of Fisher's book (O'Donald 1962), sexual selection by female mate choice did not gain wide acceptance by the scientific community until the 1970s. Moreover, though both Darwin and Wallace outlined experiments to test the idea that females choose mates based on expression of ornamental traits (see also Finn 1907 quoted at the beginning of chapter 6), and Noble anticipated modern studies of behavioral ecol-ogy by fifty years with his studies of mate choice in fish and lizards (Noble 1934, Noble and Bradley 1933, Noble and Curtis 1939), virtually all empirical tests of female mate choice relative to ornamental traits have been performed since 1982. Why did it take so long to establish the importance of female mate choice and intersexual selection, particularly related to bird plumage coloration?

I believe that female choice of mates based on expression of ornamental traits in general and colorful plumage in particular remained untested for so long primar-ily because the leading evolutionary theorists of the early twentieth century (with the exception of Fisher 1915, 1930, 1958) either ignored or discounted the idea of trait evolution driven by female choice. Einstein once remarked that "theory guides observation," and I know of no better illustration of the correctness of this statement than the lack of effort to test the idea of female mate choice. Whereas Darwin, Wallace, and contemporary evolutionary biologists seemed to lack the ability to test the idea, biologists in the early and mid-twentieth century lacked the incentive. Very few biologists made any attempt to test mate preferences of females, and the work of those who studied mate choice was largely ignored. It was only after theoreticians became interested in population genetics models of sexual selection and showed that, theoretically, female choice could lead to the evolution of ornamental traits (Kirkpatrick 1982, Lande 1981, O'Donald 1962, 1967, 1980), that empiricists began to look for mate choice in the field and lab.

In addition, a rarely mentioned reason that so few empirical tests of female choice were conducted in the hundred years following the publication of *Origin of Species* is that studies of female mate choice are technically difficult. For instance, the idea of using colored leg bands to individually mark a population of birds did not come about until the 1930s (Nice 1939) and did not gain wide application until several decades later. Dozens of products used routinely and taken for granted in manipulations of ornaments in recent studies of sexual selection such as fast-drying superglues, hair dyes in bright monochromatic colors, and art markers in bright colors have only been available for a few decades. Moreover, the statistical tests that are used routinely to distinguish patterns of choice from patterns of random behavior were not yet being used by animal behaviorists in the early part of the twentieth century. It appeared to be the technical difficulty of such work that stopped Darwin himself from conducting tests of his own idea.

I mark the beginning of the modern era of empirical studies of the role of female mate choice in the evolution of ornamental traits at the publication of Malte Andersson's study of tail length in Long-tailed Widowbirds (*Euplectes progne*) in 1982. In proportion to their body length, Long-tailed Widowbirds have among the longest tails of any birds in the world. While all male Long-tailed Widowbirds have very long tails, some males have longer long tails than others. Andersson used this species to address the fundamental question debated by Wallace and Darwin: does female mate choice select for longer tails in males? Andersson cut the tails of males and glued them back together using superglue. Some males had sections inserted, making their tails longer; some had sections removed, making their tails shorter; and some had their tails cut and glued back together with no change in length (these served as controls). Andersson found that the males with elongated tails attracted more females than males with shortened tails (Andersson 1982a). This line of research on elongated tails was picked up and greatly expanded by Anders Møller in his work on the elongated tail of the Barn Swallow (*Hirundo rustica*) (Møller 1988); summarized in Møller (1994).

Soon after Andersson's study of Long-tailed Widowbirds, experimental studies convincingly demonstrating female choice for bright integumentary coloration in fish were published (Houde 1987, Kodric-Brown 1985). However, during this exciting time of renewed interest in sexual selection and female mate choice, studies of female mate choice relative to ornamental plumage coloration became sidetracked into the study of why males of some species delay attainment of adult nuptial plumage for one or more years even though they are sexually mature

(Rohwer et al. 1980). This phenomenon, known as delayed plumage maturation, was the subject of a half-dozen detailed studies in the 1980s (see Thompson 1991 for a review). This is a fascinating topic well worth study, but in retrospect, it seems counterproductive to have invested so much effort into the study of delayed plumage maturation when the basic debate of Darwin and Wallace regarding the role of female choice in the evolution and maintenance of colorful plumage was yet to be resolved. As it turned out, little real progress in understanding delayed plumage maturation was possible until a better understanding of the role of female mate choice in the evolution of plumage coloration was achieved.

The study of female choice relative to male plumage coloration was also slowed in the late 1970s and early 1980s by poor choice of study species (with respect to the study of the function and evolution of ornamental plumage) by key research groups. In North America, much effort was focused on testing sexual selection theory on the Red-winged Blackbird (*Agelaius phoeniceus*), while in Europe, similar effort was focused on the Pied Flycatcher (*Ficedula hypoleuca*). For both of these species, adults were variable in plumage coloration, but they also had delayed plumage maturation. So, there was a yearling-specific plumage that complicated tests of female choice based on color that was independent of age. Moreover, both species were territorial, making it very difficult to disentangle mating advantages due to male-male competition from advantages gained through female choice. To this day, after dozens of studies, the role of female choice in the maintenance of colorful plumage is poorly understood for either Red-winged Blackbirds or Pied Flycatchers. It would be an injustice to neglect to add, however, that although Red-winged Blackbirds and Pied Flycatchers proved to be less-than-ideal species for the study of the role of female mate choice in the maintenance of plumage coloration, studies focused on these two species have added immensely to our understanding of a dozen other topics in evolutionary and behavioral ecology (Lundberg and Alatalo 1992, Searcy and Yasukawa 1995)

The first convincing experimental tests of the hypothesis that female mate choice selects for brightly plumaged males were my tests of female mate choice relative to male plumage coloration in the House Finch that I will describe in this book. It took 131 years after the publication in *Origin of Species* for the hypothesis that female choice drives the evolution of colorful plumage to be tested. And yet, the experiments that I conducted were simple and straightforward. While the basic demonstration of female choice for bright male plumage coloration was simple, the full implications of choice for colorful males and an understanding of what controls expression of coloration in males and what benefits accrue to females through their choice of mates has proven to be a challenging field of study.

# 2 A Red Bird in a Brown Bag

## An Introduction to the House Finch

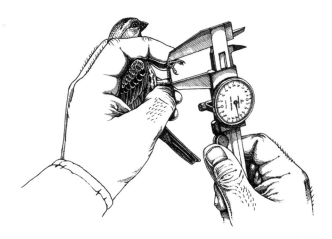

Most wild animals live in wild places. But we house finches also like living in cities, towns, and suburbs.
— Freddy Finch (1997), *Ranger Rick Magazine*

. . . this pretty little Finch [was] the most common and characteristic of the local birds . . . and was a lively and most agreeable feature in the dirty towns which it honored with its presence.
—S. F. Baird, T. M. Brewer, and R. Ridway (1874), describing the House Finch in New Mexico

*To anyone even casually familiar with the diversity of avian life on this planet, the House Finch has to be considered among the most mundane of birds. It is the bird singing from the porch at home or from the ledge outside one's office. It is common or abundant wherever it is found, and it is generally found near people. It is of average size, socially monogamous like most birds, builds a cup nest, and molts once per year—nothing that would raise the eyebrow of an ornithologist. When I meet with fellow field biologists and hear tales of cotinga leks in the lowland rain forests of Guyana or a new species of barbet from an unexplored mountain in Peru, it is almost embarrassing to admit that I catch my birds in the parking lot a hundred feet from my office and band them at my desk.*

*I chose House Finches as the focus of my research specifically because of their abundance and their accessibility. For a field ornithologist, however, conducting a study on a college campus rather than in some beautiful and fascinating natural setting is a real sacrifice. One of the great paybacks for devoting one's life to the study of the natural world is the opportunity to spend many hours surrounded by nature. Conducting a nest watch on the third level of a parking garage in the middle of Ann Arbor does not provide the same quality of natural experience as would a similar nest watch in a rain forest in Peru.*

*I approached my doctoral studies at Michigan with the goal of conducting convincing, if not definitive, tests of Darwin's hypothesis that colorful male plumage exists because females prefer to mate with colorful males. I knew from the start that the clock was my greatest enemy. To accomplish what I saw as critical experiments, I needed simultaneously to run laboratory mate choice trials and field experiments. On any given morning, I had to set traps, band birds, manipulate the plumage color of captured males, spend a few hours observing marked birds in the wild population, catch specific birds in aviary flocks to use in mate choice trials, and set up and take down two or three mate choice trials. This ambitious approach to my graduate research was only possible with aviary and mate-choice facilities and a study population all in close proximity. As it worked out, my aviaries were on the roof of the museums building, a room for my mate choice trials was on the third floor, and my finch study site was the University of Michigan campus, centered on the museum's building. My facilities could not have been in closer proximity. So I spent my days running up and down the stairs of the museum literally doing three things at once.*

*I've had a fair amount of success in accomplishing my goals of testing female choice relative to male coloration and unraveling the proximate control and information content of male coloration. One could conclude that I made the right choice in not going to the Canary Islands to study Canaries, or not looking harder for a potential study bird in Costa Rica or Africa. There are days, however, when spring migration has just passed with me too busy to get away from campus and enjoy the spectacle, or when a colleague recounts the excitement of a field camp in Panama, on which I feel the price I paid for my spreadsheets of data was a bit too high. But then a study will reveal a fascinating pattern related to plumage coloration in the House Finch, or I will get an idea for a new study that would only be possible with an abundant species that does well in an aviary. And I'm reminded that, while there can be adventure in the process of getting to one's bird, there is adventure of a different sort in scientific discovery.*

House Finches are often referred to as "small" songbirds; my students and I are as guilty of this as anyone. From the perspective of a 100- to 300-pound mammal, they are small birds. They weigh about 20 g, and from their bills to the tip of their tails, they are about 14 cm long. In a pinch, I can hold four finches comfortably in one hand. However, within the Order Passeriformes—the songbird order that includes about half of all birds—House Finches are actually of average size. They are larger than most birds in passerine families like tyrant flycatchers, chickadees, wood warblers, and kinglets, but they are smaller than most birds in passerine families like thrushes, jays, blackbirds, or starlings.

I will describe differences among subspecies of House Finches in chapter 10. For the general description that follows, I will be referring only to the subspecies *Carpodacus mexicanus frontalis*, which is the House Finch subspecies that occupies virtually all of the United States (including the Hawaiian Islands) and southern Canada (Hill 1996b). Most of the other subspecies are found in Mexico and

Pacific Islands (Hill 1996b). *C. m. frontalis* is the House Finch with which most readers will be familiar.

## Physical Characteristics

### General Appearance

House Finches are sexually dimorphic, meaning that males and females differ in physical characteristics. More specifically, they are sexually dichromatic, meaning that males and female differ in coloration. Females are drab, cryptically colored songbirds. They have brown to gray coloration on their dorsal surfaces and dirty-white plumage with heavy gray/brown streaks on their undersides. Novice bird watchers often have a difficult time identifying female House Finches because they have no bright colors or bold patterns to distinguish them easily from a dozen other cryptically colored species with more or less similar plumage markings. Males are a different story. Over much of their surface they have the same cryptic plumage coloration as females, but in three plumage regions—the rump, the under-side from the bill to the belly, and the head including the fore-crown and eye stripes—melanin (brown and gray) coloration is replaced by colorful carotenoid pigmentation. These patches of ornamental plumage are typically red or at least reddish, but they vary among males from brilliant scarlet red to very pale yellow (Grinnell 1911, Hill 1990, 1992, 1993a, Michener and Michener 1931; see also this book's cover). It is this variation in male plumage color that has been the focus of my research and that is the focus of this book. In the chapters that follow, I will discuss in detail the variation in male plumage redness and brightness, the prox-imate or physiological basis for this variation in male coloration, and what this variation means to the social lives of these birds. For the remainder of this chapter, I will present a brief overview of the behavior, ecology, and breeding biology of the House Finch. This will provide critical background for discussion of the control, function, and evolution of colorful plumage. For a detailed description of the breeding biology and life history of House Finches, see Hill (1993b).

### Age and Sex Determination

House Finches have one complete (prebasic) molt of their plumage each year in the late summer or early fall from July to October (Hill 1993b, Michener and Michener 1940, Stangel 1985) (Figure 2.1). Using plumage coloration alone, one can deter-mine the sex of individuals that have completed their first prebasic molt with about 99% accuracy. Females sometimes have a wash of red over their crown, undersides, and rumps, but in females carotenoid pigmentation overlays heavy melanin streak-ing in all patches (Hill 1993d). In contrast, in males, melanin pigmentation is lacking in the portions of plumage that have heavy carotenoid pigmentation. Thus, even the drabbest yellow male looks very different in plumage pattern than a female, because melanin streaking is lacking from the fore-crown, throat, and upper breast and rump. In an experiment in which we implanted males with testosterone (a study that will be discussed in detail in chapter 8), my student,

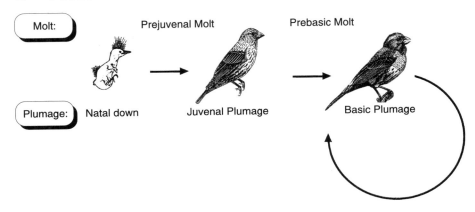

Figure 2.1. The sequence of molts and plumages of an individual House Finch. Once an individual completes its first prebasic molt in the late summer of its hatch year, it then undergoes one complete molt of its feathers each year in the late summer.

Andrew Stoehr, and I once "created" a male with virtually no carotenoid pigmentation in its plumage. My students nicknamed this bird the "Gray Ghost" because every patch of feathers that should have been colored with carotenoid pigments was instead a pale grayish white. But even with no carotenoid pigment to fill in the patches, this male did not look like a female—it looked like a male lacking carotenoid coloration. The reason that distinguishing the sexes by coloration is not 100% accurate is that a very few males grow a female-like plumage in their first year (Hill 1996b). These males have heavy melanin streaking where they should have only carotenoid pigmentation, and are indistinguishable from females (Hill 1996b). This topic of delayed plumage maturation in House Finches will be taken up in detail in chapter 10.

When young House Finches grow their first (juvenal) plumage after hatching, they have a plumage pattern that is very similar to that of adult females. Both males and females wear this female-like juvenal plumage until they undergo their first prebasic molt in late summer (Figure 2.1). Distinguishing young birds in juvenal plumage from adult females is more difficult than distinguishing between the sexes in basic plumage. With practice, however, females and juvenals can still be separated dependably. The problem of distinguishing adult females from juvenal-plumaged individuals exists from about March or April, when young of the year first appear in the population, until about mid-September when most birds have completed their prebasic molt. If one focuses on the gross similarity of adult females and hatch-year birds, they seem impossible to distinguish. If, on the other hand, one examines adult females and hatch-year birds carefully, conspicuous differences in plumage coloration and pattern become evident. By spring or early summer, the plumage of adult females is quite worn and gray with very narrow or no light margins on the tertials or greater secondary coverts (Figure 2.2). In contrast, the plumage of hatch-year birds is fresh and rich brown in color, with relatively wide buffy margins on the greater secondary coverts and especially the tertials. The rump color of adult females is flat gray often with a wash of yellow, orange, or red. The

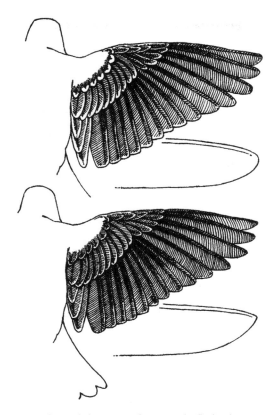

**Figure 2.2.** The upper surface of the wing of a recently fledged House Finch in juvenal plumage (top) and an adult female House Finch in basic plumage (bottom) in July, one of the summer months when juvenal finches can be difficult to distinguish from adult females. House Finches in juvenal plumage have wide buffy edges to their greater secondary coverts (the feathers covering the base of the flight feathers) as well as their tertials (three innermost flight feathers). House Finches in basic plumage in the late spring and summer, when young birds enter the population, have very narrow pale margins to their greater coverts and tertials but lack the wide buffy edges. The flight feathers of adult females tend to be more pointed than the flight feathers of juvenals.

rumps of hatch-year birds, in contrast, never have a wash of yellow, orange, or red and are rich brown in coloration. With experience, one can confidently distinguish hatch-year birds from adult females in the hand during banding or even at a distance through binoculars.

I know of no morphological features that will distinguish males from females when they are in juvenal plumage. Moreover, once birds have completed enough of their prebasic molt to obscure their previous plumage, the only way to tell hatch-year from older birds is by looking at the extent of skull ossification (Pyle et al. 1987), the process in which cartilage hardens into bone. Until about the first of November, most hatch-year birds have portions of the skull that are not yet ossified, while adult birds always have completely ossified skulls. The degree of skull ossification can be seen by parting the feathers of the crown and looking through the transparent skin on top of a bird's head. The problem with using skull ossification to age House Finches is that some of the earliest fledging birds have skulls that are completely ossified by October 1 and perhaps even earlier. Thus, skull aging must be used with caution. A bird with an incompletely ossified skull is certainly a hatch-year bird, but a bird with an ossified skull in late summer or early fall may be either an individual that is more than a year old or an advanced hatch-year bird.

## Plumage Coloration and Age

An important consideration for much of the research that I present in this book is the effect of age on plumage coloration. I was particularly interested in whether males in their first breeding season (yearling males) were less brightly colored than males in their second or later breeding season (older males). I will present details for how I quantified the plumage coloration of males in the next chapter. For now, all that is important is that in my plumage color scores redder and more brightly colored males received higher scores than yellower and more drably colored males. For this comparison, I considered only *frontalis* House Finches, which do not have delayed plumage maturation. By definition, males in these populations molt into definitive basic plumage in their first prebasic molt (Figure 2.1), and there is no yearling-specific plumage (Hill 1996b). Lack of delayed plumage maturation, however, did not mean that the age groups did not differ in average plumage color. I used plumage color scores of known-age males to test the hypothesis that yearling males are less colorful than older males.

I had plumage-color data for known-age birds from populations in California, Michigan, and Alabama. The yearling males used in these comparisons were first captured in juvenal plumage and then recaptured after their first prebasic molt but before their second prebasic molt. The adults that were included were banded as adults of unknown age and returned in the next breeding season when their color was scored. The ages of California birds came from the banding records of the late Dr. Dick Mewaldt, in whose yard I trapped finches in 1990. In all three populations, I found that two-year-old and older males were significantly brighter than yearling males (Figure 2.3). The difference was most striking for the California population, in which males displayed highly variable plumage coloration. In the two eastern populations, there was complete overlap in the plumage redness of adult and yearling male finches; only the central tendency of the distributions was different among the age groups. So, although yearling males in these populations do not have delayed plumage maturation, they are duller on average than older males. The cause of this difference in plumage coloration between yearling and older male House Finches appears to be that individual males tend to get brighter with age (Figure 2.4) and that more brightly plumaged males are more likely to survive than less brightly plumaged males (see chapter 7 and Figure 7.5).

I recorded not just the plumage coloration of males but also the extent of carotenoid pigmentation in the ventral plumage, which I will refer to as patch size. I used the same set of known-aged males to look for age effects on patch size. Interestingly, patch size did not differ significantly between yearling and older males (Figure 2.3).

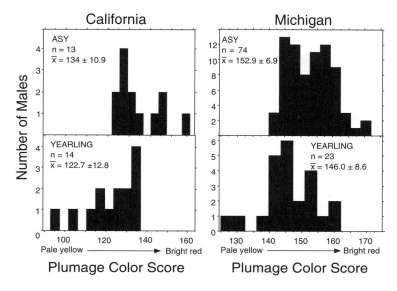

**Figure 2.3.** The relationship between age and plumage brightness in two populations of the House Finch. There is no age-specific plumage coloration or pattern in populations of the familiar northern subspecies of the House Finches, *C. m. frontalis*. Individual males tend to get brighter between their first and subsequent breeding seasons (see Figure 2.4), and bright males tend to survive better than drab males (see Figure 7.5). Consequently, males in at least their second breeding season (ASY) averaged brighter in plumage coloration than yearling males (California: d.f. = 25, $t = -2.56$, $P = 0.02$; Michigan: d.f. = 95, $t = 3.96$, $P = 0.0001$; Auburn, Alabama (not illustrated): d.f. = 150, $t = 3.87$, $P = 0.0001$; two-tailed $t$-test) (from Hill 1992; Auburn data not previously published). Interestingly, I found no significant differences in the patch sizes (measured as proportion of ventral surface with carotenoid pigmentation) of yearling and ASY males (Michigan: SY males $x = 0.71 \pm 0.09$, ASY males $x = 0.73 \pm 0.08$; d.f. = 95; $t = -1.43$, $P = 0.15$; not illustrated).

## House Finch Ecology

### Habitat

The distribution of House Finches in North American has changed dramatically in the last 150 years. Details of the range expansion of House Finches will be given in chapter 10. Briefly, when European people colonized North America, House Finches were found only in western North America from Oregon and Colorado south through most of Mexico. They were purposefully introduced to the Hawaiian Islands in the mid-nineteenth century, and are now common on all major islands in the chain. They were introduced accidentally to the eastern United States around New York City in 1940, from where they spread throughout eastern North America, from southern Canada to Florida. The western population also underwent a natural range expansion. The result is that House Finches are now found in most places in the United States (excluding Alaska) and southern Canada.

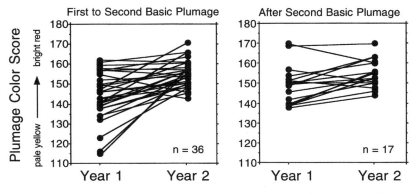

Figure 2.4. Change in plumage coloration between years for male House Finches in Ann Arbor, Michigan. Illustrated are data from the 1989–1990 field season. For statistical comparisons, data from four field seasons, 1988–1991, were included. The mean increase in plumage coloration between the first and second basic plumages (left panel) was significantly greater than the mean change between subsequent basic plumages (e.g. second to third, third to fourth etc.; right panel) ($t = 2.04$, d.f. $= 84$, $P = 0.04$, two-tailed $t$-test). The greater increase in plumage coloration between the first and subsequent years and the tendency of redder males to survive better than drabber males (Figure 7.5) likely accounts for the average difference in plumage coloration between yearling and older males (see Figure 2.3).

Across most of the lower forty-eight states and in southern Canada, the House Finch is now among the most common backyard birds. They are resident in most parts of the United States and most individuals in most populations do not migrate, but in some of the recently established populations in the eastern United States individuals have apparently evolved migratory behavior (Able and Belthoff 1998, Belthoff and Gauthreaux 1991b). In western North America, away from towns, farms, and ranches, House Finches inhabit open savanna, scrubland, and desert. They prefer areas with scattered trees, canyons, rock outcrops, and open space. They are not found in the interiors of forests, but they will nest at forest edges up to about 2000 m of elevation. They are absent from featureless open grasslands or dry desert flats. Although they commonly dwell in dry habitats, they must have daily access to water (Bartholomew and Cade 1956).

What humans have created over the last few hundred years throughout vast areas of the United States and southern Canada in the form of towns, suburbs, campuses, industrial parks, and shopping malls, is ideal finch habitat, and finches have taken to it readily. Virtually all House Finches in eastern North America and through the Great Plains live in close association with people. Although many House Finches in western North America live in natural habitats, a large proportion— perhaps now the majority—of western finches also lives in human-constructed environments. Because their habitat requirements are met so nicely in urban and suburban areas, finches would be common in these habitats throughout North America even if people paid no attention to them. But, people do pay attention. A significant proportion of the population in the United States and Canada enjoys putting out seed and watching the birds that the seed attracts (even if they don't know the identity of the birds they are observing). So, to an ideal habitat we have added a vast and essentially inexhaustible food supply. It is no wonder that House Finches exist in such abundance in developed areas of North America.

## Competition with Other Species

Because they have increased markedly in abundance in the last hundred years or so and have moved into backyard settings throughout North America, one could reasonably ask whether the increase in House Finches has negatively impacted other bird species. I have seen no convincing evidence that the increase in House Finches has come at the expense of any other bird species. House Finches are primarily occupying a food-and-space niche—consuming buds and seeds and nesting around buildings—that no native bird was using. The species that seems to overlap most with House Finches in terms of food and nesting space is the introduced House Sparrow (*Passer domesticus*). House Sparrow numbers have fallen rather dramatically in the eastern United States since the middle of the twentieth century, and several authors have suggested that the increase in House Finches caused the decline of House Sparrows (Kricher 1983). However, if one looks at Breeding Bird Survey data, the start of the decline of House Sparrows predated the arrival of House Finches by fifteen to twenty years in states such as Michigan, Alabama, and Missouri (Figure 2.5). In addition, despite the fact that House Finches arrived in these states at different times in the late twentieth century, the decline in House Sparrows in the three states is remarkably similar. Moreover, House Sparrows are dominant to House Finches at feeders (Brown and Brown 1988, Kalinoski 1975), and House Sparrows commonly usurp active House Finch nests when the two species occur together (Bergtold 1913, Evenden 1957, Hill 1993b). So it appears that the decline of House Sparrows was not caused by the increase in House Finch abundance in the eastern United States.

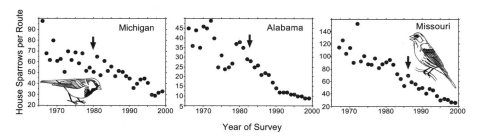

**Figure 2.5.** The decline of House Sparrows in relation to the spread of House Finches in eastern North America. Shown are the mean numbers of House Sparrows per Breeding Bird Survey Route in Michigan, Alabama, and Missouri from 1966 to 1999. The date on which House Finches first bred in each state is shown with an arrow. Note that the decline in House Sparrows preceded the introduction of House Finches in all three states. The same general pattern can be seen for all eastern states. Adapted from Sauer et al. (2001).

The primary effect of House Finch abundance on other species of birds is likely to be at feeders. Finches commonly patronize feeders in such numbers that they overwhelm other species. Certainly at specific feeders the presence of House Finches can mean the absence or greatly reduced presence of other species of birds. Over the years, I have spoken with dozens of bird-feeding enthusiasts who are dismayed at the loss of favored feeder visitors like Carolina Chickadees or White-breasted Nuthatches when large groups of House Finches begin monopolizing their feeders. It should go without saying, however, that no population of native birds is dependent on feeders for survival. Hamilton and Wise (1991) used Christmas Bird Count data from Indiana specifically to test the hypothesis that the increase in House Finches is associated with a decrease in American Goldfinches (*Carduelis tristis*) or Purple Finches (*Carpodacus purpureus*). They found no evidence that the increase in House Finches in Indiana over the last twenty years has had any impact on native cardueline finches. Thus, while it might be annoying for feeder watchers to have swarms of House Finches drive other species away from their feeders, there is no evidence that increasing House Finch numbers negatively impact any native species of bird.

## Population Studies

### Study Sites

Over the last fifteen years, I've studied House Finches primarily at two sites: on the main campus of the University of Michigan in Ann Arbor and on the campus of Auburn University, in Auburn, Alabama. My Michigan studies were conducted over four breeding seasons from 1988 until 1991. In Auburn, I banded the breeding population for three years (1993 to 1995) without monitoring pairing or nesting activity. In 1996, with my graduate students, I began an intensive study of the Auburn breeding population, which has now been conducted over five breeding seasons. At various times, I have also sampled House Finches in Ontario, New York, California, Hawaii, and Guerrero, Mexico, but studies away from Michigan and Alabama have been relatively brief collecting trips. Observations of populations away from Michigan and Auburn will be discussed in chapters 10 and 11. In this chapter and for the first two parts of the book, I will focus on my studies of breeding populations in Michigan and Alabama.

The breeding population of House Finches on the University of Michigan campus was substantially denser than the population on the Auburn University campus. Although both study sites comprised about 100 hectares (250 acres), I followed 66 to 72 breeding pairs in each of my four years in Michigan, but only 30 to 45 pairs per breeding season in Alabama. The Alabama population has had its advantages, however. Female House Finches build open, cup nests and use a wide variety of nest sites both on human structures and in vegetation. The primary requirement for a nest site seems to be a solid base on which to place the nest and some sort of overhang, preferably within about 0.25 m of the nest to protect the nest from rain. In Michigan, early in the season (from March until early May) finches nested primarily in small (less than 8-m tall) spruce trees, which were planted abundantly on campus. These nests in small spruce trees tended to be relatively easy to find and reach, although finches built quite a few nests in larger spruce trees that were not accessible. In early May when leaves started to appear on the ivy that covered most buildings on campus, many pairs of finches shifted to nesting in the ivy, particularly where it grew under eaves and overhangs of buildings. Unfortunately, on the Michigan campus, ivy commonly grows to the third or fourth story of buildings, not only putting many nests out of reach but also making many nests very hard to find and monitor. My Michigan study site also had a number of large, old buildings with wooden rafters that extended under their eaves, twenty or more meters above the ground. Many pairs of finches used gaps in these rafters as nest sites, and these nests were totally inaccessible. Thus, although the Michigan population was ideal for generating large datasets on plumage variation among male and female House Finches and on pairing success (which was the focus of my dissertation), it was not good for studies of nest success or for attempting to calculate annual reproductive success. The University of Michigan campus also proved to be a hard place to conduct a study of paternity. It was easy enough to establish whether or not a male fathered the chicks in the nest that it was attending (Hill et al. 1994b), but there were simply too many potential fathers, including some unbanded males, to have a chance of finding the sire(s) of extra-pair young.

The Auburn campus, in contrast, had no spruce trees and no ivy growing on any building. What trees were present were generally not suitable for House Finch nesting—they were too large and sparse. Compared to the University of Michigan campus, the Auburn campus had few choice nest sites for finches. I did not intensively study breeding House Finches in my first three years at Auburn (1993–1995), but I did notice that nests were located primarily in various light fixtures, nooks, and ledges on buildings mostly within reach of a ladder.

It was the apparent lack of suitable nesting sites on the Auburn University campus that gave me the idea of putting up nesting boxes for finches. The boxes I had in mind were not the sort of closed wooden boxes with a small entrance hole that are put up for cavity nesters like the Eastern Bluebirds (*Sialia sialis*) and Pied Flycatchers (*Ficedula hypoleuca*). House Finches, like all cardueline finches, are cup nesters and do not use cavities. Rather, I envisioned shallow wooden boxes with open tops that would be placed under the eaves of buildings around campus. Moreover, I was thinking that the nest boxes could be constructed following a box-within-a-box design. The idea was to get the birds to build their nest within an inner box that could then be lifted out and placed on an electronic balance (Figure 2.6). Such boxes would allow my students and me to control the placement of nests so that they would be accessible and observable. Moreover, the ability to insert an electronic balance would allow us to measure the amount of food brought to nests and hence parental care. Quantification of parental care was crucial for tests of some hypotheses related to the function of colorful plumage.

Andrew Stoehr, my former graduate student and collaborator, came up with the idea of using the bottom of plastic half-gallon milk and juice containers for the inner boxes. These proved to be ideal. They were cheap, abundant, light, and cleanable (or disposable). We found a type of small electronic scale (Acculab Pocket Pro 150-B) that we could insert between the milk-carton bottom and the

Figure 2.6. The finch nest boxes and video camera set-up used on the Auburn University campus study site. Boxes were designed so that the plastic inner box could be lifted out and placed on an electronic balance. In this way, we recorded not only the number of provisioning visits by parent birds but also the mass of food delivered. This box, which contained a House Finch nest, was 3.5 m off the ground under a covered pedestrian walkway on the Auburn campus.

outer wooden box. Moreover, we placed the boxes in locations that would allow us to set up a video camera nearby, so the nest attentiveness of parents could be filmed at each nest. This nest-box design worked wonderfully in allowing us to keep track of reproductive success of pairs and to make detailed observations of provisioning by both males and females.

When we first started erecting nest boxes on the Auburn University campus, we were extremely cautious about defacing buildings with our boxes. We mostly placed the boxes around the athletic facilities in spots where they were inconspicuous. We put up 180 boxes in January of 1997 and over 80% of the House Finch pairs that nested on our study site nested in our boxes that first year. Because the boxes were so successful and the campus community seemed supportive, we put up more boxes in 1998 becoming more brazen in where we would place boxes— including the back entrance of the campus police station itself. I've found that when conducting bird studies on a university campus, the best approach is not to ask for permission. It is better to adopt the attitude that it is inherently appropriate for a university faculty member or student to conduct research on university grounds. Asking an authority figure (e.g., head of facilities, occupational safety, campus police) for permission forces them to make a decision, and the easy decision for them is usually "no." If you request permission and the answer is "no" you can either abandon the project or begin what might be a long appeal process. If, on the other hand, you start the project and then get told not to do it (which has never happened to me), you have the powerful argument of time and resources already invested. The best policy, in my opinion, is to use good judgment, keep a low profile, and do what you need to do.

As we added more boxes to the study area, we started putting up boxes in places that presented technical challenges for filming. To record the display of the scale and hence the weights of food provided by parents at nests, we could not simply place the camera on a tripod on the ground and shoot up toward nests; we had to place the camera at the same level as or above the nest. By the second year, we were gluing nest boxes to the brick sides of buildings and under beams 10 m above the ground, so getting cameras up to the level of the nests became increasingly difficult. My graduate students, Paul Nolan and Andrew Stoehr, and in the third and fourth years, Kevin McGraw, Anne Dervan, and Renee Duckworth, took it as a personal challenge to find a way to film every nest, no matter how awkward the location. They ended up filming many nests by strapping a tripod that held the camera to the top of an extension ladder. For nests in parking garages, we parked a departmental van in the parking space adjacent to the nest and placed the video camera on the roof of the vehicle. In other situations, cameras were clipped into straps wrapped around concrete beams 10 or more meters above the ground. In the end, we got almost all birds nesting in our boxes, and we found a way to film every nest. It was and is a great set-up for a study of a bird population. It is a virtually unique system for monitoring the nesting biology of a species that does not nest in cavities.

## Capturing Birds

One of the first technical difficulties that I had to overcome at the beginning of my study was catching House Finches on my campus study sites. Mist nets, the most widely used method for capturing small birds and the technique with which I was most familiar when I began my studies, were impractical on a busy university campus. Box traps placed on the ground also proved ineffective because House Finches preferred not to feed on the ground and especially because of the huge populations of Fox Squirrels (*Sciurus niger*) on the University of Michigan campus and Gray Squirrels (*Sciurus carolinensis*) on the Auburn University campus that constantly set off or got captured in ground traps. I needed a trap that I could suspend above the ground and that squirrels could not easily enter. As a boy I had built traps to catch fish that were basically baskets of wire with funnel entrances. I decided to try the same basic trap design for trapping finches (Figure 2.7).

House Finches are by nature clingy birds. By this I mean that, unlike seed-eating species such as Northern Cardinals (*Cardinalis cardinalis*), House Sparrows, or Song Sparrows (*Melospiza melodia*), which prefer to stand on the ground and feed, House Finches are like Carolina Chickadees (*Parus carolinensis*) or Great Tits (*Parus major*) in preferring to perch off the ground while they eat. The best feeder design for these clingy bird species is a tube feeder, which is basically a cylinder filled with seeds to which birds cling as they extract seeds through small holes. This type of feeder became the focus for my trap design. I placed tube feeders at out-of-the-way nooks and corners around campus and kept them filled with seed. I soon had many finches regularly visiting the feeders. The basket traps that I built were

Figure 2.7. The type of basket trap used to capture House Finches on the campuses of both the Auburn University and the University of Michigan. After birds began regularly using a feeding station, the feeder was placed inside the trap. Finches passed through small entrance tubes to reach the feeder and could not find their way back out. Over twenty birds could be captured at a time in this type of trap.

designed so that the tube feeder could be hung inside the trap and the trap/feeder rehung where the birds were used to visiting the feeder. To get to the feeder inside the basket, the finches had to pass through one of four small (5 cm × 5 cm), hardware-cloth tubes that projected about 10 cm into the trap. No doors shut, and a bird had no idea that it was trapped as it moved to the feeder. It was not until a finch was ready to leave that it would find that it could not. These basket traps proved to be very effective at capturing House Finches, allowing me to band virtually every bird in both study populations each year.

When I removed birds from the traps, I placed them in brown, paper (unwaxed) sandwich bags. I found this to be the ideal way to hold small birds temporarily. Many banders use cloth sacks or portable wooden boxes to hold birds temporarily. Both of these techniques, however, have many disadvantages. Cloth sacks hold the bird in an uncomfortable and stressful position. When fully opened, brown paper bags have a flat bottom and create a space in which a finch can stand upright in a comfortable position. Most banders use cloth bags or wooden boxes for several or many birds before washing them, which undoubtedly leads to the transmission of parasites. Indeed, one could argue that wooden boxes can never be effectively disinfected. Brown paper bags, in contrast, are thrown away after being used once. If seed is placed in brown paper bags, House Finches will eat, and one can collect fecal samples from the bottom of the bag if such samples are needed for research. You can even write notes on the outside of the bag, which is very handy when a team of researchers is processing birds. I had to endure a lot of ribbing from my friends in the Bird Division at the University of Michigan for having brown sandwich bags full of birds bouncing around my office all the time, but I still have not found a better way to hold small birds temporarily.

## Breeding Biology

### Annual Phases

For typical migratory passerines breeding in temperate regions, individuals pass through six major phases in the course of a year: spring migration, breeding, prebasic molt, fall migration, wintering, and prealternate molt (Figure 2.8). For a typical resident House Finch, this list of six annual events simplifies to three: breeding, prebasic molt, and wintering. For House Finches, the "winter" period runs from October through December and seems to be a time of energetic recovery from the previous breeding season and the prebasic molt. Finches are generally inconspicuous during this winter season, and I have spent relatively little time watching finches from October through December. This book will focus on the breeding and molting periods, both of which appear to be energetically demanding and critically important to the reproductive success of individuals.

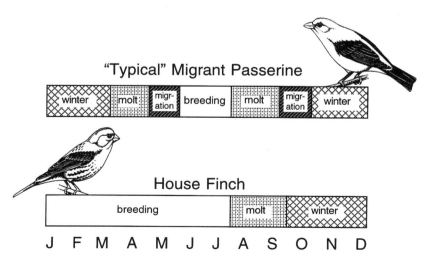

Figure 2.8. The breeding cycle of a typical Neotropical migrant passerine and the breeding cycle of the House Finch. Because most House Finches do not migrate and have only one molt per year, they have a relatively simple annual cycle with a long period for breeding (including pre-nesting courtship activity). This book will focus on periods of breeding and molt.

*Pair Formation, Pair Stability, and Divorce*

The breeding season for House Finches is among the longest of any North American bird (just ask my exhausted graduate students in August). In Alabama and probably in most of the warmer parts of the House Finch's range, the breeding season begins about the first of January as males start to sing daily and the first associations between male and female finches become obvious. In colder parts of the House Finch's range, such as Michigan, cold and inclement weather may keep birds from singing regularly for several weeks past the New Year. By February, even in the coldest parts of the House Finch's range, most males are singing throughout most of the day. The rise in male song marks the beginning of the six weeks or so between the onset of courtship and the start of nest building and egg laying. This is a dynamic time for House Finches. As they have done since the end of the previous breeding season, House Finches spend most of the day during this courtship period in flocks ranging in size from a half-dozen to a hundred birds. Within these flocks, males display and try to associate with females; females assess potential mates and allow preferred males to associate with them. Squabbles involving displacements from perches occur incessantly, but more serious fights that involve pushing, biting, or grabbing or that last more than a second or two are rare.

Males are not territorial, meaning that they do not defend any particular space or resource. They defend only access to their female. This lack of territorial behavior appears to be a function of their diet. House Finches are primarily granivores, but they will eat a variety of vegetable matter including fruits and buds (Beal 1907, Hill 1993b). They do not eat insects as a regular part of their diet. Consequently, their food appears not to be economically defensible: it is either widely dispersed, such as the seed heads of dandelions (*Taraxacum officinale*), which appear and disappear sporadically and unpredictably across a landscape, or it is extremely clumped (seeds in feeders). Lack of territorial behavior is central to the mating system of House Finches, and it is one of the features that makes the House Finch an ideal species for the study of female mate choice.

In the absence of territories, male-male interactions play only a minor role in pairing. During the pre-mating period, males sometimes displace one-another from perches, and occasionally one sees males chasing each other. However, once a female has decided with which male she wants to associate, the members of the pair work together to drive off intruding males and females. When teamed with a female, even the least dominant male appears able to drive off aggressive intruders. The male and female of a pair also drive off female intruders, so a female can only pair with a male if he chooses to pair with her. Thus, pairing in the House Finch is a cooperative venture driven by mutual mate choice (see chapter 9). Like any cooperative venture, though, it is not without its conflicts.

I've always been struck by how closely the mating system of House Finches parallels the mating system of humans. Just as in humans, the duration of pair bonds in House Finches can last from days to years. Early in the season it is common to observe a male and female associating as a pair during a morning's observation but then never to see them together again. My colleagues in animal behavior, who work diligently not to corrupt their observations of animals with anthropomorphic biases, will cringe, but to me these short associations seem equivalent to dates in humans. Birds are attracted to one another initially and associate for a time, but either the male or the female switches to a different partner before a strong pair bond is established. As will be discussed in detail in chapter 6, male plumage coloration plays an important role in mate attraction, and presumably initial pair bonds are often based on the quality of male ornament display.

Males provide food to females during the period of pre-nesting association, a behavior called "courtship feeding." This behavior could be used in conjunction with cues such as plumage coloration in finalizing mate choice—if a male fails to provide adequate resources during courtship feeding, the female might end her association with him and search for another mate. In other species of birds, particularly fish-eating gulls and terns, courtship feeding appears to play a central role in mate choice (Nisbet 1973). Courtship feeding in House Finches remains unstudied, however, so exactly what sorts of assessments are made during the period of initial association between prospective pairs is unknown.

As in humans, divorce is relatively common in House Finches. I define divorce as the termination of a pair bond during a breeding season, after the female has laid her first egg of the season and before the female has laid her last clutch of eggs. In Michigan, of ninety-two banded pairs that I followed through a first nest attempt between 1988 and 1990, eleven (12%) ended in divorce (Hill 1991). In each of these cases, both members of the original pair were observed after the nest failure, so the divorce was not due to the death of one member of the pair. Invariably, such divorces followed a nest failure, not a fledged brood (Hill 1991; see Figure 7.4). Moreover, the nest failures that preceded a divorce were not caused by predation, but rather, in virtually all cases, they were associated with the desertion of a clutch of eggs. Thus it appeared that the female abandoned not just the male but her entire reproductive effort with that male. In no cases were chicks abandoned, only eggs (see chapter 7 and Figure 7.4 for further discussion of divorce in the House Finch in the context of sexual selection).

Although divorce occurs in about 12% of pairs, the most common pattern of association between a male and female House Finch is for the pair to stay together for an entire breeding season. Returning to my observations of Michigan House Finches, all eighty-one pairs that stayed together for a second nesting attempt stayed together for the remainder of that breeding season. Divorces only occurred after a failed first nest attempt by the pair. At the start of the next year, however, some of these birds were nesting with different mates. Out of the eighty-one pairs that remained stable during a breeding season, forty-nine had both members of the pair return the next year. Of these forty-nine pairs that had the potential to maintain the pair bond to the second year, thirty-eight (76%) did so. The remaining birds nested with new partners.

Once birds were observed to be paired for two breeding seasons, they remained together for life. None of the birds from the thirty-eight pairs listed above that were stable for two years was seen paired to a different partner in that or subsequent years. The male and female of the most stable pair of House Finches that I observed were already banded (by Robert Payne) when I started my study in Michigan. Based on banding records, this male and female had been caught in a trap together in the spring of 1987, suggesting that they were paired in that year, although no one was collecting pairing data at the time. I recorded them as a stable pair in 1988 through 1991—four certain and five likely breeding seasons. Two other pairs stayed together for four years, and several other pairs at both Ann Arbor and Auburn remained together for three years. Thus, it appears that some House Finches pair for life.

An interesting question is whether a male and female House Finch that nest together in consecutive years stay together as a pair throughout the fall and winter. As I mentioned earlier, my observations from October to December are rather limited, but my impression is that House Finches do not remain paired throughout the year. From the period of fall molt through December, finches gather in flocks of variable size with no evident associations between males and females. Most of the local breeding populations leave the study site in both Michigan and Alabama during this period, so I do not have extensive observations of individuals with known-pair history. Those marked individuals that I have observed during the fall and winter have not been with their mates from the previous breeding season.

Hooge (1990), who studied House Finches on a small island off the California coast, where thirty-nine pairs bred and lived throughout the year, suggested that his birds did remain as pairs throughout the year. However, he based his conclusion on the association between males and females in the "non-breeding season," without defining the observation dates. Observations could have been made in January or February when pairs began to form. Even if one assumes that Hooge (1990) was right and that the House Finches he observed do remain as pairs throughout the year, one has to suspect that this behavior was affected by the fact that these birds were constrained to remain in proximity to one another by the small size of the island. For continental populations, I believe that birds form pair bonds anew each year.

## Nest Construction and Egg Laying

Relative to other species of songbirds, nesting begins early for House Finches. Even in Michigan, where many Neotropical migrants do not begin nesting until the first or second week of June and many resident species don't begin nesting until late April, the first House Finch eggs appear in nests each year in mid-March (Figure 2.9). March is not exactly a balmy month in Michigan, and on several occasions I have observed females incubating eggs in nests that would eventually fledge young when the highest air temperature for the day reached only 10°F (−12°C). In the much more temperate climate of Alabama, the first egg has appeared in a nest on our study site in February each year. In 1998, we were surprised to find an egg in a nest on 10 February, thirteen days earlier than our previous early nesting record. This nest fledged five young, and the pair that got such an early jump on breeding ended up fledging four broods of young—seventeen chicks in all.

Nesting begins with the selection of a nest site through a conspicuous behavior that I call "prospecting." Prior to beginning construction of a nest, pairs will explore an area, examining each potential nest site. A pair of finches may begin prospecting weeks before the female actually starts to build its first nest, and the male or the female of a pair commonly places a single twig at sites that they have visited. I did not appreciate the extent of this twig-placement behavior until we started monitoring nest boxes on the Auburn University campus. The plastic inner

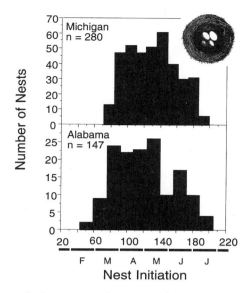

Figure 2.9. Dates on which House Finches initiated nests in Ann Arbor, Michigan (1988–1989) and Auburn, Alabama (1997–1999). Julian date (January 1 = day 1) and corresponding calendar month is shown.

cups of our nest boxes are cleaned out or replaced every winter so that any sign of nest building is obvious. We start checking all boxes on the study site in early February, and starting in early to mid-February, we commonly observe single twigs placed in boxes. Often a first twig is followed by a second or third twig placed days or weeks apart. Most of the boxes that receive single twigs early in the season are eventually occupied and have full nests built in them, but some are never used. Whether the pair that placed the twigs in a site is generally the pair that nests in the box is unknown, as is the significance of this twig-placing behavior.

Early in the season, individuals are observed only occasionally prospecting for nest sites. In the days immediately preceding the start of nest building, however, pairs spend a large part of their day searching prospective nest sites. Some pairs of finches concentrate all of their searching on a small area—sometimes one ivy-covered wall or one clump of trees. Other pairs move more widely as they search for nest sites. On the University of Michigan campus, I once observed a pair prospecting for nest sites at two locations more than 1 km apart on the same morning.

Nests are constructed primarily, if not exclusively, by the female. During nest building, males closely shadow their females—a behavior that behavioral ecologists call mate guarding. Mate guarding is essentially unstudied in the House Finch, but in other species of songbirds researchers have shown that mate guarding is a strategy by males to ensure that they fertilize the eggs produced by a particular female (Birkhead and Møller 1992). Males work to maintain exclusive sexual access to their mate during the period when her eggs are most likely to be fertilized. In other passerine birds, mate-guarding behavior reaches a peak a few days before the first egg of a clutch is laid during nest building (Birkhead and Møller 1992). Without having carefully quantified mate-guarding behavior, it appears to me that such behavior also peaks in male House Finches during nest building, just before the first egg is laid.

Typically, as a female works on the nest, her mate remains a short distance (about 10 m) away, often singing from a nearby perch from which he can clearly see his mate. The female can move around an area, for instance to the ground below the nest, without any response by the male, but as soon as she moves so as to take her out of view, the male immediately follows. Also, if any male comes between the mate-guarding male and its mate, it is immediately challenged. This is the one period in the breeding cycle in which I have seen aggressive encounters that escalate beyond a quick displacement or a few pecks. Occasionally, I have seen a male intrude on a pair in which the male is intently mate guarding and then not immediately give way when challenged by the guarding male. A chase or very rarely an actual fight will ensue, but in every case I have observed the intruding male ends up retreating. Males clearly vary in how persistently they guard their mates, and mate guarding is a topic that deserves more careful study in the House Finch.

Like most passerines, female House Finches lay one egg per day until the clutch is complete. When females are laying their eggs, they give a loud continuous call that can last for tens of minutes at a time. The purpose of this call is unknown. Montgomerie and Thornhill (1989) proposed that such egg cries are given during female fertile periods to incite competition among males and allow females to make a better choice for the father of their offspring. Extra-pair copulations are infrequent in the House Finches (Hill et al. 1994b; see chapter 6) and dominance does not appear to be a criterion in mate choice (Hill 1990, McGraw and Hill 2000c; see chapter 8), so it seems unlikely that egg cries in the House Finch are meant to incite competition. However, as I just described, males do fight to keep intruding males away from their mates, and it is possible that females encourage such aggressive interactions. Alternatively, egg cries may be a signal to males that incubation is starting and that male provisioning must begin.

## Incubation

Females do all of the incubating of eggs beginning with the last one or two eggs of a clutch. By the time the female begins incubating, her male ceases mate guarding almost entirely and spends his time foraging for himself and her (and probably pursuing extra-pair matings). As I mentioned above, and as is true of most cardueline finches (Newton 1972), males return to the nest at regular intervals during incubation and provide the female with food that is regurgitated from their crops. Using scales placed under nests on the Auburn study site, my students and I determined that the mass of food brought by males during incubation feeding was relatively constant, and that counting feeding visits was a reliable way to quantify the food resources brought to a female by a male (Nolan et al. 2001). Female are typically fed by the male either on the nest or at a nearby site. Some females returned directly to the nest (if they weren't fed on the nest) after a brief period of preening, fluffing, and wiping their bills. Other females left the area for a period, and I observed some of these females foraging during these absences from the nest. Whether male attentiveness during incubation affects the time that females spend away from the nest is untested.

## Nest Success and Number of Broods

Like many temperate songbirds that begin nesting early, House Finches have multiple broods each year. Although individuals vary in the duration of their nesting activity, as a population, finches are breeding from February until August. This allows females that are in good condition and that avoid nest failures to raise as many as four broods of young per year. The mean (± standard deviation) number of clutches for thirty-five pairs over two years on the Auburn University campus was 2.81 ± 0.85 (McGraw et al. 2001). Some pairs attempt one nest and are done for the season. Presumably female and perhaps male condition dictates how many nesting attempts are made per season, but this has not been formally studied in the House Finch.

Predation is the primary cause of nest failure, followed by nest abandonment. Because House Finches nest in urban settings, nest predation is relatively low. Predation of nests is sporadic, however, and in some circumstances it can be quite high. In Ann Arbor, 43% of nests were destroyed by predators over the four years that I monitored the population. In Auburn, 8% (13/164) of nests were lost to predators in 1997 and 1998, but the rate of nest predation rose to 21% (28/133) in 1999 and 2000. The difference in predation rate between 97/98 and 99/00 was that Blue Jays (*Cyanocitta cristata*) discovered that our nest boxes were a good place to look for eggs and chicks. A few or perhaps just a single Blue Jay more than doubled the predation rate, and in a small area of campus, virtually no finches fledged. A decade earlier I observed a similar change in predator behavior on the University of Michigan campus. Nest success in freestanding lamps was very high for the first two years on my Michigan study site. In the third year, however, Blue Jays, and again perhaps even one Blue Jay, learned to look in the lamps for eggs and chicks. Virtually no nests placed in lamps fledged thereafter.

## Provisioning

Both parents feed chicks in the nest. Beal (1907) analyzed the crops and stomach contents of House Finch nestlings in California and found that they were fed almost exclusively plant matter. My students and I have never conducted a careful analysis of what parents feed to nestlings, but the crops of chicks are transparent so we have seen the meals of hundreds of chicks. Chicks appear to be fed primarily green seeds and buds. Despite the fact that adults eat sunflower seed at our feeders, we infrequently see sunflower seeds in the crops of chicks.

House Finches never carry food for chicks in their mouths. Rather, both males and females regurgitate food to chicks from their crops. My students and I used video cameras to record the provisioning behavior at nests on our Auburn study site six and eleven days after eggs hatched. We found that, compared to many songbird species, House Finches feed chicks infrequently. Pairs made 2.53 ± 0.58 feeding visits per hour on day 6 and 2.83 ± 0.50 feeding visits per hour on day 11. Males provide slightly more food to chicks than females. On average, males make 51.6% of the feeding on day 6 and 53.0% of feeding visits to the nest on day 11. (See chapter 7 and particularly Figures 7.1 and 7.2 for how male provisioning relates to plumage coloration, and chapter 8 and Figure 8.5 for how hormones affect male provisioning rate.)

Most pairs attempt at least a second nest after a first brood of young has fledged. In most cases, the female focuses exclusively on the first brood until young fledge from the nest. After young leave the nest, however, the male alone feeds and cares for the fledglings while the female builds another nest and lays another clutch of eggs. In some cases, the female leaves chicks to the exclusive care of the male before chicks have fledged, and she begins building a new nest. We calculated re-nesting speed for twenty-two pairs in Auburn in 1997, and the average number of days it took pairs to lay an egg after fledging young from a previous nest was 5.9 ± 5.8 days. The longest it took a female to initiate a second nest was 24 days, while the minimum was −4 days. This latter female had four eggs in a second nest before young fledged from the first; the male alone fed the young during their last few days in the nest. It appears that even after the last nest of the year, the male alone feeds fledglings, but this assertion is supported by only a few anecdotal observations. In general, the behaviors of House Finches during the fledgling and very late nesting stages of development have not been adequately studied.

## Molt

For adults, prebasic molt begins soon after breeding has ended. Hatch-year finches begin molt as early as May, but most hatch-year and adult House Finches molt in July, August, and September (see Figure 2.1). During these late summer/early fall months, adult birds become scarce at our feeding stations. We only occasionally capture adults during banding from July to September, and I have also had trouble capturing adult House Finches at feeding stations in August in California and Ontario. In contrast, hatch-year birds come into feeding stations in large numbers from July through September. It appears that hatch-year birds from a local area stage at feeding stations for several weeks before they disperse to the area in which they will breed.

## What Makes the House Finch a Good Study Species?

In this chapter I have provided some background on the natural history and breeding biology of House Finches that will help put the studies of plumage coloration outlined in subsequent chapters into context. After much thought and consideration of many species of birds, I chose to focus my studies on House Finches because they appeared to be the ideal bird species for a study of the function and evolution of plumage coloration. When choosing a study species with minimal information and no personal experience, one's idealistic and optimistic impressions are usually dashed to bits after the first few weeks of work. There is generally a very good reason why the bird is not well studied, and all of the anticipated advantages of studying the species are overwhelmed by unforeseen difficulties (e.g., catching the birds, finding their nests, or observing the birds without disturbing them).

House Finches, however, in almost every case, proved to be easier to study and more appropriate for various tests than I anticipated even in my most optimistic mood. They are common and widespread throughout the United States and southern Canada, and they nest on college campuses, so wherever I have moved, they have been waiting for me. They can easily be captured in basket traps, making it possible to color mark a large population in relatively little time and with little help. Their nests are very easy to find. If you miss a female building a nest—which usually doesn't happen because they make no effort to conceal their nest-building activity—the female will cry out from the nest as she lays eggs. Males are tremendously variable in color display, so even with crude color measurement systems, I was still able to capture adequate variation in male coloration to conduct meaningful studies. Finches adapt very well to captivity, and in late summer when I have been most interested in establishing captive flocks, young birds come into feeding stations in huge numbers. With relative ease, year after year, I have been able to establish large captive flocks for experiments.

As a focal species for a study of mate choice and sexual selection, the House Finch was ideal. Males are not territorial and defend no resources other than the female. This meant that if a female chooses a mate, she must base the choice on characteristics of the male itself and not on the quality of resources defended by the male. As a consequence, I was able to avoid completely the confounding effects of territory quality that made the study of sexual selection so difficult in species like the Red-winged Blackbird. House Finches are also resident in the area in which they breed, and they are present in their nesting area for weeks before the start of nesting. This meant that there was time to conduct manipulations on the birds (such as changing the coloration of males) during the mate-choice period before the start of nesting. With many migrant species there is so little time between arrival and nesting that such manipulations are not possible.

When I first started working on House Finches in Michigan, I worried about the fact that it was an introduced population. However, I've become convinced that the inferences that I've drawn from introduced eastern populations of finches are as valid as inferences drawn from any populations of finches. Whether they are recently introduced or not, female House Finches still have to find a mate, and in both long-established and introduced populations the fitness consequences to females of making a good choice of mates are large. Moreover, the existence of many introduced populations of House Finches in a diverse array of habitats has provided tremendous opportunities for me to conduct comparative studies and to ask evolutionary questions that would have been possible with few other species.

## Color Vision and the Quantification of Color

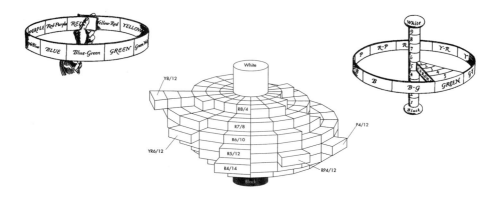

> It is no longer necessary to rely on subjective assessments of colour and conspicuousness, nor on assessments which rely upon human vision. This is important because the vision of many animals is different than that of ours, to say nothing of within-population variation in human colour vision.
>
> —J. A. Endler (1990)

> It would be quite facetious to say that we did all this [developed tristimulus color notation] just to satisfy mankind's yearning for TV. But it makes an interesting theory.
>
> —Anonymous, from *Colortron User's Manual*

*In my experience, successful science is the result of grasp of theory, hard work, and luck. On more than one occasion luck has played a disproportionate role in determining the direction and outcome of my research. Ever since Thornhill and Møller first proposed the use of fluctuating asymmetry as a measure of condition in studies of ornamental traits in the late 1980s, I had been interested in testing whether symmetry of carotenoid pigmentation might predict female mate choice. Unfortunately, in the late 1980s when I first tried to quantify carotenoid pigment symmetry in House Finches, I focused only on the pigmentation of the breast—the largest and, to my way of thinking, most important ornament of male finches. The long body feathers in the breast region are easily ruffed and repositioned during handling, and without pondering the problem sufficiently, I decided that it was too difficult to judge pigment symmetry accurately. Years later when I was processing House Finches that I had caught with my Advanced Ornithology class at Auburn, I came upon a male that had the most peculiar head coloration. The left side of its forecrown was yellow, but the right side was bright red. It looked like the character "Twoface" from "Batman" comics. The moment I viewed that male, it struck me that the crown plumage was the place to look at pigment symmetry. There is an obvious midline to the crown coloration, the feathers are short and cannot be repositioned with handling, and it is fairly easy to score the degree of bilateral symmetry of pigmentation. With that bird, I started recording the pigment symmetry of all male House Finches that we captured, and the study of pigment symmetry has turned out to be an important addition to my studies of condition-dependent trait expression in the House Finch. I have yet to see another male with such complete red/yellow asymmetry. If we hadn't happened to capture that male that day, I likely would have never integrated symmetry measures into my studies of plumage coloration. It really makes you wonder about missed opportunities in a research program.*

Male House Finches are more variable in expression of colorful plumage than virtually any other species of bird in the world. At one extreme, male House Finches can be brilliant, scarlet red on the crown, rump, and underside. But color expression ranges across a continuum from bright red through drabber reddish orange, to orange, to pale yellow (see this book's cover). It is critical to point out that these are not discrete color morphs. Rather, males vary continuously from pale yellow to bright red and, as I will discuss later, the mean coloration of males varies among locations. In some areas it is hard to find a male that is red; at other locations, it is hard to find a male that is not red.

Variation in ornamental plumage coloration was not only the focus of my interest in House Finches, it represented the first major obstacle that I had to overcome in beginning my study of plumage coloration. How does one measure the coloration of a bird? One can use a ruler and a scale to measure accurately the wing length, tarsus length, and mass of a bird. But what are the units of color? To begin my study of plumage coloration I had to devise a means to quantify the plumage coloration of males by a repeatable and detailed means. In addition, I had to establish that female House Finches could perceive the plumage color displayed by males. If females couldn't see the coloration, the coloration could not be functioning as a signal used in mate choice.

## Tristimulus Versus Tetrastimulas Vision

As I quickly learned when I began my study of plumage coloration, the color of an object is generally described in what is called "tristimulus" notation. This system of color description was not developed for bird coloration or even for general biological descriptions. It was developed for the many industrial uses of careful color quantification.

Humans have color vision that is based on the stimulation of three types of cones in the retina—red cones that are maximally stimulated by light at wavelengths of about 565 nanometers (nm), green cones that are maximally stimulated by light at about 530 nm, and blue cones that are maximally stimulated by light at about 420 nm (Tovee 1995). Together, these three types of cones give humans a range of visual perception from about 400 nm (violet) to 700 nm (red) (Figure 3.1). Below 400 nm lies ultraviolet, and above 700 nm lies infrared. Within the bounds of human vision lies red, orange, yellow, green, blue, indigo, and violet. To describe accurately how an object appears to the human eye, therefore, one must basically describe the pattern of cone stimulation associated with viewing the object. Because there are three cone types, three variables (tristimulus scores) are required to fully describe color for a human observer. Thus, tristimulus color notation is a means to quantify the color of an object as it is perceived by the human eye.

A few other relevant points come to bear, some of which I ignored, and was fortunate to get away with ignoring, when I started my study of House Finches. Tristimulus color notation describes the coloration of an object as a human sees it, not as a bird sees it. Because my study focused on the evolution of ornamental plumage in the context of intraspecific signaling, how a human perceived plumage coloration was not nearly as important as how it was perceived by a House Finch.

My assumption, and the assumption of many biologists, was that human color vision and avian color vision were basically equivalent. However, it was known at the time I began studying House Finches (had I done a better job of researching the topic I would have known it), and it is better established now, that the avian visual system and the human visual system are fundamentally different. (See Bennett et al. 1994 for a discussion of the issue of human color perception versus avian color

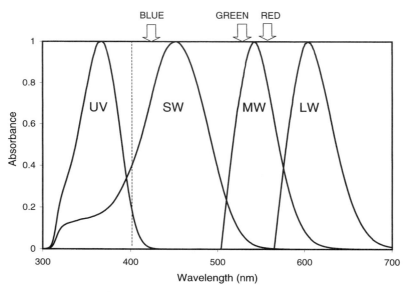

Figure 3.1. The normalized spectral sensitivity of retinal cone cells of the European Starling (*Sturnus vulgaris*). Unlike the human visual system, which is based on three types of cones, songbirds have four cone types, one of which is UV-sensitive. Note also that the red, green, and blue cones of humans (indicated by arrows) are maximally stimulated at different wavelengths than the red, green, and blue cones of songbirds. Humans perceive light from about 400 to 700 nm, while birds perceive light from about 340 to 700 nm. Reproduced with permission from Cuthill et al. (2000).

perception in the context of sexual selection.) Rather than being based on three cones, the typical avian visual system is based on four cones, the fourth being an ultraviolet cone (Chen et al. 1984, Chen and Goldsmith 1986, Tovee 1995) (Figure 3.1). A four-cone visual system is now commonly stated as the visual system of all diurnal birds. Thus, most birds have a range of visual perception from about 340 nm to about 700 nm and see a color dimension that is invisible to humans. Birds also have substantially greater density of cones in the retina (Walls 1942), and consequently see the world in far greater detail than humans, but that is perhaps less relevant to my attempts at measuring House Finch color than the range of avian visual perception.

Having just explained the importance of a fourth, UV-sensitive cone to avian vision, I am going to ignore UV coloration and color perception for the rest of this book. The House Finch was among the first bird species tested for spectral sensitivity, and the study showed that they perceive UV light with a lower peak sensitivity at around 370 nm (Chen and Goldsmith 1986). Although House Finches can see UV wavelengths (Chen and Goldsmith 1986), carotenoid pigments reflect light primarily in the visible spectrum and variation in light reflection by feathers pigmented with carotenoids occurs primarily above 500 nm, the portion of the visible spectrum that lies opposite the UV region (Figure 3.2). Consequently, variation in expression of carotenoid-based pigmentation is visible to both a finch and a human, and tristimulus color description can be reasonably used to describe variation in carotenoid pigmentation among male House Finches (Hill 1998a).

In addition to my work on House Finches, I have conducted research on the blue plumage coloration of Blue Grosbeaks (*Guiraca caerulea*) and Eastern Bluebirds (*Sailia sialis*), both of which have structural plumage coloration that varies in how it reflects light primarily in the violet/UV portion of the spectrum

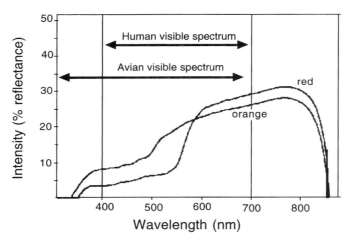

Figure 3.2. The reflectance spectra of the breast plumage of two male House Finches with red and orange coloration, respectively. House Finch plumage reflects minimally in the UV, and in my studies of plumage coloration in the House Finch I focused on light reflected in the human visible spectrum. (See Keyser and Hill 1999 for details of how reflectance data were collected.)

(Keyser and Hill 1999, 2000). For these species, how birds and humans perceive a plumage display are substantially different (most variation in coloration is invisible to a human but conspicuous to the birds), and tristimulus scoring of color is obviously inappropriate for these species. Whether or not tristimulus color description is appropriate for a study of plumage coloration clearly depends on the type of plumage coloration under study (Hill 1998a).

## Quantification of Color

### Visual Quantification of Color

Opting to measure plumage coloration with a tristimulus system still left me with the problem of how to derive a tristimulus plumage score for the birds that I was studying. There are many good and interchangeable systems of tristimulus color descriptors. The standard tristimulus system that is the basis for computer monitors, television screens, scanners, and digital cameras is red/green/blue, or RGB. This is the simplest system to understand because it closely mimics the way the human visual system perceives color. A computer or TV screen works by flashing millions of red, green, and blue phosphors at the human eye. These phosphors stimulate red, green, and blue cones in the retina in complex patterns that are sent to the brain and translated into images that blend the three primary colors into millions of colors. A similar but subtly different color system is cyan/magenta/yellow, or CMYK. The RGB tristimulus system is based on the idea of starting with black (no color) and adding colors to create the desired appearance. The CMYK system, in contrast, is based on the idea of starting with white (all colors) and subtracting unwanted colors to create the desired appearance. The CMYK system is the basis for color printing and copying on white paper. Neither of these widely used tristimulus systems was ideal for describing bird coloration, however. The system of color description that I have found most useful for quantifying plumage coloration in House Finches was developed by an artist, Albert Munsell.

In developing his color quantification system, Munsell wanted a color system that was intuitive. He recognized the three basic components of color that one person would use to describe an object to another person. First is the hue or what in common usage is typically referred to simply as the "color" of an object. This is the component of color description that we first learn as children. "What color is the crayon?" prompts a child to describe the hue of the crayon: for instance "Red" or "Green." Munsell took the visible spectrum of hues and bent them into a circle so that red abutted violet (Figure 3.3). He then divided the hues into five primary hues (red, yellow, green, blue, and purple), and five intermediate hues (yellow-red, green-yellow, blue-green, purple-blue, and red-purple). Assigning numbers to the ten hues allowed for a numeric description of color hue (although Munsell used an alphabetical code for hue).

The next descriptor in the Munsell tristimulus system is saturation, also known as chroma or intensity. This describes how much color is present. An easy way to think of variation in saturation is to think of red ink on a white piece of paper. If the paper is covered completely by red ink, the color is maximally saturated. If, on the

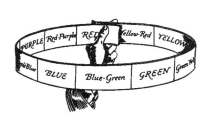 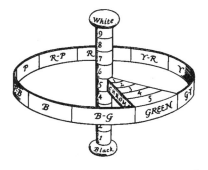

Figure 3.3. A diagram of the tristimulus color-scoring system devised by Albert Munsell. Hue is a 360-degree color wheel. Saturation (chroma) is shown as the spokes of the wheel. Tone or blackness is the axis on which the wheel can slide up and down. Illustrations from *A Grammar of Color* (Cleland 1921).

other hand, ink is applied as tiny dots evenly spread across the surface of the paper and covering half of the paper's surface, the color will appear only half as saturated (more pink). Less color is present. If only one fourth of the surface of the paper is covered in red dots, the color is less saturated still. In the Munsell system, the range of possible saturation is from white to maximally saturated color and this range is divided into six uniform steps, with white getting a score of 0 and a maximally saturated object getting a score of 5.

The final color descriptor in the Munsell system is tone (also known as brightness or value). It deals with the black/white component of color. Tone is the darkness of the object, with the range of possible darknesses divided into ten intervals from black (score 0) to white (score 10). Shades of gray lie between.

With these three color descriptors—hue, saturation, and tone—it is possible to describe numerically the color of an object just as with length, width, and height it is possible to describe the volume of a box. Because my entire research program would depend on detailed comparison of plumage coloration among males, accurate color description was fundamental. The few studies of plumage coloration conducted in the 1970s and 1980s, including my own studies of Black-headed Grosbeaks, *Pheucticus melanocephalus* (Hill 1988a, b, 1989), either ranked the color of males based on an observed range of variation or compared birds to color chips from a paint store. However, these techniques created a problem in that they were not generally repeatable and subtle variation in coloration was lost. The benefits of careful and standardized measurement of coloration in a study where coloration is the focus of research are the same as the benefits of careful measurement of length, width, and mass in a study of size. Imagine that instead of measuring wing length or bill width to the nearest 0.1 mm, as is the standard in field ornithology, one simply ranked males into one of five size categories. Clearly a fine-scale ruler is needed for study of size, and a standardized and detailed color measurement system is needed for a study of coloration.

Once I had decided to use Munsell-type tristimulus scoring as my means of quantifying plumage coloration, I had to decide how to generate the appropriate numerical descriptors for a patch of feathers. Basically, there were two ways that I could obtain tristimulus scores for a colored object. I could have used a reflectance spectrophotometer, or I could have compared the object to color chips in a standard color reference and recorded the tristimulus descriptors of the chip that most closely matched the object's color. This latter method uses one's own visual system as a color-measuring device. As a graduate student in a zoology museum, I did not have access to a reflectance spectrophotometer; I didn't know how to get access to one; and, I didn't have necessary funds required to buy one. As a matter of fact, even the high-quality color chips produced by the Munsell Company were beyond my financial means. Luckily, I discovered what was basically a poor man's Munsell—a small hardback book called the *Methuen Handbook of Colour* (Kornerup and Wanscher 1983).

The *Methuen Handbook of Colour* was directly comparable to Munsell color chips, except that in the *Methuen Handbook* hues were divided into thirty intervals. In the Methuen system there were eight saturation intervals (designated with numbers) and there were eight tone intervals (designated with letters). I had a preconceived notion, which I have since been able to test and verify (Hill 1996a), that more ornamented finches had redder (as opposed to yellower) plumage, that that they had more saturated coloration, and that they had less black in their plumage coloration. Therefore, when I scored the coloration of a finch, I made sure that higher scores were given to colors that were redder, more saturated, and less black. To do this, for each male, I recorded the color plate number that most closely matched the hue of its plumage. The ornamental plumage coloration of male House Finches conveniently matched the color plates from 2 to 11, with 2 being yellow and 11 being purplish red. When I scored the saturation of a male's plumage, I recorded the saturation number given on the page in the book, and finches varied from 3 (least saturated) to 7 (most saturated). Finally, I scored the tone of a male's plumage by converting the letters given in the book to numeric scores. Finches varied from C = 4 (most black) to A = 6 (least black). Thus, the plumage coloration of a male House Finch received three numeric color descriptors according to which color chip in the *Methuen Handbook of Colour* it seemed to match most closely.

To make things a little more complicated, male House Finches have colorful plumage not on one spot on their body, but in three distinct patches—the dorsal portion of the head, the ventral portion of the torso, and the rump (Figure 3.4). Coloration is sometimes different among these colored patches, so I thought that it was important to record the color of different patches separately. Thus, for each bird, I recorded the hue, saturation, and tone of seven plumage regions: the crown, the eyestripe, the rump, and four divisions of the large ventral patch. I did not divide the ventral patch evenly, but rather, I divided it in a manner that I thought would help distinguish among differently colored patches of feathers. The end result was that I recorded tristimulus descriptors for seven patches, giving me twenty-one total plumage scores.

For many analyses of finches that I published based on these book scores, I simply summed the twenty-one plumage scores. Because of the way I set up the scoring system, higher scores corresponded to brighter plumage, and so higher additive scores equated to brighter overall plumage. Thus, for a male to get a very high plumage score it had to have bright red plumage in all of the colored

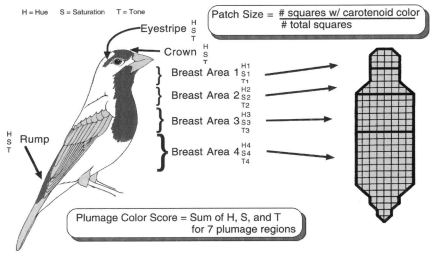

**Figure 3.4.** An overview of the system used to quantify expression of carotenoid-based plumage ornamentation of male House Finches by visual assessment. The body plumage of males was divided into seven plumage regions, and each region was given a hue (H), saturation (S), and tone (T) score based on comparison to plates in the *Methuen Handbook of Colour* (Kornerup and Wanscher 1983). A plumage color score was then calculated by summing the twenty-one individual plumage scores. The extent of ventral carotenoid pigmentation (patch size) was measured using a transparent grid overlay in the pattern shown. The number of squares covering feathers with carotenoid pigmentation was divided by total squares.

patches of its body. If a male House Finch had, for instance, a drab rump but an otherwise colorful plumage the lower hue, saturation and tone scores from its rump would slightly lower its overall plumage score. This additive plumage score was virtually identical to a First Principal Component descriptor from a Principal Components Analysis run with all plumage variables (Hill 1990). I used my additive score rather than Principal Component variables because the additive scores were positive integers that I could remember and interpret more easily than the Principal Component descriptors. I called the sum of all twenty-one color descriptors the "Plumage Color Score" for a male (Figure 3.4).

Although an occasional male might have a bright-red crown and breast and a drab-yellow rump, there were generally strong correlations between the plumage color scores of the crown, breast, and rump (Figure 3.5). Moreover, although hue, saturation, and tone could vary independently, there was a relatively strong correlation between plumage hue and saturation—redder males tended to have more saturated color (Figure 3.6). Tone did not vary much among males, but there were still significant relationships between tone and redness and between tone and saturation.

## Quantification of Patch Size

Male House Finches of the familiar *frontalis* subspecies have red pigment that extends, on average, from the base of the lower mandible to cover about 65% of the ventral surface (Figure 3.4). I refer to this extent of feathers with carotenoid coloration as patch size. The patch size of breast coloration varies considerably among males (see color plate in Hill 1992; Figure 3.7). At the time I began my studies of plumage coloration, almost no work had been conducted on the quality of the color display by birds. Rather, studies of plumage coloration were focused primarily on patch size and social interactions (see Butcher and Rohwer 1989 for a review). So, I certainly wanted to include patch size in my studies of coloration in male House Finches, and I quantified the patch size of the breast plumage using a transparent grid overlay (Figure 3.4). When I was processing a male, I would align the top of the grid with the base of the lower mandible and count the number of squares in the grid that overlapped feathers with carotenoid pigmentation. Patch size was significantly positively related to plumage color score (Figure 3.7), but there was sufficient variation in patch size independent of plumage color score for patch size to be an important component of plumage coloration to include in my studies.

Male House Finches also vary in the patch size of rump coloration (Yunick 1987), but I have never recorded rump patch. I assumed that rump and breast patch size would be correlated and that breast patch size would be more important, but these remain untested assumptions.

## Quantification of Pigment Symmetry

The degree of perfection of bilateral symmetry of ornamental traits has been used by behavioral ecologists as a measure of ornament quality and of the condition of the individual that produced the ornament (Møller 1990b, Thornhill 1992, Watson and Thornhill 1994). Beginning in 1995, I recorded the degree of pigment symmetry of crown feathers. I focused on crown feathers because they are short and cannot be repositioned during handling. The crown has a relatively small patch of feathers with carotenoid pigmentation and a clear midline, so the feathers on the right and left side can be easily compared. I scored symmetry as the percent of feathers that had dissimilar pigmentation between the right and left side. To simplify the process, I decided to use a zero-to-five symmetry scale based on the percent of asymmetrically colored feathers on the crown: (5) 0% (perfect symmetry); (4) 1% to 5%; (3) 5% to 50%; (2) 50% to 95%; (1) 95% to 100%. Feathers were counted as asymmetrically colored when the corresponding feather on the opposite side of the crown was obviously a different hue. Males varied substantially in the symmetry of their crown pigmentation, and the symmetry of crown pigmentation was significantly correlated with the hue and saturation of overall plumage coloration (Hill 1998b) (Figure 3.8). As with patch size, even though pigment symmetry was significantly related to plumage coloration, there was variation in pigment symmetry independent of variation in plumage color score.

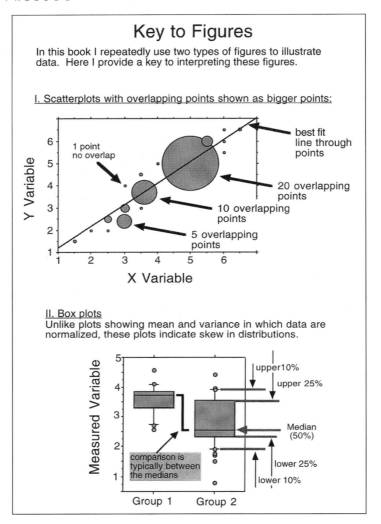

# Key to Figures

In this book I repeatedly use two types of figures to illustrate data. Here I provide a key to interpreting these figures.

I. Scatterplots with overlapping points shown as bigger points:

- best fit line through points
- 1 point no overlap
- 20 overlapping points
- 10 overlapping points
- 5 overlapping points

Y Variable

X Variable

II. Box plots
Unlike plots showing mean and variance in which data are normalized, these plots indicate skew in distributions.

Measured Variable

- upper 10%
- upper 25%
- Median (50%)
- comparison is typically between the medians
- lower 25%
- lower 10%

Group 1    Group 2

## Mechanical Quantification of Color

I quantified coloration using visual comparison of plumage to the *Methuen Handbook of Colour* from 1987 until 1995. The biggest drawback to this book scoring system was that, although it was repeatable for me (Hill 1990), it was not very repeatable between different observers. This meant that if I wanted dependable color data on birds, I had to score the color of every bird personally. After 1993, when I took a faculty position at Auburn University, I was involving increasing numbers of graduate and undergraduate students in my research projects, and it was frustrating and constraining for both the students and me to have to have me handle every bird. I had to find a way to automate the process of color scoring.

In 1995, my friend and former postdoc mentor Bob Montgomerie, who is very good about staying abreast of the latest technological innovations, told me about a

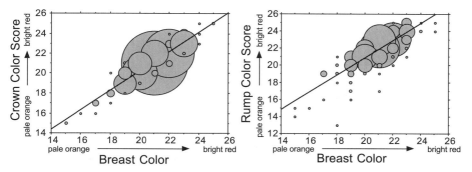

Figure 3.5. The relationship between the color score (hue + intensity + tone) of breast plumage and the color score of crown plumage (left scatterplot) and rump plumage (right scatterplot). Shown are 1990 Michigan data only ($n = 214$). Note that, although both relationships were strong, the relationship between crown and breast ($r_s = 0.87$, $n = 548$, $P = 0.0001$) was higher than for rump and breast ($r_s = 0.76$, $n = 548$, $P = 0.0001$) (statistical comparisons for four years of Michigan data 1988–91). If the coloration among patches was mismatched, the most common state was for a male to have a drab rump with an otherwise bright plumage. See the "Key to Figures" box on page 48 for an explanation of the figure.

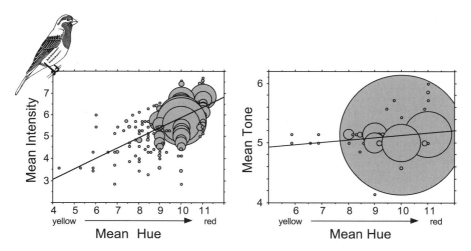

Figure 3.6. The relationship between the three color axes (hue, saturation, and tone) used to quantify the plumage coloration of male House Finches. Illustrated are the mean scores for seven plumage regions (see Figure 3.4). Redder males tend to have more intensely pigmented plumage, hence the positive correlation between hue and saturation ($r_s = 0.58$, $n = 548$, $P = 0.0001$). Tone varied little among males, but there was still a significant relationship between tone and hue ($r_s = 0.21$, $n = 548$, $P = 0.0001$), and tone and saturation (not shown; $r_s = 0.39$, $n = 548$, $P = 0.0001$).

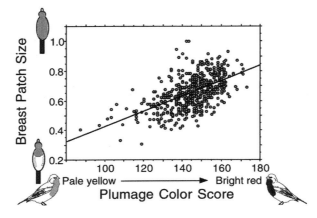

**Figure 3.7.** The relationship between carotenoid-based plumage coloration and the proportion of ventral plumage with carotenoid pigmentation (patch size) for male House Finches captured in Ann Arbor, Michigan. Males with brighter plumage coloration also tended to have larger patches of color ($r = 0.56$, $n = 547$, $P = 0.0001$). Adapted from Hill (1992).

new Macintosh-compatible color-measuring device. I found an ad for it in a computer catalogue, and I ordered one that same day. The device is called a Colortron (X-rite, San Francisco), and for me it was revolutionary because it simultaneously overcame the two arguments I had for not measuring coloration with a spectrometric device: it was inexpensive and it was very easy to use. At the time I bought my first Colortron unit, it was about one-third the price of a personal computer or about one-tenth the price of the fiber-optic reflectance spectrophotometers that were, up to that point, the only portable color-measuring devices available.

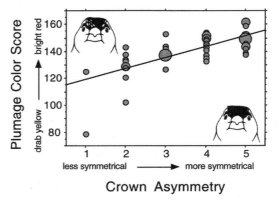

**Crown Asymmetry**

**Figure 3.8.** The relationship between the bilateral symmetry of crown pigmentation and the plumage color score of male House Finches in Auburn, Alabama. Brightly colored males also tended to have symmetrically pigmented crowns ($r_s = 0.70$, $n = 52$, $P = 0.0001$). Adapted from Hill (1998a).

Moreover, it was very easy to use. I was taking color measurements and generating interpretable tristimulus color scores within ten minutes of carrying the package of the new machine into my office. You literally plugged it into the computer, opened the software, and placed the small spectrophotometer against an object. Here was a machine that even a technophobic person like myself could use.

When the Colortron was pointed at an object, it generated a reflectance spectrum showing the intensity of light reflection at 10-nm intervals from 400 nm to 700 nm. It also calculated tristimulus descriptors for the coloration of the object. Before I could transfer the color scoring of plumage done in my studies to the Colortron, however, I had to know how Colortron tristimulus scores compared to the book scores that I had been using for eight years. By scoring a set of birds with both my book method and the Colortron, I found that the numbers generated by the Colortron were directly comparable to the book scores (Hill 1998a). The Colortron's tristimulus notation, however, was different from that used in the *Methuen Handbook of Colour*. Instead of dividing color space into thirty hue intervals, the Colortron divided the range of hues into a 360° color wheel, with the zero point arbitrarily and inconveniently located between red and purple. Male House Finches ranged in hues from 355 (purple red) to 32 (yellow). We converted hue scores of 355 to 359 to scores of −5 to −1 to make the hue scale linear. Saturation was measured by the Colortron as a continuous range of proportions from 0 (no color) through 0.50 (half of possible saturation), to 1.00 (maximally saturated). The saturation of the plumage of male House Finches ranged from 0.27 to 0.85. Tone was similarly measured as a range from 0 (no black) through 0.50 (midway toward maximum blackness) to 1.00 (fully black). The tone of the plumage of male House Finches ranged from 0.18 to 0.68.

One important difference between the color scores generated by the Colortron and color scores generated using the Methuen, Munsell, or other tristimulus scoring systems was that the units assigned to hue were reversed. Hue values from the Colortron increased as one went from red to yellow. This meant that redder males were given *lower* hue scores, and it meant that if there was a relationship between redness and, for instance, patch size, then the correlation would be negative not positive. For this reason, switching from book scores to Colortron scores could get a bit confusing, but in presenting the results of research in this book I will work to state the patterns clearly.

With the Colortron, we suddenly had an automated means to take repeatable color measurements from birds. This meant that my students did not have to rely on me to score the plumage of every bird. Moreover, because the Colortron divided color space into more intervals, it did a better job of capturing subtle variation in coloration. The Colortron has become an indispensable tool in my studies of plumage coloration in the House Finch.

## Summary

For my study of the function and evolution of plumage coloration, it was essential that I find dependable means to measure coloration. For a century, the standard means of color description has been tristimulus notation, in which color is described using three color axes—hue, saturation, and tone. A potential problem with tristimulus color description is that it is based on the three-cone human visual system, which is fundamentally different than the four-cone avian system. However, for carotenoid-based color displays, tristimulus color scoring captures the pertinent variation in coloration. In my studies of House Finches, I first used visual comparison of colored feathers to a standard color reference to generate hue, saturation, and tone scores for plumage. I combined these three color axes into a single plumage color score. I later switched to generating tristimulus scores of colored feathers using a reflectance spectrophotometer. By carefully quantifying male plumage coloration, I have been able to compare not only the plumage coloration of individual finches within a population, but also the plumage coloration of males across years and among populations. Although plumage coloration was the focus of most of my research with House Finches, I also recorded the extent of carotenoid pigmentation (patch size) and the symmetry of carotenoid pigmentation. Plumage color, patch size, and pigment symmetry are all intercorrelated, but some variation in patch size and pigment symmetry persisted independent of variation in plumage coloration.

The development of a repeatable and standardized means of scoring the size, symmetry, and coloration of carotenoid ornamentation of male House Finches set the stage for my studies of the function of ornamental coloration.

# PART 2

## The Proximate Control and
## Function of Red Plumage

# 4 You Are What You Eat

## Plumage Pigments and Carotenoid Physiology

It has been supposed that change from red to yellow in caged birds is in some way caused by change in food, or by general deterioration in bodily vigor, or perhaps due to a lack of a normal amount of muscular activity. . . That food can be the prime cause of the color-modification is possible; but the following facts [vegetable diet and wide geographic range] do not give this explanation more than a remote possibility.
—J. Grinnell (1911), speculating on why male house finches turn yellow when held in cages

Any male which consumes more carotenoids during foraging will have brighter carotenoid colors. Thus male brightness, at least with respect to carotenoid colors, is a direct indicator of feeding success.
—J. A. Endler (1983), describing carotenoid coloration in guppies

Researchers seem to think that, because carotenoids cannot be produced endogenously, animals have little control over their carotenoid loads and must be limited in the wild. However, there is little evidence of carotenoid limitation in the wild.
—J. Hudon (1994), in a critique of my paper showing that dietary access to carotenoid pigments affects expression of plumage coloration in the House Finch

*Since I began conducting experiments with aviary birds in 1987, an annual tradition for me has been, each summer, to search for suitable locations where I can capture birds for captive flocks. I don't like to remove birds from my study population, so I've had to rely on the backyard feeding stations as a source for my captive birds. This has put me at the mercy of the kindness as well as the lunacy of the bird-feeding public. The ideal set-up is when the owner of a feeding station grants me access to his or her yard while he or she is away for the day. I then have a peaceful morning collecting my birds. All too often, however, the homeowner is home.*

*One of the most memorable interactions that I've had with a homeowner involved a dentist in southern Michigan who, I was told, had feeders with lots of finches. I don't want to use his real name, so I will just call him Dr. Jones. I contacted him and made arrangements to work in his yard one Saturday morning in July.*

*I have to admit that dentists have always made me a little nervous. No matter how much I try to rationalize and focus on the good that they do, it is hard for me not to associate a dentist with pain and suffering. Unfortunately, my dealings with this dentist did nothing to foster my trust in the profession.*

*The day started off well enough. I set up a couple of mist nets near Dr. Jones' feeding stations, a wave of birds hit the nets, and together the dentist and I went out to extract the birds that had been caught. Dr. Jones had some banding experience, so he was enthusiastic about helping me remove birds from the net. They were mostly House Finches, but a few individuals of other species had also been caught.*

*As we were working on extracting the birds Dr. Jones asked in a rather disinterested voice, "What are you doing with the House Finches?"*

*"I'm studying the effect of diet on plumage color" I answered in a chipper voice, expecting a follow-up question like "How could diet change their color?"*

*Instead, Dr. Jones leaned forward and asked, "No experiments?"*

*"Experiments?" I replied, not sure what he meant.*

*"Yea. Aren't you going to do any dissections—cut em up?" he asked with a strangely misplaced grin that made me a little uncomfortable.*

*"No, I won't need to conduct any invasive experiments on any of these birds. I'll release them after they molt this fall." I replied, with the feeling that it was not the answer that my host was hoping to hear.*

*Having thrown four or five hatch-year House Finches into paper bags, the dentist came to one of the House Sparrows that had been part of the wave of birds. House Sparrows are introduced in North America, so they have no legal protection, and many people dislike them because they compete with native birds for nesting cavities.*

*"These little trash birds don't deserve to live" he muttered, and before I could grasp what he meant, he hurled the bird onto the pavement with a full throwing motion. On impact the bird made a loud POP. The dentist grinned at my look of shock. "It's the most humane way to kill 'em," he said.*

*I'm not a squeamish person. I've had to put down birds on occasion for the sake of science, but there was something about this brutal execution, and the obvious joy that the dentist seemed to derive from popping birds on the driveway, that made me shudder.*

*I had seen enough. I made up some excuse about appointments and a busy schedule and within the hour I had all my stuff packed and was ready to leave. I half expected to find the remains of an ex-wife as I was removing my net poles.*

*"Who works on your teeth?" Dr. Jones asked as I was loading up my car to leave.*

*"Ah, I don't have a regular dentist,"* I replied, regretting it immediately.

*"You and your wife should come by my office next week. I could clean your teeth and give you a checkup,"* he said. As he was making the invitation I could see three corpses plastered to the driveway behind him.

*"Sure, sure,"* I said. *"I'll call and make an appointment with your receptionist."*

That's one appointment I never made. As a matter of fact, it was several years before I could muster the courage (or the pain drove me) to see another dentist. Too bad, because it was a great yard for collecting birds.

The ornamental coloration of male House Finches results from carotenoid pigments deposited in growing feathers (Brush and Power 1976). Carotenoids are among the most common and widespread pigments in nature (Fox and Vevers 1960, Goodwin 1984). They are responsible for the red, orange, and yellow coloration of leaves in the fall, for many of the bright colors of fruits and vegetables (including carrots from which the pigments get their name), and for the yellow, orange, and red integumentary coloration of many invertebrates, fish, amphibians, reptiles, and birds (Goodwin 1984, Stradi 1998).

Carotenoids are large hydrophobic (water-fleeing) molecules. Each carotenoid molecule is a chain of forty carbon atoms, joined with alternating single and double bonds, with six-carbon rings at both ends. Carotenes are the simplest carotenoids, containing only carbon and hydrogen atoms. Xanthophylls, which are the oxygenated derivative of carotenes, have one or more keto- or hydroxy- functional groups on their terminal cyclohexene rings (Figure 4.1). The hue of a particular type of carotenoid results from the absorption of light by the unsaturated carbon backbone (known as the chromophore) as well as any functional groups. The specific hue

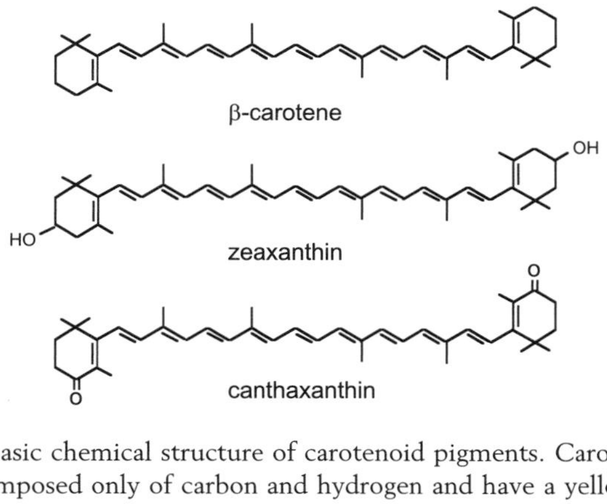

β-carotene

zeaxanthin

canthaxanthin

Figure 4.1. The basic chemical structure of carotenoid pigments. Carotenes, such as β-carotene, are composed only of carbon and hydrogen and have a yellow hue. Carotenes are modified into xanthophylls by the addition of oxygen-containing functional groups, such as the two hydroxyl groups added to β-carotene to produce zeaxanthin, a yellow pigment. Addition of a ketone to the fourth position on each side of the molecule produces one of a group of xanthophylls known as 4-keto-carotenoids, such as canthaxanthin. 4-Keto-carotenoids have a red hue and feature prominently in producing red coloration in bird plumage. For an introduction to carotenoid structure and plumage color, see Stradi (1998).

produced by a type of carotenoid is also dependent on its concentration. For instance, β-carotene dissolved in vegetable oil can appear red-orange if highly concentrated but appears yellow at lower concentrations. Various carotenoid pigments, however, have characteristic hues, and for the sake of describing plumage pigments in this book, I will refer to some carotenoids as "yellow" pigments (even though they can perhaps produce an orange hue if highly concentrated), and to others as "red" pigments (even though they may appear orangish or even yellowish at low concentrations). The colors that I will use to describe carotenoids will be consistent with the colors that the pigments generally appear on chromatography plates and the colors that they typically produce in feathers (e.g., Brush and Power 1976, Stradi 1998).

Carotenoids are especially interesting as biological colorants because animals cannot synthesize them (Völker 1938, Goodwin 1984). All animals are heterotrophs and, by definition, heterotrophic organisms cannot make their own saccharides, amino acids, nucleic acids, or fatty acids from carbon dioxide and water as autotrophic plants can. Animals have to obtain these basic building blocks from the food they eat. So, all components of an animal's body ultimately come from the diet. However, scientists recognize two different classes of macromolecules needed by animals. The first group of compounds includes substances like DNA, alcohol dehydrogenase, and cholesterol, which can be manufactured by animal cells from one of the four basic biological precursor molecules. Other macromolecules, which are generally called "vitamins," cannot be manufactured from basic biological precursors and must be ingested as macromolecules. Carotenoids fall in the latter group. Some animals can modify carotenoids once they are ingested, such as by adding functional groups that might change their color (Stradi 1998), but no animals can make carotenoids from basic biological building blocks (Goodwin 1984). Thus, expression of ornamental coloration that relies on carotenoid pigments depends ultimately on access to dietary sources of carotenoid pigments.

Carotenoid pigments are unique among biological pigments in being diet-dependent. Melanin pigments, which provide the gray and brown coloration of the wings, tail, and body plumage of House Finches and are found in the plumage of virtually all birds, are made from basic biological precursors—the amino acids tyrosine, tryptophan, and phenylalanine (Brush 1978, Fox 1976). These amino acid precursors for melanin production are part of the basic diet of all animals. Some bright green, blue, purple, and iridescent coloration found in the plumage of some birds, but not House Finches, is produced by the microstructure of the feathers, which differentially reflects and absorbs wavelengths of light to create a color display (Finger 1995, Finger et al. 1992). Again, structural coloration may be dependent on general nutrition (Keyser and Hill 1999), but it is not dependent on the consumption of specific pigment molecules. Only carotenoid-based coloration has a tight link to specific dietary components.

In this chapter I will first review what is known about the pigmentary basis for carotenoid-based plumage coloration and the physiological process by which carotenoids move from food to feathers. I will focus particularly on aspects of this process that might be dependent on the physiological condition of a bird and that might help us to understand the condition-dependent nature of carotenoid-based plumage coloration. I will end the chapter by focusing specifically on what has been learned of the pigmentary basis for plumage coloration in the House Finch and the physiological control of this coloration.

## A Review of Carotenoid Physiology

To obtain red plumage, a male House Finch has to move specific carotenoids from the environment into its feathers. This is a long and complex process, and I find it helpful to think of the process that determines individual expression of carotenoid pigmentation as having two distinct phases: acquisition and utilization (Hill 1999b) (Figure 4.2). Acquisition refers to the accruement of carotenoids from the environment into the gut. It includes all behaviors and structures that facilitate obtaining carotenoids. The role of carotenoid acquisition in determining plumage brightness was the early focus of my studies on the proximate control of plumage coloration. Utilization, on the other hand, refers to the manner in which acquired carotenoids are used by the physiological systems of the bird. It includes the absorption of carotenoids from the gut, the transport of carotenoids from the gut to target organs in the body, the modification of carotenoids (e.g., the oxidation of yellow carotenes to produce red xanthophylls), and finally the deposition of carotenoids in feathers.

Both the acquisition and the utilization of carotenoids are complex processes. Acquisition involves, among other things, choice of foraging habitat, choice of foraging patch, recognition of food items, choice of food items, ability to handle and ingest food items, and competition for food items. Virtually none of these

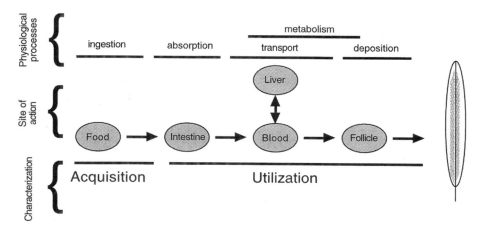

Figure 4.2. The pathway of carotenoid pigments from the gut to the feather. All carotenoids that are used as colorants by birds must be obtained in the diet. Acquisition refers to all behavioral and physiological processes that affect the type and quantity of carotenoids that reach the gut. Utilization refers to all processes that affect how ingested carotenoids are used. Utilization involves four processes: absorption, transport, metabolism, and deposition. Each of these processes plays an important role in the quality of feather coloration that is produced.

components of carotenoid acquisition has been studied in wild birds. A few of these components have been studied in captive House Finches, and I will present these studies in chapter 5 when I discuss how environmental variables can affect expression of carotenoid coloration. In this chapter, I am going to focus on how carotenoid resources are utilized within the body.

Carotenoid utilization in birds is even more poorly understood than carotenoid acquisition. The majority of studies on absorption, metabolism, and transport of carotenoids has been conducted on humans and other mammalian (e.g., rat, mice, hamster) systems. The mammals that are the focus of recent studies of carotenoid physiology lack carotenoid-based ornamental displays and generally have diets with small amounts of carotenoids (Hill 1999a). Consequently, we can expect the manner in which these mammals utilize carotenoids to be fundamentally different than the way vertebrates with high-carotenoid diets and bright carotenoid displays utilize carotenoids (Hill 1999a). Recent work by Hudon (1991, 1989, 1996) and Stradi (Stradi 1998, Stradi et al. 1995, 1996, 1997), particularly on carotenoid metabolism, has improved our understanding of the carotenoid physiology of songbirds. By necessity, in reviewing the physiological control of carotenoid pigmentation, I will use the human/mammalian literature to describe some basic aspects of carotenoid utilization, but whenever possible I will focus on the bird literature for insight on how carotenoid-based ornaments might function as honest signals of individual condition.

## Carotenoid Absorption

The process of carotenoid utilization begins during digestion with the liberation of carotenoids from the matrix, generally protein, to which they are bound in food. Bioavailablity of the carotenoids in food is a topic of interest in the human carotenoid literature, where variables such as the preparation of food (e.g., if and how it is cooked) and the fiber content of the diet that contains carotenoids, are studied relative to the efficiency of carotenoid uptake (Parker 1997). To my knowledge, bioavailability of ingested carotenoids has never been considered in discussions of how carotenoid utilization could affect expression of carotenoid-based integumentary coloration of vertebrates. One could imagine, however, that many variables could affect the efficiency with which an individual is able to extract carotenoids from food including, for birds, the efficiency of mastication, the type, amount, and size of grit in the crop, and the biochemistry and flora of the gut. Carotenoids are hydrophobic, so, as carotenoids are released from food in the gastro-intestinal tract, they are concentrated in lipid droplets. In the small intestine, these relatively large lipid droplets are emulsified by bile salts, leaving very small lipid/carotenoid droplets called micelles. It is in these micelles that carotenoids are thought to be absorbed by intestinal mucosal cells (Parker 1997). The effect of variation in how birds release carotenoids from food and prepare them for absorption by mucosal cells is unstudied.

The process of carotenoid absorption is generally described as passive (i.e., not requiring energy to move molecules) (Parker 1997), but, paradoxically, in birds it is highly selective. In general, birds absorb xanthophylls (which tend to be redder/ oranger pigments) almost to the exclusion of carotenes (yellower pigments) (Brush

1981, Goodwin 1984), prompting Goodwin (1984:165) to call birds "xanthophyll accumulators." Yellow carotenes and xanthophylls are far more abundant in the food of most birds than are red xanthophylls (Hill 1996a, Stradi 1998), so by absorbing xanthophylls preferentially over carotenes, birds are choosing a rarer group of pigments over a more common group of pigments. Moreover, the few observations that are available suggest that birds absorb carotenoids that can be used for plumage pigments more readily or to the exclusion of carotenoids that serve as neither plumage pigments nor precursors of plumage pigments (Hill 1996a).

The most extensive studies on the absorption, metabolism, and deposition of carotenoid pigments in a bird species with ornamental plumage coloration were conducted by Denis Fox and colleagues on several species of flamingos. One of the research foci of Fox and colleagues was the selectivity of dietary carotenoids used by flamingos (Fox 1962, 1976, Fox and Hopkins 1966, Fox and McBeth 1970, Fox et al. 1967, 1969, 1970). They fed American Flamingos (*Phoenicopterus ruber*) carotenoid-deficient diets for several weeks, until the levels of carotenoids in the flamingos' blood were very low. The diet of the flamingos was then supplemented with either lutein, zeaxanthin, lycopene, γ-carotene, α-carotene, β-carotene, or astaxanthin. The blood concentration of carotenoids remained unchanged after dietary supplementation with lutein, zeaxanthin, lycopene, and γ-carotene. Apparently, American Flamingos do not assimilate these carotenoids from food (Fox and McBeth 1970, Fox et al. 1970). Moreover, lutein and γ-carotene appeared in quantities in the feces, further supporting the assertion that these carotenoids were not absorbed. When β-carotene was added to the diet, increased concentrations of echinenone and canthaxanthin appeared in the blood, indicating that β-carotene had been absorbed and metabolized into these pigments. Likewise, astaxanthin in the diet caused an increase in canthaxanthin in the blood. When α-carotene was fed, it appeared in low concentrations in the liver and caused phoenicopterone to appear in the blood (Fox and McBeth 1970, Fox et al. 1970). Thus, all of the carotenoids assimilated by American Flamingos from the diet were plumage pigments or precursors for plumage pigments (Fox 1962, Fox and Hopkins 1966) (Figure 4.3); those that were not assimilated were neither pigments nor precursors.

Similar selectivity of absorption of carotenoids has also been observed in other birds. In one of the first studies looking at how various carotenoids were absorbed, Weis and Bisbey (1947) showed that, in chickens, xanthophylls are absorbed from the diet more readily than carotenes. The degree of specificity of carotenoid uptake in chickens was demonstrated by Tyczkowski and Hamilton (1986). They fed chickens a mash that contained three carotenoids—cryptoxanthin, lutein, and zeacarotene. They found that cryptoxanthin, which is not used as an integumentary pigment, was not absorbed from the gut, while lutein and zeacarotene were both absorbed but at different sites. Lutein, which is used as an integumentary pigment by chickens, was absorbed quickly in the upper small intestine; zeacarotene, which is a vitamin A precursor, was not absorbed until later and in the lower small intestine. Thus, chickens are selective in both the type of carotenoids that they absorb and in the site of absorption. Yet another example concerns the African Oriole (*Oriolus auratus*). Lutein, which is the

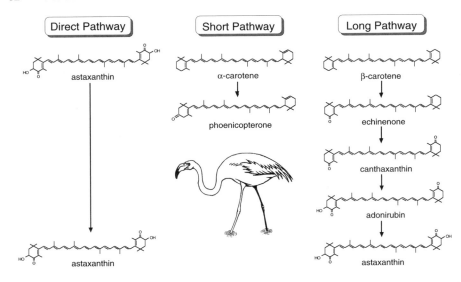

Figure 4.3. The metabolic pathways of carotenoid pigments in the Greater Flamingo proposed by Fox et al. (1967). To obtain red carotenoid pigments needed to fully color their plumage, flamingos can convert β-carotene to astaxanthin through a four-step pathway, convert α-carotene to phoenicopterone through a one-step pathway, or use dietary astaxanthin directly as a plumage pigment. Presumably, the cost of pigmentation is proportional to the number of metabolic steps involved.

primary pigment of African Oriole (*Oriolus auratus*) plumage, is absorbed efficiently from food, while α-carotene, which serves as neither pigment or pigment precursor, is not absorbed (Thommen 1971). Similarly, Völker (1938) found that Canaries (*Serinus canarius*) absorbed lutein and zeaxanthin and used these pigments to color their feathers, but that violaxanthin, β-carotene, and lycopene, which Canaries do not use as feather pigments, were not absorbed from food.

Selective absorption of carotenoid pigments from those pigments available in the gut seems to be a common characteristic of pigment physiology for the species of birds that have been studied. There are several important implications for the information content of carotenoid displays in the observation that birds are selective in their absorption of carotenoid pigments. First, there must be a cost associated with carotenoid absorption. If carotenoid absorption were completely passive and free of costs, then there would be no reason for carotenoid absorption to be selective. Second, the selectivity of carotenoid absorption suggests that birds have sophisticated physiological adaptations that enable them to use dietary carotenoids efficiently and to maximize carotenoid display. Birds appear able to efficiently absorb those pigments that are needed for ornament display and to void those pigments that are not useful in ornament display. To an evolutionary biologist this is perfectly obvious—under the pressure of intense sexual selection (documented in chapter 6), physiological mechanisms have evolved that maximize ornament display. To my knowledge, however, no physiologist has approached the study of avian carotenoid pigmentation with such an evolutionary perspective. I suggest that such a perspective will allow for better understanding of the pigment physiology of species with ornamental carotenoid displays.

There are as yet no data on carotenoid absorption by House Finches.

## Carotenoid Transport

After absorption from the gut, carotenoids are transported through the blood to the site of metabolism or deposition. In the late 1950s and early 1960s, it was discovered that carotenoids are associated with lipoproteins in human blood (Cornwall et al. 1962, Krinsky et al. 1958), and shortly thereafter it was demonstrated that birds also used lipoproteins to move carotenoids through their blood. In the first study of carotenoid transport in birds, Fox et al. (1965) found that 83% of total plasma carotenoids (primarily canthaxanthin and astaxanthin) of Roseate Spoonbills (*Ajaia ajaja*) was associated with high-density lipoproteins (HDL) in plasma; the remaining 17% was associated with low-density lipoproteins (LDL). Lipoproteins are molecules with a surface protein layer that encases a lipid core, in effect "hiding" the lipids and making the molecule hydrophilic (water-seeking). These molecules are known to play a critical role in the transport of non-polar substances like cholesterol through the blood. The descriptors "high-density" and "low-density" associated with lipoproteins refer to where these molecules occur in a centrifuge gradient. "High-density" and "low-density" are used to broadly classify lipoproteins, but within each of these general classes there may be many different lipoproteins.

In a more detailed study of carotenoid transport, Trams (1969) reported that in White and Scarlet Ibises (*Eudocimus ruber* and *E. albus*) about 92% of carotenoids in the plasma were carried by HDLs and the remaining 8% by LDLs. He proposed that pigments associated with HDL were hydroxy- and 4-keto-carotenoids that had been metabolized from dietary precursors by the birds, while pigments associated with LDL were dietary pigments. In more recent studies using mammalian models, it has been noted that hydrocarbon carotenoids (entirely carbon and hydrogen) tend to associate with LDL while more polar carotenoids (containing oxygen) tend to associate with HDL (Parker 1996). Thus in mammals, which commonly transport hydrocarbon carotenoids like β-carotene, most carotenoids are associated with LDL.

No research on carotenoid transport in birds has been conducted since the work of Trams (1969), despite the fact that Trams made several suggestions in his paper which should pique the interest of biologists interested in the signaling properties of carotenoids. First, HDL constituted about 17% of plasma proteins in Scarlet Ibises compared to about 1.5% in rats. Trams suggested that this might be an adaptation for greater lipid transport in birds, perhaps to aid powered flight, but one could also imagine that more HDL is an adaptation for more efficient movement of carotenoids that are needed to pigment feathers used in ornamental display. Moreover, Trams showed that the half-life for carotenoid pigments in plasma of Scarlet Ibises is about five days, which he also mentions is the plasma half-life for HDLs. Because carrier proteins have such a short half-life, a large amount of carrier protein has to be made by a bird every day. It is not hard to imagine how a wide range of condition factors, including nutrition and parasites (discussed in chapter 5), might have an impact on the ability of a bird to maintain production of HDLs and hence the capacity to maximally pigment plumage. Finally, Trams detected a specific HDL in the blood of Scarlet but not White Ibises that he proposed was a carotenoid-specific carrier protein unique to the Scarlet Ibis. As with absorption, specificity in carotenoid carrier proteins could perhaps increase the efficiency and speed with which carotenoids are moved from the site of absorption to the site of deposition (just as specialized carrier proteins utilize small quantities of steroid hormones such as testosterone very efficiently (Wingfield 1979)). Unfortunately, no study of carotenoid transport specificity has been conducted. Carotenoid transport remains among the least studied aspects of carotenoid pigmentation, but it is a field of study that holds great potential for improving our understanding of how carotenoid pigments serve as signals of individual quality.

## Carotenoid Metabolism: Selective and Efficient Use of Resources

As I discussed earlier, animals cannot synthesize carotenoids *de novo*; they must obtain them either directly or secondarily from plant tissues (Goodwin 1950, 1984, Völker 1938). Once ingested, however, carotenoids can be modified substantially by birds. Thus, birds can obtain the oxidized xanthophylls that color the plumage of many species either directly from their diet or, more commonly, through the metabolic conversion of precursors, such as lutein, that are obtained from the diet (Brush 1978, 1990, Rawles 1960) (Figure 4.4). From feeding experiments, and from a very few studies of dietary and feather pigments in wild birds, it is clear that many avian species obtain most or all of the pigment that they use to color their plumage from the metabolic conversion of dietary carotenoid precursors.

The metabolic pathways leading from dietary precursor to fully oxidized end-product can sometimes be inferred from the carotenoid composition of the blood,

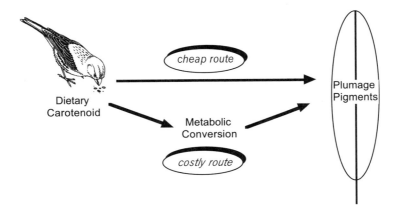

Figure 4.4. Like all vertebrates, House Finches are incapable of synthesizing carotenoid pigments. All pigments must be obtained in their diet. However, some dietary carotenoid pigments can be deposited unchanged in plumage, which is an energetically cheap pathway to plumage color. Other dietary carotenoid pigments must be biochemically modified one or more times, which requires energy and enzymes and hence is an energetically more expensive route.

liver, skin, and feathers (Fox et al. 1967). For example, the tissues of Chilean and Greater Flamingos (*Phoenicopterus chilensis* and *P. antiquorum*) contain five carotenoids (Fox et al. 1967). From the structure, concentration, and distribution of these pigments, combined with controlled feeding experiments, Fox et al. (1967) proposed two metabolic pathways by which feather pigments were derived from dietary pigments: (1) a one-step short pathway involving the oxidation of α-carotene to phoenicopterone (4-keto-α-carotene; Figure 4.3); and (2) a four-step, long pathway in which β-carotene is oxidized to echinenone (4-keto-diketo-β-carotene), thence to canthaxanthin (4,4′-diketo-β-carotene), adonirubin (3-hydroxy-4,4′-diketo-β-carotene), and finally to astaxanthin (3,3′-dihydoxy-4,4′-diketo-β-carotene) (Figure 4.3). Interestingly, these two pathways, like many of the known metabolic pathways in birds, change a yellow precursor molecule to an end-product that is red. Thus, flamingos potentially have three ways to obtain red carotenoid pigments that vary in their metabolic costs: (1) direct ingestion of red pigments (metabolically cheapest); (2) ingestion of pigments that can be altered into red pigments through a short metabolic pathway (more expensive); or (3) ingestion of pigments that can be altered into red pigments through a long metabolic pathway (most expensive) (Figure 4.3). One would predict that birds would utilize the cheapest pathway available, but this prediction has never been formally tested in any species (see "Energetic costs," p. 73 below).

Some support for the hypothesis that males will use dietary carotenoids in the most efficient manner comes from the observation that if a male can obtain a metabolic end-product in its diet, it will bypass metabolic pathways and use the dietary carotenoids directly. (Recall the quote from Goodwin 1984 that birds are "xanthophyll accumulators.") For instance, American Flamingos will convert β-carotenes to hydroxylated xanthophylls such as astaxanthin (pathway 3 in Figure 4.3) when necessary, but they will more readily absorb astaxanthin from the diet and use it to pigment their plumage directly (Fox and McBeth 1970).

Although birds are capable of substantially modifying the carotenoids that they ingest, not all carotenoids are suitable precursors for a particular end-product. There is strong evidence that α- and β-carotenes follow independent pathways (Fox and Hopkins 1966, 1967, Lee 1966) and that birds cannot convert between α- and β-carotenes and their derivatives (Brush 1968). α- and β-carotene and their derivatives are the most widespread and abundant carotenoids in plant tissue (Goodwin 1980, Goodwin and Britton 1988, Gross 1987, Mangels et al. 1993). Restrictions on the pathways available for the synthesis of various carotenoids are potentially of great importance in understanding the evolution of bright plumage. Commitment to a particular pigment or set of pigments as plumage colorants makes a portion of carotenoids in the diet unusable. Female preference may select for the color or color saturation of male plumage, but female choice cannot select directly for the metabolic pathway that produces the pigments. Thus, one would expect males to adopt the metabolic pathway or pathways that enable them to maximize color display. In other words, if β-carotene is the primary carotenoid available in the diet, then β-carotene or a metabolic derivative should be used to pigment the plumage. Likewise, if metabolic conversion of carotenoids is energetically costly, then the birds should use the most energetically efficient pathway that yields a required end-product (Figure 4.4). The examples of selective absorption of carotenoids that I reviewed appear to support this idea. However, the hypothesis that birds have physiologies adapted to optimize the use of the carotenoid pigments that are available in their diets remains to be tested.

### Carotenoid Metabolism: Studies of Cardueline Finches

The studies of carotenoid metabolism most relevant to plumage coloration in House Finches are the studies of Eurasian cardueline finches, including congeners of the House Finch, conducted by Stradi et al. (1995, 1996, 1997). Studies of carotenoid composition of plumage by Stradi and associates, extended and corrected earlier studies on plumage pigments of cardueline finches (Brockmann and Völker 1934, Lönnberg 1938, Völker 1934, 1938) including the House Finch (Brush and Power 1976). Prior to Stradi's work, quantification of plumage pigments was based on analytical procedures that relied on harsh extraction techniques (see House Finch pigment section, p. 74) and that did not allow positive identification of carotenoids. Erroneous information on plumage pigments can lead to incorrect deduction regarding the primary metabolic pathways of carotenoids in cardueline finches. In all species of cardueline finches with red plumage coloration that Stradi et al. studied, they found a similar mix of plumage pigments (Figure 4.5). The primary pigment of red species and the basis for their red coloration was the red pigment 3-hydroxy-echinenone, with smaller amounts of the red pigments astaxanthin, adonirubin, and canthaxanthin. Also present in small amounts were the isomers of 3-hydroxy-echinenone: 4-oxo-rubixanthin and 4-oxo-gazaniaxanthin. Although careful analysis of diet was not conducted, based on other published studies Stradi et al. (1996, 1997) proposed that the primary dietary carotenoids of cardueline finches with red plumage coloration were lutein, zeaxanthin, β-cryptoxanthin, rubixanthin, and β-carotene. Based on this assumption, they proposed the following metabolic pathways for the production of red carotenoid pigments in cardueline finches:

β-cryptoxanthin ⟶ 3-hydroxy-echinenone ⟶ adonirubin
(yellow/orange)　　　　　(red)　　　　　　　　(red)

zeaxanthin ⟶ astaxanthin
(yellow)　　　(red)

β-carotene ⟶ echinenone ⟶ canthaxanthin
(yellow/orange)　(red)　　　　　(red)

lutein ⟶ α-doradexanthin
(yellow)　　　(red)

rubixanthin ⟶ 4-oxo-rubixanthin
(yellow)　　　　　(red)

gazaniaxanthin ⟶ 4-oxo-gazaniaxanthin
(yellow)　　　　　　　(red)

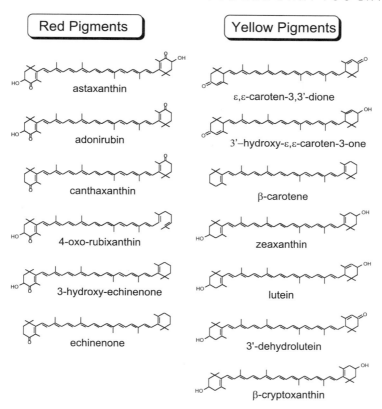

| Red Pigments | Yellow Pigments |

Red Pigments:
astaxanthin
adonirubin
canthaxanthin
4-oxo-rubixanthin
3-hydroxy-echinenone
echinenone

Yellow Pigments:
ε,ε-caroten-3,3'-dione
3'-hydroxy-ε,ε-caroten-3-one
β-carotene
zeaxanthin
lutein
3'-dehydrolutein
β-cryptoxanthin

Figure 4.5. The molecular structure of the thirteen carotenoid pigments found in the plumage of male House Finches collected in coastal California and Guerrero, Mexico. The six 4-keto-carotenoids listed on the left are red pigments. The remaining seven pigments are yellow. Adapted from Inouye et al. (2001).

Some yellow carotenoids found in plumage were presumed to be incorporated directly from diet, including lutein and zeaxanthin, but others were proposed to be metabolic derivatives from the following pathways:

lutein ⟶ 3-hydroxy-ε,ε-carotene-3,3'-dione (canary xanthophyll 1)

lutein or zeaxanthin ⟶ ε,ε-carotene-3,3'-dione (canary xanthophyll 2)

lutein or zeaxanthin ⟶ 3'-dihydro-lutein

All of the metabolic conversions of carotenoid pigments that contribute to the coloration of plumage are presumably enzymatically mediated (Brush 1990). However, little is known about the enzymes that are involved in the metabolism of carotenoids. At one extreme, one could propose that all of the enzymes that catalyze changes in carotenoid structure are generalist enzymes that catalyze a range of molecular conversions within the organism. For instance, Brush (1990) suggested that addition of an OH-functional group to carotenoid molecules might be the result of activity of such a generalist enzyme. At the other extreme, one could propose that each metabolic conversion of each carotenoid pigment that is involved in feather pigmentation requires a specialist enzyme that catalyzes only that reaction and that must be synthesized for that purpose. There has been too little detailed characterization of enzymes that function in carotenoid metabolism to definitively support or reject either of these hypotheses, but the studies that have been conducted suggest that there is at least some level of enzyme specificity in birds. For instance, female and basic-plumaged male Scarlet Tanagers (*Piranga olivacea*) have yellow body plumage pigmented by isozeaxanthin, while alternate-plumaged males have bright red body plumage pigmented by canthaxanthin (Brush 1967). This difference is the result of a single metabolic conversion of isozeaxanthin to canthaxanthin (Brush 1967). Presumably, males in the spring have an enzyme that facilitates this conversion, but this enzyme is lacking in females and males in the fall. Such an enzyme obviously cannot be a generalist enzyme with many functions in the organism, or it could not be lacking (or turned off) in females or in males for much of the year. Similarly, males of the finch genera *Loxia* and *Pinicola*, which have red plumage, have 4-keto carotenoids in their plumage that are lacking in females, which have drab yellow plumage. The 4-keto carotenoids found exclusively in males are metabolic derivatives of yellow dietary pigments, and Stradi et al. (1996) proposed that the presence of the metabolic end-products in the plumage of males but not females indicated that there was a 4-oxygenase enzyme found only in males. Again, such a sex-specific enzyme could not be of very general function or it could not be lacking in females.

The latest work by Stradi et al. (1997) suggests that the enzymes that catalyze carotenoid metabolism in red cardueline finches are neither generalists nor precisely substrate-specific. Rather, the enzymes that catalyze carotenoid metabolism appear to be specific to particular carotenoid end-groups and hence to groups of structurally similar carotenoid pigments. For instance, all of the red pigments found in the plumage of cardueline finches result from the oxidation of the β-end group of dietary carotenoids. Thus, a single enzyme might be responsible for all of the metabolic conversions from yellow pigments to red pigments listed above. The fact that α-doradexanthin is lacking from the plumage of these finch species, however, indicates some level of enzyme specificity. α-Doradexanthin is the oxidative derivative of lutein, which is found in relative abundance in the food and feathers of these birds. Despite having a suitable molecular structure, however, lutein is not metabolically converted into β-doradexanthin (Stradi et al. 1997).

The site of metabolic transformation of carotenoids in birds remains a matter of debate. The liver has long been implicated as the primary site of metabolism (Erdman et al. 1993, Schiedt et al. 1985), although others have proposed that metabolism occurs in the feather follicle itself (Schlinger et al. 1989). In a study of plasma color in relation to plumage coloration, my colleagues and I found support for the hypothesis that metabolic conversion of carotenoids occurs before carotenoids reach feather follicles (Hill et al. 1994a). We compared the coloration of plasma from male House Finches (scored by visual assessment of spun blood in hematocrit tubes) undergoing molt in the wild to the coloration of breast feathers that they were growing. We found a strong positive correlation—males growing redder feathers had redder plasma, presumably indicating greater quantities of red carotenoids being transported in blood (Figure 4.6). Given the finding of Stradi and others that red carotenoid pigments result primarily from metabolism of yellow dietary pigments, we are left with the conclusion that carotenoid metabolism must occur before the pigments reach their final destination. Why would redder males be transporting red pigments unless they were created at a site distant from the point of deposition?

Despite the fact that we found evidence that metabolic transformation of carotenoid occurs before carotenoid reach the follicle, Inouye (1999) found evidence that carotenoid metabolism does not occur in the liver. She extracted the carotenoid pigments from the livers of male House Finches just prior to molt, and she found no 4-keto (red) carotenoids. Only yellow apo-carotenoids were detected. So, the site of carotenoid metabolism in male House Finches and in birds in general remains unknown.

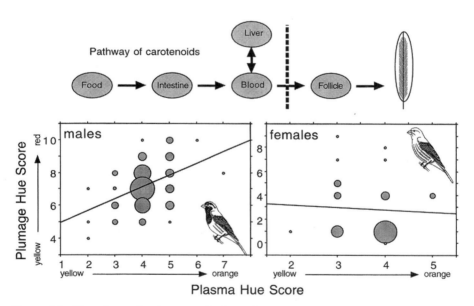

**Figure 4.6.** The relationship between the hue of growing feathers and the hue of plasma for House Finches captured in August in coastal California. Hue was quantified by visual comparison to a standard color reference (see text). Plasma color was assumed to be proportional to the concentration of circulating carotenoid pigments. All birds were undergoing prebasic molt at the time of capture. Plumage hue was scored for breast feathers for males and rump feathers for females. There was a strong and significant correlation between plasma hue and feather hue for males ($r_s = 0.42$, $n = 85$, $P = 0.0001$), but not females ($r_s = -0.05$, $n = 42$, $P = 0.74$). The observation that brighter males had redder plasma suggests that differences in color expression among males are due to processes that occur before deposition of carotenoids (i.e., due to processes to the left of the dashed line in the figure above the scatterplots). Adapted from Hill et al. (1994a).

*Carotenoid Deposition*

Once carotenoids reach the feathers, they are deposited in lipid droplets within the cells of feather follicles. As the maturing feather keratinizes, the lipid solvent disappears, leaving the carotenoid pigment dispersed in the keratin (Lucas and Stettenheim 1972). Unlike melanin pigmentation, which can form intricate bars, stripes, and other patterns within feathers, carotenoid pigmentation is rarely found to create any sort of specific pattern within individual feathers. The lack of pattern in carotenoid pigmentation within individual feathers is no doubt a result of lack of precise depositional control of carotenoid pigments by follicle cells. Carotenoids are added to growing feathers as amorphous substances (Brush 1990), and carotenoid-based plumage patterns result from different coloration of entire feathers, not parts of individual feathers. The role of deposition in controlling expression of plumage coloration, if any, is unknown.

*Carotenoproteins and Coloration*

An interesting but almost entirely neglected topic related to the physiology of carotenoid pigmentation of feathers is the binding of carotenoids to carotenoproteins and the effect that such association has on the coloration produced by the pigment. Virtually all carotenoids used as feather pigments bind to the keratin (protein) structure of the feather (Stradi et al. 1995), but this relatively weak binding of carotenoids to the protein surface of feathers is different than the formation of true carotenoproteins. Cheesman et al. (1967) define "true carotenoproteins" as compounds in which carotenoids are bound to proteins in stoichiometric proportions (i.e., with the proteins and carotenoids associated in specific ratios that reflect specific binding sites for the carotenoids on the proteins). Carotenoids bound to proteins in this fashion can have a spectral reflectance curve that is quite different than the spectral curve of the carotenoids in an unbound state (Thommen 1971).

The topic of carotenoproteins as it relates to feathers is virtually unstudied. The best known case of a color display resulting from a carotenoprotein is the blue carapace color of lobsters of the genus *Homarus*, which results from a carotenoprotein comprised of astaxanthin bound to a simple protein (Thommen 1971). When not bound to protein, as it usually appears in feathers, astaxanthin is a red carotenoid pigment. More recently, Stradi et al. (1995) discussed how the same mixture of carotenoids can produce both the yellow body plumage and red face of the European Goldfinch (*Carduelis carduelis*). They proposed that differences in the attachment of carotenoids to keratin caused a shift in the reflectance spectrum of the pigments and accounted for the different color displays. Whether this latter example involved the formation of true carotenoproteins is unknown. Because carotenoid-protein binding can change the hue of the color display, and, as we will see in later chapters, plumage hue determines female mating response, carotenoid-protein binding is a topic that holds huge potential for improving our understanding of the proximate control of plumage coloration in birds. It is a topic on which few biologists seem to be focused.

## From Food to Feathers

The process from absorption in the gut to deposition in the feather can be very rapid. Test (1969) found that when he supplied raw, grated carrots to Northern Flickers (*Colaptes auratus*) that had been deprived of carotenoids for several weeks, bright carotenoid pigments appeared in developing feathers in a few hours. As a result, there was an abrupt transition between the distal portion of the feather, which had keratinized before carrots were added to the diet and lacked pigments, and the proximal portion, which was brightly pigmented. The speedy assimilation of carotenoids is potentially of great significance in interpreting plumage variation and the function of bright plumage. Rapid deposition suggests that the diet during the period of molt can be very important to the pigmentation of feathers. The need for large quantities of carotenoids within a narrow time period that is more or less synchronized among males (fall or spring molt) presents a situation in which intrasexual interactions and foraging skills could influence expression of plumage brightness. In addition, the period of molt is an energetically stressful period for birds (Lindstrom et al. 1993, Murphy and Taruscio 1995). The added caloric requirements that molt imposes on a bird may make it more difficult for a male to accrue and metabolically alter sufficient carotenoids to fully pigment its plumage.

The need for daily or even hourly access to dietary carotenoid pigments could explain the patchy appearance of carotenoid-based plumage coloration seen in many species of birds, but particularly conspicuous in many male House Finches. Within the plumage of a male it is not uncommon to observe small groups of feathers that are a different hue or brightness relative to other feathers in the patch (e.g., an area of yellow in an otherwise red breast patch). One explanation is that these patches are the result of short-term limitations of dietary carotenoids. I found support for this hypothesis during a feeding experiment when a group of males being maintained on a low-carotenoid diet and growing yellow feathers were accidentally supplied with canthaxanthin (a red pigment) for one morning. This single pulse of carotenoids caused males to grow variable patches of red feathers in their otherwise yellow plumage and they looked remarkably like some of the wild birds that we handle. Alternatively, the different hue or brightness of a patch of feathers could be the result of utilizational effects such as parasites or access to energy, as described later in chapter 5. By recording the pigment symmetry of crown feathers, my lab group has begun to study the evenness of pigmentation in House Finches.

## Carotenoid Storage

The argument that access to carotenoids during the period of molt is critical to the expression of plumage coloration hinges on the assumption that stored carotenoids are relatively unimportant in pigmenting feathers. At least some species of birds under some circumstances, however, can store significant quantities of carotenoids and use the stored carotenoids to pigment their plumage. For instance, after feeding three species of African weaver finches (*Euplectes afra*, *E. nigroventris*, and *Ploceus cucullatus*) a carotenoid-deficient diet for three months, Kritzler (1943) found that

*E. afra* grew normally colored, yellow feathers, *P. cucullatus* molted into a slightly dulled, yellow plumage, and *E. nigroventris* grew pale, yellowish feathers in place of normal, bright orange plumage. All pigmentation was apparently drawn from carotenoid stores. Brush and Power (1976) found that male House Finches deposited small quantities of canthaxanthin in their plumage three months after feeding experiments involving this carotenoid had been terminated. In his study of Northern Flickers, however, Test (1969) found that two months on a carotenoid-deficient diet was sufficient to cause flickers to molt virtually unpigmented feathers. Furthermore, Thommen (1971) could find no evidence for carotenoid storage in Ring-necked Pheasants (*Phasianus colchicus*).

Despite the evidence for carotenoid storage in some species of birds, including House Finches fed canthaxanthin, I found no evidence for storage of carotenoid pigments that could be used as feather pigments in the House Finch (Hill 1992). I captured bright-red adult males just before the onset of their prebasic molt, and placed them on a diet containing few carotenoid pigments. These birds had more than eleven months in the wild to store carotenoid pigments. I predicted that if stored carotenoid pigments were used by males to color their feathers, then males that had the opportunity to store carotenoid pigments in the wild would grow redder plumage than males that had been held on low-carotenoid diets throughout the year. I found, however, that the males caught just prior to molt grew pale yellow plumage just like males held throughout the year on a low carotenoid diet (Figure 4.7). There is no evidence for significant storage of carotenoid pigments that can be used to color feathers in the House Finch or in any species of bird in the wild.

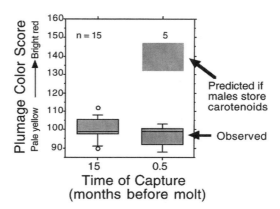

Figure 4.7. A test to determine whether male House Finches use stored carotenoids to pigment their plumage. Twenty males were held on plain-seed (low-carotenoid) diets during molt. Fifteen of these males had been held on low-carotenoid diets for 15 months prior to the experiment. Five males were captured just prior to molt and had essentially one year to store carotenoid pigments. There was no significant difference in the color of plumage grown by these two sets of males, suggesting that House Finches do not use stored carotenoids as an important source of plumage pigments ($Z = -0.08$, $P = 0.40$). See the "Key to Figures" box on page 48 for an explanation of this figure. Adapted from Hill (1992).

Inouye (1999) further confirmed that male House Finches do not use stored carotenoid pigments as an important source of plumage pigments during molt. She extracted and identified the carotenoid pigments in the livers of males just prior to fall molt. When males were held on low-carotenoid diets, the quantities of carotenoids in the livers of males dropped to nearly zero in seven days. These observations contradict predictions of the hypothesis that the liver serves as a source of stored carotenoid pigments during molt for male House Finches.

## Energetic Costs

Throughout this chapter and throughout this book, I invoke a non-trivial energetic cost for the creation of bright carotenoid-based color displays. I propose that there is a significant cost to metabolically converting dietary carotenoid pigments into pigments used to color plumage (apo-carotenoids into 4-keto-carotenoids). Consequently, I propose that it is significantly cheaper for birds to use dietary carotenoids directly as feather colorants than to make plumage pigments from dietary precursors (Figure 4.4). An obvious and important question is what specifically is energetically costly about putting red pigment molecules into feathers?

While there is no doubt that a small amount of energy is needed to drive the chemical reactions that change yellow pigments to red pigments, the activation energy required for the relatively few carotenoid molecules that are metabolized is certainly going to be trivial compared to the overall energy budget of the bird. Such a small energy investment is not likely to be limiting. I propose, however, that it is not the cost of driving chemical reactions that imposes an energetic burden on birds pigmenting their plumage with carotenoids. Rather, it is the creation and maintenance of the "machinery" needed for carotenoid absorption, transport, and metabolism that is energetically costly.

In the previous review of carotenoid physiology, in example after example, we saw how the processes involved in the production of plumage pigmentation were highly specialized and specific to certain pigments or groups of pigments. Thus, it appears that in absorbing, transporting, and metabolizing carotenoid pigments used in plumage, birds cannot rely on preexisting physiological systems that serve general functions. Rather, entirely new physiological systems must be created and maintained for the sole purpose of getting the right carotenoids into feathers in sufficient concentration to create an effective ornamental display. For instance, carotenoid absorption is selective, and such selective carotenoid absorption likely requires specific membrane proteins (Furr and Clark 1997). Similarly, carotenoid transport requires high- and low-density lipoprotiens that have a short half-life and have to be continuously synthesized (Trams 1969). Carotenoid metabolism requires enzymes that are specific to groups of carotenoids with similar molecular structure (Stradi 1998, Stradi et al. 1997) and because the carotenoid substrate needed for the reactions is likely present in small quantities, a relatively high concentration of enzyme will be needed if carotenoids are to be metabolized efficiently (Hill 1996a). It is the maintenance of these physiological systems needed to handle carotenoid pigments that I propose competes for energy that could be used for body maintenance and that imposes an energetic cost on birds. Testing this idea should be relatively straightforward.

## The Current State of Our Knowledge of Carotenoid Physiology

Carotenoid physiology as it relates to ornamental color display stands as a vast unexplored research frontier. To really understand how individual condition can affect expression of plumage coloration, we must have a better understanding of the mechanisms used by birds to absorb, transport, metabolize, and deposit carotenoid pigments. As summarized above, we know very little about any of these processes of carotenoid pigmentation in birds, and particularly passerine birds. Two of the most basic physiological processes in carotenoid utilization that remain fundamentally unstudied are the means by which carotenoids are transported from the gut to feathers, and where and by what means ingested carotenoids are altered metabolically. Is there one general lipoprotein carrier used to move all carotenoids? Is there a unique carrier for each type of carotenoid that is transported? Or are there a few different types of carrier lipoproteins for general classes of carotenoids? A parallel line of questions can be formulated by substituting "enzymes for carotenoid metabolism" in place of "lipoprotein carrier" in the previous sentences. Understanding the processes by which birds transport and metabolize carotenoids will help us to understand basic observations, such as the specificity of carotenoid uptake, and to better estimate the cost of carotenoid utilization. I have no doubt that a better fundamental understanding of how birds utilize carotenoid pigments will lead to a better fundamental understanding of how carotenoid-based plumage coloration functions as a signal in birds.

## The Pigmentary Basis of Ornamental Coloration in House Finches

The pigment mix that gives a male House Finch its ornamental coloration is complex. Brush and Power (1976) originally reported that ornamental plumage coloration in the House Finch resulted from three primary carotenoid pigments: the yellow/orange pigment β-carotene; the orange/yellow pigment isocryptoxanthin; and the red pigment echinenone (Figure 4.8). In addition to these primary pigments, Brush and Power also noted smaller quantities of pigments that they could not identify positively. Most interesting to me, Brush and Power reported that the plumage coloration of a male was a function of the relative proportion of β-carotene, isocryptoxanthin, and echinenone in its plumage. In the feathers of males with pale yellow plumage, they found only β-carotene and some unidentified xanthophylls—isocryptoxanthin and echinenone were not present; in the feathers of males with orange coloration they found β-carotene, unidentified xanthophylls, and isocryptoxanthin but no echinenone; and, in the feathers of males with bright red coloration they found β-carotene, unidentified xanthophylls, isocryptoxanthin, and echinenone. Thus, the study by Brush and Power (1976) indicated that the plumage hue of male House Finches was a function of the mix of three pigments, with plumage redness being determined largely by the concentration of the red pigment echinenone. Moreover, they proposed that the three primary carotenoids that they found in the plumage represented steps in a metabolic pathway with

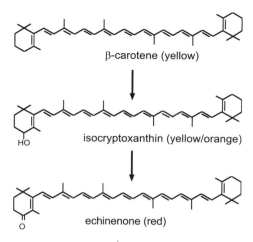

β-carotene (yellow)

isocryptoxanthin (yellow/orange)

echinenone (red)

Figure 4.8. The metabolic pathway of carotenoid pigments used by male House Finches to pigment plumage as proposed by Brush and Power (1976). This hypothesis has now been modified substantially (see Figure 4.11), but it influenced my approach to the study of plumage coloration in the House Finch.

dietary β-carotene being converted first into isocryptoxanthin, and then isocryptoxanthin being converted into echinenone.

As I mentioned in the preface, the study by Brush and Power (1976) was among the first to consider the pigmentary basis for variation in plumage coloration among individuals within a species, and it was instrumental in stimulating my research on plumage coloration. It was, however, conducted with analytical methods that have been much improved in recent decades. Brush and Power (1976) used thin-layer chromatography plates to separate carotenoids. The carotenoids were then identified by comparison to the spots produced by known carotenoid standards. They did not directly identify the carotenoids extracted from feathers, as is now possible, and because of the limitation of their analytical techniques, Brush and Power (1976) had to list many carotenoid plumage pigments as unidentified xanthophylls. So, although the study of Brush and Power (1976) was an important first step in the determination of the pigmentary basis for the variation in expression of ornamental plumage among male House Finches, it was not the definitive work on the topic.

In a series of three papers, Stradi and his co-workers at Universitá Degli Studi di Milano revolutionized the study of carotenoid pigments in feathers and, in particular, the feathers of cardueline finches (Stradi et al. 1995, 1996, 1997; see previous section). They used new gentle extraction techniques to release carotenoids from feather keratin, to which carotenoids apparently bind quite tightly. According to Stradi et al. (1995), the extraction techniques that had been used in previous studies of feather pigmentation were likely to cause chemical modification of the carotenoids being studied. They cautioned that previous studies may have identified extraction artifacts rather than the pigments found in the feathers of birds. Using their new gentle extraction technique, high performance liquid chromatography

(HPLC) to separate the carotenoids, and mass spectroscopy analysis to identify the carotenoids, Stradi et al. analyzed the carotenoid content of twenty-one species of cardueline finches from Eurasia.

They found that species with yellow plumage coloration, for example, Greenfinches (*Carduelis chloris*), European Siskins (*C. spinus*), or Serins (*Serinus serinus*), had primarily the yellow pigments ε,ε-carotene-3,3′-dione and 3-hydroxy-ε,ε-carotene-3,3′-dione (sometimes referred to in the literature as "canary xanthophylls") (Stradi et al. 1995). A variety of other pigments were present, but the canary xanthophylls were the most common pigments in all yellow species that were studied. In red species of cardueline finches from the genera *Loxia, Pinicola, Carduelis, Carpodacus*, and *Uragus*, they found fourteen carotenoid pigments (thirteen of these pigments are illustrated in Figure 4.5), but in all red individuals from these species they found that 3-hydroxy-echinenone was the most abundant plumage pigment (Stradi et al. 1996, 1997). In these studies, Stradi et al. focused on Eurasian species and did not analyze the plumage of House Finches.

In the summer of 1992, with Bob Montgomerie and Caron Inouye, I traveled to central California and southern Mexico to collect the feathers and gut contents from wild House Finches (Inouye et al. 2001). The House Finches collected at these two sites belonged to two different subspecies: C. m. *frontalis* in California and C. m. *griscomi* in Mexico (Hill 1996b, Moore 1939). The importance of the subspecies status of these populations and the important ways in which they differed will be discussed in detail in chapter 10. Here I will focus on individual variation in plumage pigments within these populations. Our goals in this study were to determine what pigments were in the diets and plumages of House Finches, and to gain a better understanding of the link between diet and expression of ornamental plumage coloration. We chose to conduct this study in late summer so that we could capture birds that were growing their ornamental plumage. Thus, we collected birds at the time when an individual's diet would have a direct effect on the coloration of its growing feathers. We also collected finches at a time of year when it was easy to distinguish hatch-year (HY) birds (birds that had hatched in the summer in which they were collected) and older (AHY) birds (birds that had hatched in a previous summer). Thus, we ended up comparing individuals within four groups of males: HY males and AHY males from each of the two populations.

We collected sixty-two HY and sixty-nine AHY males in San Jose County, California and fifty-nine HY and thirty-two AHY males in the vicinity of Chilpancingo, Guerrero, Mexico. Because these birds were collected three years before I started using a reflectance spectrophotometer for color quantification, we scored the color of male plumage by comparison to the *Methuen Handbook of Colour*. We sacrificed the House Finches that we captured and preserved their entire plumage and contents of their guts. Caron Inouye had the job of determining the carotenoid content of the feathers and gut contents of these birds. She ended up enlisting the help of Stradi and his techniques for precise identification of the carotenoid pigments of feathers as described above.

In the plumages of males from both age groups in both populations, we found the same basic mix of thirteen carotenoid pigments (Inouye et al. 2001; Figure 4.5), although not every male had all thirteen carotenoid pigments in its plumage, and the relative abundance of the various pigments differed among the populations and age groups. Overall, what we found in the plumage of male House Finches was very similar to what Stradi et al. (1996, 1997) found in the plumage of other red cardueline finches. The most abundant pigment in the feathers of HY and AHY males from both populations was the red carotenoid 3-hydroxy-echinenone, and the second most abundant pigment was the yellow carotenoid lutein (Inouye et al. 2001).

We used the results of these carotenoid analyses to answer three basic questions: (1) What is the relationship between the pigment composition of ornamental plumage and the redness and saturation of the plumage? (2) How do young and old males differ in the composition of carotenoid pigments in their plumage? (3) How do males from the two different subspecies differ in the composition of carotenoid pigments in their plumage? In the remainder of this chapter I am going to focus on the comparisons within populations, and hence the first two questions posed above. A comparison of the different subspecies of House Finches will be taken up in chapters 10 and 11.

## The Pigmentary Basis of Plumage Redness and Saturation

The thirteen carotenoid pigments that combine to produce the coloration of a male House Finch can be divided into those that have a ketone functional group at the fourth position at the β-end of the molecule (4-keto carotenoids) and those that lack this functional group (see Stradi 1998 for an overview of carotenoid biochemistry). The reason for grouping carotenoids as "4-keto-" or "other" is that the six 4-keto-carotenoids found in the plumage of House Finches are red pigments, while the other seven pigments are yellow or yellow/orange (Figure 4.5). Comparing the abundance of 4-keto and other carotenoids allowed us to investigate the importance of red pigments versus yellow pigments in determining the hue and saturation of plumage.

We considered plumage hue and plumage saturation separately in our study, and I will consider these two color dimensions in turn, starting with plumage hue. Our basic prediction going into this study was that we would verify with more detailed biochemical data what Brush and Power (1976) had reported twenty years before: the redness of plumage was a function of the proportion of red versus yellow pigments in the plumage. And indeed, that is what we found. In both populations, the abundance of most 4-keto (red) carotenoids was positively correlated with plumage redness; whereas, the abundance of most yellow carotenoid pigments was negatively associated with plumage redness (Inouye et al. 2001; Figure 4.9). Not surprisingly, the proportion of total red pigments versus total yellow pigments in male plumage was a good predictor of the hue of male plumage (Inouye et al. 2001; Figure 4.10). The only exception among age groups and populations was the plumage of adult males from the *griscomi* subspecies; in that group, there was no significant relationship between plumage redness and proportion of red pigments in plumage. However, virtually all adult *griscomi* males were bright red, so there may have been too little variation in plumage hue for a relationship between pigment composition and plumage redness to have been detected. The contribution of specific red pigments varied among populations and age groups

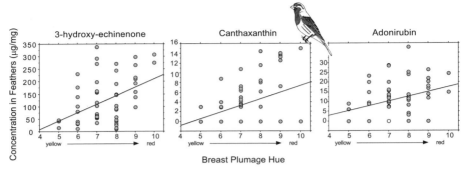

Figure 4.9 The concentration of three 4-keto-carotenoids (red pigments) in the plumage of male House Finches collected in coastal California in relation to the hue of their breast plumage. Hue was quantified by visual comparison to a standard color reference (see text). Not surprisingly, plumage redness is a function of concentration of red 4-keto-carotenoids deposited in plumage (3-hydroxy-echinenone: $r = 0.39$, $n = 27$, $P = 0.04$; canthaxanthin: $r = 0.57$, $n = 27$, $P = 0.002$; adonirubin: $r = 0.49$, $n = 27$, $P = 0.04$). Six of the six 4-keto-carotenoids found in the plumage of male House Finches collected in coastal California were positively associated with plumage hue, including the three illustrated above (see Inouye et al. 2001).

(Inouye et al. 2001), but as the most abundant red pigment, 3-hydroxy-echinenone consistently had a significant positive influence on plumage redness.

Our prediction regarding the relationship between the saturation of carotenoid-based coloration and the pigment composition of plumage was equally simple. We predicted that the concentration of total carotenoids would be positively correlated with color saturation. Unlike our results with plumage hue, however, our predictions regarding plumage saturation were not as clearly supported by our observations. Under the best of conditions, saturation is hard to measure by visually comparing feathers to a standard reference like the *Methuen Handbook of Colour*. Among 548 males for which I scored plumage coloration in Ann Arbor, 92% had saturation scores of 5, 6, or 7, and 72% had scores of 6 or 7 (Figure 3.6). Thus, with book scores, there is relatively little observable variation in plumage saturation to which comparisons can be made. To make matters worse, in our study of plumage pigments in California and Mexico, males were captured during molt and for many males too few new feathers had grown to allow for an accurate assessment of color saturation. For most males, we scored only the hue of incoming feathers. This left a small sample of males with saturation scores, and most of these had scores of 6 or 7. For these reasons, we decided to group males into a high-saturation group (score 7 or 8) and a low-saturation group (score 5 or 6) and to use a paired comparison to look for effects of pigmentation on plumage saturation. For males sampled in Mexico, the high-saturation group had a significantly higher concentration of total carotenoids, as was predicted. For males sampled in California, however, concentration of total carotenoids was not significantly different among high- and low-saturation groups, but concentration of red pigments was significantly higher in the high-saturation group. Thus, while it seems generally to be the case that greater pigment concentration means more saturated coloration, in our study of House

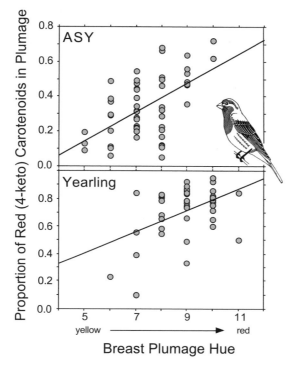

**Figure 4.10.** The hue of breast plumage of male House Finches collected in coastal California in relation to the proportion of 4-keto-carotenoids (amount of red carotenoids versus total carotenoids) in their feathers. Hue was quantified by visual comparison to a standard color reference (see text). The upper figure shows the relationship for ASY males ($n = 19$, $r = 0.48$, $P = 0.04$), and the lower figure shows the relationship for yearling males ($n = 39$, $r = 0.60$, $P = 0.0001$). Adapted from Inouye et al. (2001).

Finches the link between pigment concentration and plumage saturation was not nearly as strong as the link between proportion of red pigments and plumage redness. It would be instructive to repeat the study by comparing pigment concentration to saturation scores measured with a reflectance spectrophotometer that provides more accurate assessment of plumage saturation.

## Age and Pigment Composition

There were no differences in the pigment composition of the plumage of HY and AHY males in the *griscomi* population (the population from southern Mexico), but there were significant differences between HY and AHY in the *frontalis* population (see chapter 10 for a detailed description of these populations). This is a paradoxical result because *griscomi* males have delayed plumage maturation; yearling males have female-like plumage with relatively little ornamental coloration (Hill 1996b; see chapter 10). In populations of the *frontalis* subspecies, in contrast, males reach definitive plumage in their first fall (prebasic) molt and only differ in mean plumage coloration (see Figure 2.1). Nevertheless, yearling males of the *frontalis* subspecies differed from adults in having higher levels of the yellow pigments zeaxanthin and lutein and lower levels of the red pigments adonirubin and astaxanthin. Adonirubin and astaxanthin are thought to be metabolic derivatives of dietary carotenoids, while zeaxanthin and lutein are likely ingested directly (Inouye et al. 2001). Thus, the differences between adult and yearling males from the *frontalis* population appear to reflect differences in the capacity for yearling males of these populations to metabolically convert dietary carotenoid pigments, or perhaps differences in their abilities to ingest red pigments that can be used directly (without metabolic conversion) as plumage pigments. Male House Finches are very young (one to four months old) when they first produce ornamental plumage, so it would not be surprising if they differed from older males in their biochemistry and hence their ability to metabolically convert some carotenoid pigments. Alternatively, yearling males may have had access to fewer or different food resources than adult males, may have been subjected to higher parasitism than adult males, or they may have been impacted by some other environmental factor (e.g., poor nutrition) to a different degree than older males, which may have caused the difference in pigment composition (see chapter 5).

### Carotenoid Metabolism in the House Finch

Because the pigment composition of male House Finches is essentially the same as for the Eurasian finches studied by Stradi et al. (1996, 1997), the metabolic pathways that have been proposed for other red species of cardueline finches (see previous section) are likely to be the same as those pathways used by male House Finches to pigment their plumage. Especially important metabolic pathways for House Finches appear to be the conversion of $\beta$-cryptoxanthin to 3-hydroxy-echineone, $\beta$-carotene to canthaxanthin, rubixanthin to 4-oxo-rubixanthin, and zeaxanthin to astaxanthin (Inouye et al. 2001; Figure 4.11). One should bear in mind, however, that all of these pathways have been deduced from what carotenoids were observed in or presumed to be in food and what carotenoids were observed in feathers. All of these pathways remain to be confirmed with studies of radio-labeled molecules.

## Summary

The overall result of the plumage pigment study is rather trivial: redder birds have more red pigments as opposed to yellow pigments in their plumage. This is essentially what Brush and Power concluded in 1976 and it is what anyone with even the most rudimentary knowledge of carotenoids pigments would have guessed. There is, however, a bit more to the results of studies of the pigmentary basis of carotenoid color display than simply that the red birds are those that deposit more red pigments.

When Brush and Power (1976) presented the first interpretation of how plumage pigments create the yellow-to-red continuum seen in male House Finches, it was a simple story. There was a yellow pigment, there was an orange pigment, and

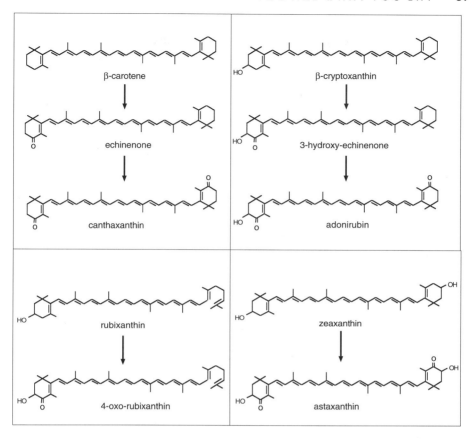

Figure 4.11. The primary metabolic pathways of carotenoid pigments used by male House Finches to pigment plumage, as proposed by Inouye (1999). This hypothesis is a modification of the single pathway proposed by Brush and Power (1976; see Figure 4.8). These pathways have been deduced from the carotenoid content of food and feathers; they have not been confirmed by tracing labeled molecules.

there was a red pigment. The red pigment was manufactured from the yellow pigment through the orange intermediate. Thus, the yellow, orange, and red pigments lay on a continuum from common (yellow) to rare (red), from accessible in the environment (yellow) to inaccessible (red), and from cheap to put in feathers (yellow) to expensive to put in feathers (red) (Hill 1996a). This simple scenario made it easy to see why birds in better condition would be redder: they would ingest more yellow precursors to produce more red pigments and they would have more energy available to convert yellow pigments into red pigments. The contrast of red (good, high quality) to yellow (poor, low quality) seemed to hold across diverse levels of analysis, from the whole bird to the individual pigment molecules.

Our more recent study of the carotenoid pigments in House Finch plumage complicates the picture considerably. First, there are many more pigments in the plumage of House Finches than Brush and Power suggested, and hence many more potential pathways to the same color end-point. Some of the yellow feather pigments are dietary (β-cryptoxanthin, β-carotene) and can be deposited unchanged; other yellow pigments appear to be the products of metabolic conversion of dietary pigments (canary xanthophylls and 3′-dehydro-lutein) (Inouye et al. 2001). Similarly, some of the 4-keto (red) carotenoids of House Finch plumage appear to be dietary pigments that have been deposited unchanged (astaxanthin and adonirubin), while others (echinenone and 3-hydroxy-echinenone) are almost surely the products of metabolic conversion of dietary pigments (Inouye et al. 2001). Thus, there appear to be many pathways for an individual House Finch to achieve any particular color display. To achieve bright red coloration, a finch can potentially ingest either the dietary precursors to red pigments and manufacture its own red pigment, or it can ingest red pigments and put them directly into its plumage. Based on the abundance of 3-hydroxy-echinenone (a metabolic derivative) in the plumage of virtually all male House Finches, it appears that all males use the pathway of carotenoid metabolism as a primary means of pigmenting their plumage. Data suggest that older males may be better or more efficient at this than yearling males. In addition, males have variable amounts of red carotenoid pigments that have been ingested and deposited in feathers unchanged. These pigments appear to have a significant positive effect on plumage coloration, so skill in acquiring these red pigments may play a role in expression of plumage coloration among male House Finches.

The costs of acquiring red plumage pigments from dietary precursors versus directly from red pigments in the diet are likely quite different. Red pigments that are ingested require no modification and hence no metabolic cost. These pigments simply need to be absorbed and transported to growing feathers (Figure 4.4). Such pigments appear to be rare in the environment (Hill 1996a), and they may invoke a substantial search time. In contrast, yellow carotenoid precursors that can be converted into red carotenoids for feather display appear to be much more abundant in food, but they have to be changed metabolically to red pigments. The cost of such conversion may not be trivial for birds.

Although we now know much more about the physiology of the carotenoid pigmentation of the feathers of passerine birds than we did just a few years ago, investigation of the topic is still in its infancy. Most aspects of the process of pigmenting feathers with carotenoids (in the House Finch or in any species of bird) remain unknown, and many fascinating questions remain to be answered. For those looking for a research topic related to visual signaling in animals, a study of the physiological costs of carotenoid absorption, transport, or metabolism would undoubtedly yield fascinating results and would add substantially to our understanding of how carotenoid-based color displays work. For instance, parasites may affect carotenoid metabolism making red dietary pigments particularly valuable to parasitized birds. Similarly, compared to older birds, first-year males may be less able to convert yellow pigments precursors to red plumage pigments, making red dietary pigments more valuable to young versus older males. These observations raise the interesting question, do birds change their pigment acquisition strategies based on their condition? I believe that studies on the strategies of carotenoid acquisition by male House Finches hold the potential to provide important new insights into how signal honesty is maintained in carotenoid signaling systems.

# 5    A Matter of Condition

## The Effects of Environment on Plumage Coloration

The very frequent superiority of the male bird or insect in
brightness or intensity of colour, even when the general coloration is
the same in both sexes, now seems to me to be, in great part, due
to the greater vigour and activity and the higher vitality of the male.
The colours of an animal usually fade during disease or weakness,
while robust health and vigor adds to their intensity.
>                               —A. R. Wallace (1878), in one of the earliest
>                        published suggestions that colorful plumage reflects
>                                               the condition of a male

A change of chemical composition will nearly always mean the
absorption of different rays of light and therefore a different colour;
but the quality of the latter, as measured by our aesthetic sense, will
bear no relation to the strain put upon the organism in producing
the pigment.
>                               —E. B. Poulton (1890), commenting on
>                         Wallace's idea that plumage color reflects the
>                                         excess vitality of males

It is difficult to say whether the change in color of the caged house
finch (*Carpodacus mexicanus frontalis*) from red to yellow is due
principally to a change in food, or to the confinement and general
deterioration of the system from captivity. Food, nevertheless, plays
some part in this, as well as in many changes in the color birds in a

wild state, which, with the present lack of experimental data, are far too complex even to be surmised.
> —C.A. Keeler (1893:229), in his rarely cited but very insightful *Evolution of the Colors of North American Land Birds*

*I'm often asked by my friends and relatives, "What do you like best about your job?" Although I could give a variety of answers—intellectual stimulation, pursuit of knowledge, interesting colleagues, and so on—my favorite and standard answer is "I get to work outside studying birds." This often leads to the follow-up: "What do you like least about your job?," to which I'm obliged to answer "I have to work outside studying birds."*

*Working with wild birds in the field and aviary is exciting, enlightening, challenging, and fun. It is also so unpredictable that to be successful and retain your sanity, you have to learn to expect impossibly disastrous setbacks. As a matter of fact, I find myself invoking Murphy's Law—anything that can go wrong will go wrong and at the worst possible moment—as a basic philosophy of field biology. I look with anxious curiosity and trepidation at what Murphy's Law will bring next. It rarely fails to disappoint.*

*Introducing the uninitiated to Murphy's Law is part of my duty in graduate training. All of my students have had to overcome Murphy and his games, but no one started out quite so naive or was pounded down quite so brutally by Murphy as Blue Brawner.*

*Blue signed up for my Advanced Ornithology course at Auburn University in 1996 as a non-thesis (coursework) master's student. He was working to increase his grade point average and biding his time as he waited to be accepted into vet school. A major requirement of this class is an independent project, and Blue was keen on conducting a study of animal disease that would tie into his impending veterinary training. I mentioned that I had wanted to study coccidia in House Finches for years, but that I had been told that coccidia were a difficult group of parasites with which to work. Blue accepted the challenge and for his class project he conducted an admirable study of coccidia in wild House Finches. Blue's class project was so impressive that I started discussing with him the idea of his switching to a thesis (research) master's program and expanding his studies of coccidia into a master's thesis. Blue had found that he really enjoyed research, and he jumped at the chance to conduct a master's thesis. The thesis, we decided, would involve a controlled infection experiment in which we could directly test the effect of coccidia on expression of plumage coloration.*

*Blue's timetable was extremely tight. Given his anticipated entry into the vet program, from the time he switched master's programs Blue had nine months to write a proposal, conduct his research, and write his thesis. In his class project Blue had made headway in establishing protocols for sampling coccidia in feces, but many techniques related to infection experiments had to be developed quickly. First and foremost he needed a drug with which to treat birds for coccidia. Colleagues at the vet school suggested amprolium, but when Blue tried this on finches it was completely ineffective against coccidial infection. The second drug he tried, pyrimethamine, turned out to be unstable in an outdoor environment. Suddenly a month was gone, and he didn't have an anti-coccidial drug that worked. He came to me dejected, relating his failures, implying that he had run into an insurmountable hurdle.*

*"Do you have other drugs to try?" I asked.*

*"Yes, several," Blue responded.*

*"I wouldn't get discouraged at this point," I said, invoking Murphy's Law. "I guarantee that you'll have bigger problems than a few failed drug trials before the project is done."*

*Within a week Blue had found a drug, sulfadimethoxine, that worked well in suppressing coccidal infection in finches, and he had the aviary full of birds ready for an infection experiment.*

*Suddenly birds started to get sick. From the grossly swollen eyes, we knew imme-diately that it was conjunctivitis caused by* Mycoplasma gallisepticum *(MG). This was a new House Finch disease that had first been reported two years earlier in Maryland and had spread southwest to our population. We had seen a few birds at feeders with the disease, but up to that point it didn't seem to be anything to worry about. Suddenly, though, it was a raging epidemic. Many of our captive birds broke out with the disease, and more than half of the wild birds that we were catching were symptomatic. It was an experimental biologist's worst nightmare. Just as we were initiating an experiment on the effect of a pathogen, a second infection was sweeping through the study population.*

*Blue made a valiant effort to keep MG out of the experiment, but we were caught in the middle of one of the worst epidemics ever recorded in a passerine bird. There was simply no way that a flock of MG-free finches could be established. After a few days of depression and lots of muttering to himself, Blue came to grips with the situation. Instead of a test of the effect of one pathogen on plumage coloration we would have a test of two pathogens.*

*In the end, the experiment was a success. We had enough birds without MG to effectively test the effect of coccidia on plumage coloration. Blue wrote a nice thesis, was accepted into vet school, and will be a practicing vet by the time these words are read.*

*The great (and terrifying) thing about Murphy's Law is that there are really no bounds. If, before Blue's study, my lab group had brainstormed about potential problems that might arise, I seriously doubt that anyone would have ventured: "Maybe a new infectious disease will jump from chickens to House Finches, sweep across the continent and into the study population, and confound all the effects of the experimental treat-ment." So we don't even try to second-guess what I consider to be one of the guiding laws of the universe. We just take Murphy as he comes.*

To understand the function and evolution of colorful plumage in male House Finches (and in birds in general), I had to understand the mechanistic basis for variation in male plumage brightness. What genetic or environmental factors enabled one male to have brilliant scarlet plumage while another male of the same age from the same population had pale yellow plumage? Did males simply inherit particular plumage coloration? It was well known that variation in plu-mage coloration in species with plumage polymorphisms like the Snow Goose (*Chen caerulescens*), the Ruff (*Philomachus pugnax*), and the Parasitic Jaeger (*Stercorarius parasiticus*), in which individuals either have dark or light body plumage, is determined by allelic variation at a single locus—individuals either inherit a dark plumage or they inherit a light plumage (Cooke and Cooch 1968, Lank et al. 1995, O'Donald 1983). In Snow Geese, which is the best studied of the polymorphic species, there are essentially no color intermediates, and the environment appears to have no effect on plumage coloration (Cooke and Cooch 1968). The plumage polymorphism of White-throated Sparrows (*Zonotrichia albicollis*), in which the crown stripes of both males and females are either black/white or tan/brown, is also under genetic control, but rather than reflecting allelic variation at a particular locus, the plumage polymorphism of this species reflects a chromosomal polymorphism (Thorneycroft 1966). If a White-throated Sparrow inherits a chromosomal type II$^m$, regardless of whether it is homozygous or heterozygous for that chromosomal type, it will display the white/black color morph (Thorneycroft 1966). It is interesting to note that all of these well-known, genetically controlled plumage-color polymorphisms involve melanin, not carotenoid, pigmentation.

Going into my study, I had reason to think that variation in plumage coloration among male House Finches was not under simple genetic control. First, male House Finches did not have discrete color morphs as is true of the species mentioned above. Rather, there was continuous variation from drab yellow to bright red (see this book's cover). Second, as I reviewed in chapter 4, I knew that the ornamental plumage coloration in the House Finch was due to carotenoid pigmentation and that Endler had shown in guppies (*Poecilia reticulata*) that individuals attained bright carotenoid-based coloration only if they had access to a high-quality diet (Endler 1980, 1983). Finally, I knew from the work of Brush and Power (1976) that the carotenoid content of the diet of House Finches during molt could greatly affect plumage coloration. Thus, I began my studies on the proximate control of plumage coloration in male House Finches with a focus on dietary access but with an understanding that carotenoid pigmentation was a complex process and that many factors were likely to influence the quality of ornamental coloration that an individual displayed.

## Carotenoid Acquisition: The Role of Dietary Access

### Feeding Experiments

I hypothesized that males with access to more carotenoids would produce brighter plumage, and I tested this hypothesis by means of aviary feeding experiments with House Finches captured in the wild. At the University of Michigan I was fortunate to have the use of large outdoor aviaries built in the early 1960s by Harrison Tordoff. These facilities were run down and in need of repair when I started my House Finch project in 1987, but with some painting, wire mending, and patching, I soon had them ready to house birds. The main cage complex in which I conducted most of my studies had six large flight cages (2.5 × 2.5 × 4 m) each connected by a window to an adjacent indoor room. There were also four smaller cages (2.5 × 2.5 × 2.5 m) that were not attached to a building. I used the larger flight cages to house both the birds in mate-choice trials (see chapter 6) and the birds in my experiments on the proximate control of expression of plumage coloration. In these latter experiments, I wanted to control access to dietary carotenoid pigments among groups of male House Finches during molt. The basic idea was to standardize an environmental variable that was hypothesized to be a key factor in determining expression of plumage coloration and then to see how much variation in plumage coloration remained.

House Finches replace all of their feathers once per year in the late summer and early fall through a process called the prebasic molt (see Figure 2.1). It is during this relatively brief period of feather growth that the plumage is pigmented. The fundamental coloration of feathers can be affected by environmental conditions only during the period of molt. After feathers are keratinized to become hardened, non-living tissue plumage color can change only by external processes such as the wearing away of feather edges that exposes different coloration beneath. Thus, my studies of the effect of carotenoid access on expression of plumage coloration were always conducted when birds were molting in the late summer/early fall.

The first feeding experiment that I conducted in 1988 was simple, and I was sure that I "knew" what the result would be. In July, about four to six weeks before finches started molting, I captured a group of male House Finches that included both yearling and older birds. I randomly divided males into two groups, with one group assigned to a low-carotenoid diet and the other group assigned to a high-carotenoid diet. All birds were provided with the same basic diet of sunflower seed and millet plus water treated with multivitamins. In addition to this basic diet, the carotenoid-supplemented group was provided with chopped carrots and sweet potatoes, both of which have abundant β-carotene but few other carotenoids (Mangels et al. 1993). To control for the potential benefits of daily fresh vegetables, the group on the low-carotenoid diet was given peeled, chopped apples, which have virtually no carotenoid pigments (Mangels et al. 1993). (See Hill 1992 for more details of the diet of birds in these experiments.) These birds were maintained on these specific diets until all birds had completed molt in mid-October. I then scored the plumage coloration of all males by comparison to the *Methuen Handbook of Colour* (Kornerup and Wanscher 1983) as described in chapter 3 (see Figure 3.4).

Brush and Power (1976) published a pathway by which they proposed that House Finches used dietary pigments to produce red coloration (see Figure 4.8). Based on the carotenoids that they found in the plumage and food of House Finches, they hypothesized that finches consumed primarily β-carotene, converted the dietary β-carotene into isocryptoxanthin, and then converted isocryptoxanthin into echinenone. The hypothesized metabolic pathways of carotenoids in red cardueline finches have been modified extensively since Brush and Power (1976) proposed their hypothesis (see chapter 4 and Figure 4.11), but when I started my feeding experiments, I had no reason to question the pathway that Brush and Power had published. Based on Brush and Power's pathway for the acquisition of red pigments, I was confident that the birds that I provided with abundant β-carotene would turn bright red by converting dietary β-carotene into echinenone. They didn't. Instead of growing feathers that were red, the males on the carrot-and-sweet-potato-supplemented diet grew feathers that were pale orange. They were significantly more orange than birds fed a plain seed diet, which were drab yellow, but they were not red (Hill 1992) (Figure 5.1). I was stumped. All the birds were very healthy with normal weight. Many of the birds would go on to live for four years in captivity. Their plumage was in great condition—strong and healthy with no fault bars. Why had birds supplemented with abundant carotenoids not turned red?

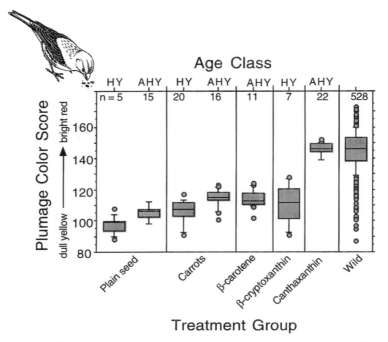

Figure 5.1. The effect of dietary access to carotenoid pigments during molt on the expression of plumage coloration in captive male House Finches. Codes above the box plots indicate whether males were molting from juvenal to first basic plumage (HY) and hence growing their first ornamental plumage, or molting from one basic plumage to another (AHY) and hence growing at least their second ornamental plumage. Below the box plots, the diet treatment of the group of males is given. Carrots provided males with β-carotene but few other carotenoids. β-cryptoxanthin was provided as tangerine juice. Canthaxanthin and β-carotene were provided as pure carotenoids in starch-gel beadlets added to drinking water. A box plot for wild males banded in Ann Arbor, Michigan, is presented on the right for comparison. For statistical comparisons of group means, see Hill (1992) and Hill (2000) from which this figure was adapted. Wild males exhibited significantly more variation in plumage color scores than did captive males after molt on any of the diets (plain seeds: $F = 0.27$, $P = 0.0007$; β-carotene: $F = 0.14$ $P = 0.0008$; canthaxanthin: $F = 1.0$, $P < 0.0001$).

The next year I repeated the experiment, but I abandoned natural food sources of carotenoids (carrots and sweet potatoes) and instead used carotenoid supplements donated by Roche Vitamin Co. As mentioned earlier, carotenoids are large hydrophobic molecules, so they float on water like oil. Moreover, they are photosensitive, making it very difficult to get them into the mouths of birds before they are degraded. Fortunately, Roche has developed a way to imbed carotenoids in starch-gel beadlets. This has the effect of making the carotenoids water-dispersible and of stabilizing the carotenoid molecules by making them less photosensitive. With the starch-gel beadlets I could simply add a standardized amount of carotenoids to water (although in the first experiment in which I used the starch-gel beadlets I also sprinkled them on chopped apples). The drawback was that only a few types of carotenoid pigments were available as starch-gel beadlets.

For this second feeding experiment, I again captured male House Finches in the summer before they began molt and divided them into three treatment groups. The first group was fed the basic seed diet plus chopped apples coated with canthax-anthin and water treated with the same. The second group was fed the basic diet plus chopped apples coated with β-carotene and water treated with the same. The third group was fed the basic diet and plain apples. As in the previous year, birds were maintained on their specific diets through the summer and fall until all indi-viduals had completed their prebasic molt. I then scored the coloration of plumage using the *Methuen Handbook of Colour*.

As described in chapter 4, canthaxanthin is a red carotenoid pigment that is found naturally in relatively small quantities in the plumage of House Finches (see Figure 4.5). Based on our subsequent study of House Finch plumage pigments (Inouye et al. 2001), one could propose that my feeding experiments would have been more robust if, rather than canthaxanthin, I had used a red carotenoid pigment that makes up a higher proportion of the red pigments naturally found in House Finch plumage, such as 3-hydroxy-echinenone. However, none of the other red pigments found in the plumage of House Finches were available in starch-gel bead-lets. Moreover, as I discussed at the end of chapter 4, one of the more interesting aspects of carotenoid pigmentation is the extent to which birds deposit red dietary pigments directly into plumage as opposed to using the metabolic derivatives of dietary pigments (Figure 4.3). From this perspective, canthaxanthin was an excel-lent choice to use in the feeding experiments.

In this second feeding experiment, males that were supplemented with canthaxanthin grew bright red plumage (although not as bright as the brightest wild males) (Figure 5.1). This observation showed that male House Finches can grow red plumage in captivity. In addition, males again grew pale yellow plumage on the plain-seed diet, and they grew pale orange plumage on the β-carotene-supplemented diet. As a matter of fact, males in the β-carotene-supplemented and carrot-and-sweet-potato-supplemented groups did not differ significantly in plumage coloration (Hill 1992). The observation that males grew the same color plumage when they were fed carrots and sweet potatoes and when they were fed a β-carotene supplement convinced me that male House Finches had not failed to achieve red plumage on a diet supplemented with carrots and sweet potato because of problems with the bioavailability of the carotenoids. Rather, when male House Finches ingested β-carotene, whether as a vegetable supplement or added directly to their diet, they appeared to deposit it unchanged in their plu-mage (but this was not confirmed by biochemical analysis). In contrast, in feeding experiments in which male House Finches were supplemented with β-carotene and then plucked to induce feather replacement, Brush and Power (1976) found isocryptoxanthin and echinenone along with β-carotene in feathers that were grown. However, the color of the feathers of these birds were still reported to be orange, so I assume that only small quantities of β-carotene were converted to red pigments.

I conducted one final variation of this sort of feeding experiment in Auburn in 1998. I caught juvenal birds in late summer before they had completed their first prebasic molt and divided them into three flocks each of twenty. Because I could not determine the sex of birds added to these groups, I ended up with approximately half the birds in each group being male, as described below. One group was maintained on plain seeds with no carotenoid supplementation, one group had canthaxanthin added to its drinking water, and one group had tangerine juice added to its drinking water. This experiment was conducted after I had read Stradi's papers showing that β-cryptoxanthin was the likely dietary precursor for the major red pigment in male cardueline finches (Stradi et al. 1997). Unfortunately, β-cryptoxanthin was not available in pure form or as starch-gel beadlets. However, in searching the extensive literature on the carotenoid content of fruits and vegetables, I found that tangerines have large amounts of β-cryptoxanthin and few other carotenoids (Mangels et al. 1993). Fresh tangerine juice was available at a local grocery store, so I added tangerine juice to the drinking water of the third group of finches (one part tangerine juice to four parts water) as a means of supplementing them with β-cryptoxanthin. I scored the plumage of males after they had completed their prebasic molt on these diets, and found that birds on the plain-seed diet turned pale yellow, birds supplemented with canthaxanthin turned red, and birds supplemented with β-cryptoxanthin were variably orange or red (Hill 2000) (Figure 5.1). None of these tangerine-juice-supplemented males was fully red, but several birds had patches of red/orange feathers.

Males fed tangerine juice during molt were more variable in plumage coloration than males in any other feeding experiment that I had conducted, and this was the first time that I had fed birds a yellow carotenoid pigment and had them grow reddish plumage. Not coincidentally, the experiment in which I provided males with tangerine juice was probably the first time I had supplemented captive birds with quantities of a suitable dietary precursor to red plumage pigments. My interpretation of the feeding experiment with tangerine juice was that, as proposed by Stradi et al. (1997) and Inouye et al. (2001), male House Finches metabolically converted dietary β-cryptoxanthin into a red pigment or pigments, probably 3-hydroxy-echinenone. Since I published the results of this study, biochemical analysis of the plumage of males from the tangerine juice-supplemented group confirmed that these males had indeed converted β-cryptoxanthin into 3-hydroxy-echinenone (R. Stradi, pers. commun.). In this captive feeding experiment, however, either too little β-cryptoxanthin was available in the diluted tangerine juice for males to manufacture sufficient red pigment to grow uniform red plumage, or some unknown environmental stressor prevented males from efficiently converting the β-cryptoxanthin that they ingested.

The most difficult part of these feeding experiments to explain remains the response by male House Finches to supplementation with large quantities of β-carotene. Brush and Power (1976), Stradi et al. (1997), and Inouye et al. (2001) proposed that β-carotene is poorly absorbed by red cardueline finches like the House Finch, but that the β-carotene that is absorbed should be metabolically converted into the red pigments echinenone and canthaxanthin. Thus, male House Finches fed large doses of β-carotene should have metabolized at least some β-carotene into red carotenoid pigments and grown a reddish plumage. We are left with the possibilities that either researchers are wrong, and House Finches are not capable of the proposed conversion of β-carotene to red pigments or that something in the captive environment kept the birds from performing this conversion (see Hudon 1994). One untested possibility is that coccidiosis, which I did not test for or treat in House Finches until 1997, prevented birds in the feeding experiments in Michigan from metabolizing β-carotene. Birds in the tangerine juice-supplementation experiment were maintained free of coccidiosis, and some of these birds grew red feathers.

There were other important observations from these feeding experiments. First, in the wild, young male House Finches average less red than older males (see Figure 2.3). In my feeding experiments, there was also a consistent tendency for young male House Finches to grow less colorful plumage than older male House Finches (Figure 5.1). None of the age differences was statistically significant (see Hill 1992), but the sample sizes in the comparisons were small. The effects of age on plumage expression in the captive experiments were about as large as the significant effects of age on plumage coloration observed among wild males. These observations provide potentially key insight into the basis for age differences in color display among male House Finches. Assuming that the age effects in the captive flocks were real, the color differences between these yearling and older males have to be attributed to differences in how the yearling and older males utilized carotenoids—their carotenoid access, after all, was standardized. This proposition that yearling and older male House Finches differ in their ability to utilize carotenoid pigments is also supported by the observations reported in chapter 4 that the pigment composition of the plumage of yearling and older males collected in the wild is different (Inouye et al. 2001). The relative roles of carotenoid acquisition versus carotenoid utilization in generating the differences in plumage expression among yearling and older males is a topic worthy of further study. Moreover, identifying the specific mechanisms of carotenoid utilization that differ between yearling and older males should help to identify those mechanisms that are most costly.

Another key result of these feeding experiments is that there was no relationship between the plumage color score of adult males at the time of capture (i.e., their natural coloration) and their relative plumage color after captive molt on a standardized diet (Figure 5.2) (Hill 1992). In other words, if males differed in their capacity to use the carotenoids that they were fed (this topic of carotenoid utilization will be discussed in more detail below), then I predicted that males that had drab plumage in the wild would grow relatively less colorful plumage than males that had bright plumage in the wild. This is not what I observed. When access to carotenoid pigments for a group of males was standardized during molt, all males converged on a very similar appearance. Bright males did not grow relatively brighter plumage, nor did drab males grow relatively drabber plumage on a fixed intake of carotenoids. As a matter of fact, with the exception of the tangerine juice-supplemented group, the variance in plumage coloration in the captive flocks was much less than the variance in plumage coloration observed in wild House Finches (Figure 5.1). Standardizing carotenoid access among males essentially eliminated variation in expression of ornamental plumage coloration.

It is important to point out, however, that in none of these feeding experiments were appropriate dietary precursors fed to males, except perhaps in small quantities in seed. Males directly deposited the canthaxanthin and β-carotene with which they were supplemented. If there are intrinsic differences among male House Finches in

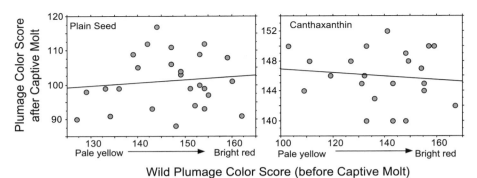

Figure 5.2. The plumage coloration of males at the time of capture in relation to their plumage coloration after undergoing captive molt on either a plain-seed diet (left) or a diet supplemented with canthaxanthin (right). There was no significant relationship between pre-manipulation and post-manipulation plumage scores for either group (plain seed: $n = 27$, $r^2 = 0.01$, $P = 0.59$; canthaxanthin: $n = 22$, $r^2 = -0.01$, $P = 0.60$). The failure of the wild plumage coloration of these males to predict the relative plumage brightness after molt on a standardized diet suggests that there are no intrinsic differences among males in their abilities to use dietary carotenoids. Adapted from Hill (1992).

their ability to use pigments to which they have access, it is likely that such differences involve the ability to modify precursor pigments into pigments for feather display. In the only experiment in which I provided males with a suitable dietary precursor (β-cryptoxanthin), only HY males were included so there were no pre-manipulation plumage scores. The critical experiment—testing how males that are bright and drab in the wild utilize suitable precursor pigments—remains to be conducted.

A result of my feeding experiments that I have not emphasized in published papers (Hill 1992, 1993a), but that is important for interpreting the response of birds to diet treatments, is that male House Finches showed a dose-dependent response to carotenoid supplementation. The pattern was most clear in experiments in which I fed canthaxanthin to birds. The quantity of canthaxanthin that I supplied to birds ranged from very high (more than 1000 mg/L), when it was supplied both on food and in drinking water, to relatively low (10 mg/L) when it was provided as a dilute solution in drinking water. The mean plumage coloration of males decreased substantially in response to the decreased supplementation of carotenoid pigments (Figure 5.3). Similarly, when I switched from a seed mix with relatively high carotenoid content to another with lower carotenoid content, the mean plumage redness of males dropped (see Hill 1992, 1994b). These results suggest that relatively large concentrations of carotenoid pigments are needed for birds to reach maximum plumage expression.

I have repeated the feeding experiments at various times using House Finches captured in Auburn, usually in preparation for other experiments. The results of these feeding experiments have been consistent with the results of my original feeding experiments. On a canthaxanthin-supplemented diet males turn red, and on plain seeds they turn pale yellow. I have not repeated the β-carotene supplementation since 1989. Taken together, these feeding experiments provide definitive

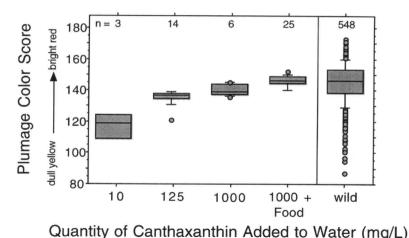

**Figure 5.3.** The response of male House Finches to supplementation with different doses of canthaxanthin during molt. Males in the right treatment group had canthaxanthin sprinkled on food in addition to the 1000 mg/L added to their water. There was a clear dose-dependent effect of carotenoid supplementation. The right box plot shows the distribution of plumage color scores for wild males in Ann Arbor, Michigan.

evidence that dietary access to carotenoid pigments can have an over-riding effect on expression of plumage coloration, at least in captive birds.

### Feeding Experiments with Other Songbirds

To see whether my observations of House Finches in captive feeding experiments would extrapolate to other species, Kevin McGraw and I conducted controlled feeding experiments with American Goldfinches and Northern Cardinals (*Cardinalis cardinalis*), songbirds in which males have bright yellow and bright red carotenoid-based plumage coloration, respectively. During fall molt, we fed male cardinals a diet of mixed seeds. During spring molt, we fed one group of male goldfinches a diet of sunflower seeds and millet, and a second group the same seed diet but with the red pigment canthaxanthin added to their water. As I had observed in House Finches, both cardinals and goldfinches that molted on plain seed diets grew feathers that were much less intensely pigmented than the plumage of wild male cardinals or goldfinches, respectively. Male cardinals that were fed only seeds converted the yellow dietary pigments β-carotene, β-crypto-xanthin, lutein, and zeaxanthin into the red carotenoids canthaxanthin, adonirubin, α-doradexanthin, and astaxanthin, and they grew a pale salmon-colored plumage (McGraw and Hill 2001). Male goldfinches that were maintained on seeds converted small amounts of the yellow dietary pigments lutein and zeaxanthin into yellow canary xanthophylls. Male goldfinches supplemented with canthaxanthin grew orange plumage, but the intensity of the coloration of their feathers was like the intensity of the yellow coloration of wild goldfinches. Thus, as in House Finches, access to carotenoid pigments during molt had a large effect on plumage coloration in these species (McGraw and Hill 2001) (Figure 5.4).

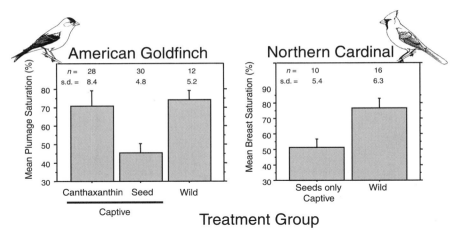

Figure 5.4. The intensity of ornamental plumage coloration of male American Goldfinches and Northern Cardinals after captive molt with standardized access to carotenoid pigments. All males were maintained on a diet of seed. Some males also had the red pigment canthaxanthin put in their water. On a plain-seed diet, male goldfinches grew pale yellow plumage that was significantly less saturated than the breeding plumage of wild males ($t > 14$, $P < 0.0001$); the plumage of male goldfinches fed canthaxanthin did not differ significantly from the plumage of wild goldfinches ($Z = 1.7$, $P = 0.1$). On a plain-seed diet Northern Cardinals grew significantly drabber carotenoid-based plumage than did wild males ($t = 10.4$, $P < 0.0001$). Interestingly, we found that the plumage variation of goldfinches fed plain seed ($F = 2.6$, $P = 0.07$) or canthaxanthin ($F = 0.83$, $P = 0.72$) or cardinals fed plain seed ($F = 0.74$, $P = 0.62$) was not significantly different than the plumage variation of wild male cardinals and goldfinches, respectively. Group sample sizes and standard deviations are given at the top of the figure. Adapted from McGraw and Hill (2001).

In other ways, however, cardinals and particularly goldfinches responded differently than House Finches when they had standardized access to carotenoid pigments during molt. First, after captive molt, male cardinals and goldfinches showed as much variation in expression of plumage coloration as is observed among wild male cardinals or goldfinches, respectively. Even when carotenoid access was standardized, and in the absence of parasites or nutritional stress, the manner in which individuals utilized carotenoid pigments generated substantial variation in expression of plumage coloration (McGraw and Hill 2001) (Figure 5.4). In contrast, after captive molt on either a high- or low-carotenoid diet, male House Finches showed significantly less variation in expression of plumage coloration than is observed among wild males in any population (Figure 5.1).

Second, the plumage coloration of a male goldfinch after captive molt was significantly positively correlated with its plumage coloration at the time of capture. Male goldfinches that were relatively bright in the wild were also the males that grew the brightest plumage coloration in the aviaries whether they were maintained on a high- or low-carotenoid diet (McGraw and Hill 2001) (Figure 5.5). In House Finches, there was no relationship between the plumage coloration of a male at the time of capture and its relative plumage coloration after captive molt (Figure 5.2).

The differences in the response of goldfinches and cardinals versus House Finches to carotenoid supplementation can likely be attributed to the types of pigments that the species were fed. Cardinals and goldfinches were fed yellow carotenoid pigments that are the precursor molecules used by wild birds. These pigments were modified metabolically into red or yellow plumage pigments, and likely it was the efficiency of these metabolic conversions that played a large role in generating variation in plumage coloration. House Finches, in contrast, apparently do not typically modify β-carotene, canthaxanthin, or the carotenoids provided by millet and sunflower seeds (see section above). Thus, the pigments used in captivity by House Finches were all routed through the energetically cheap pathway of Figure 4.4, while the pigments used by captive cardinals and goldfinches were routed through more energetically expensive pathway. This could explain many of the differences between the observed responses of goldfinches and cardinals and House Finches. However, this argument cannot be used to explain the response of goldfinches to supplementation with canthaxanthin. Like House Finches, American Goldfinches deposit canthaxanthin directly in their feathers (McGraw and Hill 2001). Nevertheless, after molt on a canthaxanthin-supplemented diet goldfinches displayed variable saturation of plumage coloration and males that had been bright in the wild grew more colorful plumage than males that had been drab in the wild (Figure 5.5).

Before we will achieve a basic understanding of the role of carotenoid access in expression of plumage coloration, we will have to have more detailed studies of the processes involved in pigmentation within individuals of a given species, and more cross-species comparisons.

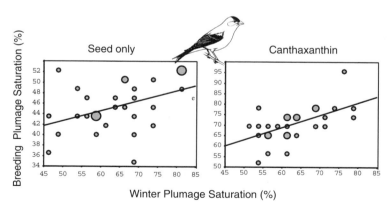

Figure 5.5. The plumage coloration of male American Goldfinches at the time of capture versus their plumage coloration after undergoing captive molt on either a plain-seed diet (left) or a diet supplemented with canthaxanthin (right). For both groups there was a significant positive relationship between a male's plumage score at the time of capture and the brightness of feathers that it grew under standardized conditions in an aviary (plain seed: $n = 30$, $r^2 = 0.17$, $P = 0.02$; canthaxanthin: $n = 28$, $r^2 = -0.32$, $P = 0.002$). Adapted from McGraw and Hill (2001).

*Patch Size, Pigment Symmetry, and Feeding Experiments*

In my feeding experiments I focused on the plumage color scores of males, but I also recorded the effects of controlled diets on patch size and pigment symmetry. As with plumage coloration, the dietary treatments significantly affected patch size—males maintained on a plain seed diet grew smaller patches of color than males fed β-carotene. In turn, males fed β-carotene grew smaller patches than males supplemented with canthaxanthin (Figure 5.6) (Hill 1992).

Variation in pigment symmetry was essentially eliminated by diet treatments. On plain-seed diets, males grew perfectly symmetrical plumage that was drab yellow, while on canthaxanthin-supplemented diets males grew perfectly symmetrical plumage that was bright red. The only feeding experiment in which I observed substantial pigment asymmetry was an experiment in which individuals were food stressed during molt. I will discuss this experiment in detail in the section on carotenoid utilization.

Thus, plumage coloration, patch size, and pigment symmetry responded differently to manipulations of access to carotenoid pigments. The observation that different components of the carotenoid-based plumage display respond in unique ways to environmental stress has important implications. It suggests that expression

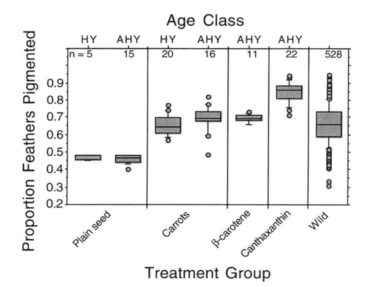

Figure 5.6. The effect of dietary access to carotenoid pigments during molt on the extent of ventral plumage with carotenoid pigmentation (patch size) in captive male House Finches. Codes are the same as in Figure 5.1. For statistical comparisons among groups see Hill (1992). Dietary access to carotenoid pigments had a large effect on expression of patch size just as it did on expression of plumage coloration. However, more variation in expression of patch size than plumage coloration persisted among groups after treatment (coefficient of variation: plain seed treatment: 6.3% plumage color, 14.8% patch size; β-carotene treatment: 4.7% plumage color, 10.3% patch size; canthaxanthin treatment: 2.6% plumage color, 16.0% patch size).

of different components of a color display might reflect different aspects of condition (Badyaev et al. 2001, Wedekind et al. 1998). By paying attention to several aspects of a color display, females may gain a better overall assessment of male condition. The manner in which females assess multiple ornamental traits is a topic of great theoretical (Iwasa and Pomiankowski 1994, Møller and Pomiankowski 1993) and growing empirical interest (McGraw and Hill 2000b), and it is certainly a topic worthy of further study in the House Finch.

## Plumage Redness and Carotenoid Access in the Wild

Until recently, I had to assume that the observations that I made of captive male House Finches in feeding experiments could be extrapolated to the wild. In the lab, carotenoid access had an overriding effect on expression of plumage coloration. Males could be changed from red to yellow and yellow to red simply by modifying their access to carotenoid pigments during molt. It seemed inconceivable to me that access to carotenoid pigments was not also affecting expression of plumage coloration in wild male House Finches.

When Bob Montgomerie, Caron Inouye, and I collected males for our studies of the pigmentary basis for plumage coloration in male House Finches (see chapter 4), we had the opportunity to test the basic idea that males with access to more dietary carotenoids grow redder plumage than males with access to fewer dietary pigments. As described in chapter 4, in San Jose, California, and Guerrero, Mexico, we collected male finches that were in the process of growing ornamental feathers. We scored the coloration of incoming feathers, and we removed the entire gut contents of the males. Later in the lab, Caron Inouye extracted the carotenoids from the gut contents and calculated the concentration of carotenoids in the samples taken from each male. We then simply correlated the redness of growing feathers to the concentration of carotenoids in the gut of each male. In California, we found a weak but positive correlation; males with a higher concentration of carotenoids in their gut tended to grow redder feathers (Hill et al. 2002; Figure 5.7). In Guerrero we found no significant relationship between concentration of dietary carotenoids and plumage hue.

One could look at the weak relationship between plumage hue and concentration of carotenoids in the gut and dismiss the pattern as unimportant or even as evidence against the idea that access to dietary carotenoids plays a role in determining expression of plumage coloration. In my view though, in this sort of analysis, the best we could have achieved is a weak relationship, even if access to dietary carotenoids is very important for wild males. Consider that plumage with ornamental coloration takes about thirty days to grow and the digestive system of a small bird like the House Finch fills and empties several times per day. That means that a hundred or more meals are ingested during the period of feather pigmentation. In our study we sampled one meal per bird and used that to characterize carotenoid access for the entire molt period. It is hardly surprising that the pattern that we observed was noisy. In addition, for this analysis we had to pool the total concentration of all carotenoid pigments—there was not enough food removed from the guts of individual birds to allow us to calculate the concentration of each component carotenoid. As I discussed in chapter 4, not all carotenoids are of equal value,

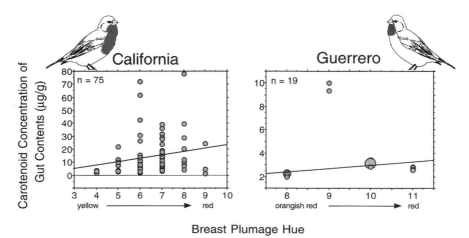

**Figure 5.7.** Dietary access to carotenoid pigments in relation to the hue of growing breast feathers for male House Finches collected in San Jose, California, and Guerrero, Mexico. Hue was quantified by visual comparison to a standard color reference (see text). There was a significant positive relationship in California ($r_s = 0.24$, $P = 0.04$) and a nonsignificant positive relationship in Guerrero ($r_s = 0.41$, $P = 0.08$). Adapted from Hill et al. 2002.

so by not looking at component carotenoids we undoubtedly lost important variation among individuals. A final reason why we should have expected a weak relationship between carotenoid concentration of food in the gut and color of growing feathers, is that a strong correlation between the carotenoid concentration of gut contents and plumage coloration would have meant that dietary access to carotenoids was the primary source for variation in plumage coloration in male House Finches. As I will discuss in the following sections, other environmental factors in addition to dietary access to carotenoid pigments are likely to have large effects on expression of plumage coloration.

The fact that we found a significant pattern despite the inherent problems with the study makes me believe that differences in carotenoid access among males really do play an important role in creating differences among males in ornamental coloration. The lack of a significant correlation between dietary carotenoids and plumage coloration in the Guerrero population could mean that access to dietary carotenoids is less important in that population, but the effect of gut carotenoid concentration on plumage coloration was at least as strong in Guerrero as it was in California—it was the small sample size that prevented this effect from being significant in the Guerrero population. Finding the same basic effect of dietary carotenoids on plumage coloration further strengthens the argument that the effect is real.

## Carotenoid Access and Color Display in Other Taxa

A few studies on species other than House Finches support the idea that carotenoid pigments are limiting resources for wild birds. The best study to date was a joint

effort conducted by field ornithologists (Slagsvold and Lifjeld 1985) and bioche-
mists (Partali et al. 1987) who combined field observations and laboratory carotenoid analyses in a study of plumage coloration in Great Tits (*Parus major*). These
authors began with the basic observation that the plumage of nestling Great Tits
from deciduous forest environments was yellower than the plumage of nestling
Great Tits from coniferous forest environments (Slagsvold and Lifjeld 1985).
They conducted cross-fostering experiments and showed that the differences in
chick plumage color were due to differences in the environments in which they
developed, not to genetic differences between Great Tits in the two habitats
(Slagsvold and Lifjeld 1985). When they analyzed the carotenoid content of the
foliage, insects, and nestling Great Tits in the two habitats, they found that the
primary pigments found in leaves and caterpillars were lutein, β-carotene, and
zeaxanthin (Partali et al. 1987). The feathers of nestling Great Tits were pigmented
only by lutein and zeaxanthin (Partali et al. 1987), and the differences in coloration
were a consequence of chicks in the coniferous forests getting fewer caterpillars and
hence fewer carotenoid pigments in their diet than chicks in deciduous forest. This
remains the most convincing field demonstration that carotenoid access can affect
expression of plumage coloration.

Another outstanding field study showing that dietary access to carotenoid pigments affects expression of ornamental coloration comes from a study of guppies—
small fish with carotenoid-based integumentary coloration. Grether et al. (1999)
sampled the carotenoid content of fish integument and algae (the primary food of
the fish) across an environmental gradient that was created by variation in the
amount of sunlight that reached the stream. They found that guppies living in
high-light environments had access to more algae and hence more carotenoid pigment and that they had brighter integumentary coloration than did guppies living in
low-light streams. Thus, the original hypothesis of Endler (1983) that access to
scarce carotenoid resources controlled expression of integumentary pigmentation
in guppies was verified. No equivalent study looking at change in plumage coloration and carotenoid access across an environmental gradient has yet been conducted
for birds.

Two studies have shown that the introduction of a new type of carotenoid
pigment to the diet of a wild population of birds can change expression of carotenoid coloration. The best known and best documented case concerns the Cedar
Waxwing (*Bombycilla cedrorum*), a species in which both male and females have a
bright yellow terminal band on the tail. In the middle part of the twentieth century,
Cedar Waxwings with orange rather than yellow tail bands began to be observed in
the midwestern and northeastern United States, and this change in plumage coloration coincided with the introduction of Morrow's honeysuckle, a shrub that produces small fruits that waxwings eat during molt (Hudon and Brush 1989, Mulvihill
et al. 1992, Witmer 1996). Biochemical analyses of waxwing tail feathers showed
that orange tails differed from yellow tails only by the presence of rhodoxanthin,
which is a primary carotenoid of Morrow's honeysuckle fruits (Hudon and Brush
1989). Feeding experiments with captive waxwings demonstrated that, when their
diet was supplemented with Morrow's honeysuckle fruits, Cedar Waxwings grew
orange tails (Witmer 1996). Thus, a change in dietary access to carotenoids in a
population of wild birds changed expression of ornamental plumage coloration. A

similar but less-well-documented case was reported for Bananaquits (*Coereba fla-veola*) on a Caribbean island in which a change in plumage coloration from yellow to orange was linked to a new dietary pigment probably produced by an introduced plant (Hudon et al. 1996).

Additional evidence that access to carotenoid pigments limits expression of plumage coloration in wild populations of birds comes from a study of Northern Cardinals (*Cardinalis cardinalis*) by Linville and Breitwisch (1997). They scored the redness of cardinal plumage over three years. Two of those years appeared normal, with typical production of fruits on which cardinals foraged and from which they obtained the carotenoids for plumage pigmentation. In the winter preceding the third year, however, an unusually severe cold spell caused a significant reduction in the amount of wild grapes (*Vitis* sp.) available to molting males. The result was that males grew significantly drabber plumage coloration in the year when fruit was scarce than in the years when fruit was available.

My studies on geographic variation in expression of carotenoid-based plumage coloration in the House Finch (presented in detail in chapter 10) also provide indirect evidence in support of the idea that access to carotenoid pigments affects plumage coloration. I found that male House Finches from different local popula-tions had very different mean color expression, with the most striking observation being that virtually no males in a population introduced to the Hawaiian islands had bright red coloration—they all had yellow or orange plumage coloration (Hill 1993a). While it is possible that a difference in parasite load or some other unex-amined variable caused Hawaiian males to be drabber than males from other popu-lations of finches, the most likely explanation was that males on the Hawaiian islands had access to fewer carotenoid pigments during molt than males from other populations (Hill 1993a). Unfortunately, I was unable to conduct the critical test of this idea by measuring the carotenoid content of the diet of males in various populations.

These studies show that, at least in some cases, dietary access to carotenoids can affect expression of plumage coloration in wild populations of birds.

## How Could There Be Differential Access?

I'm often asked by both my professional colleagues who study carotenoid pigmen-tation and the general public when they are presented with my House Finch work how some finches in a population could have greater access to carotenoid pigments than others. House Finches are highly social birds. They feed in flocks composed of males and females, young and old. They are non-territorial, and males are not particularly aggressive at food sources. They squabble to hold their place at a bird feeder, but they do not exhibit the sort of elevated aggressive behavior that one would expect if access to a particular limited resource determined the future reproductive success of males. As Zahn and Rothstein (1999:40) stated in their criticism of the hypothesis that dietary access to carotenoids can affect expression of plumage coloration in wild male House Finches: "House Finches feed in flocks, and there is no direct evidence that some males have access to certain dietary resources from which others are excluded."

I have always thought that the manner by which differential access to carotenoid pigments arises in wild populations of House Finches has not so much to do with how well individuals compete with each other for access to resources as it does with how well individuals deal with limited time and a challenging environment. Male House Finches have to grow their feathers at the same time that they pigment them. This means that during the molt period a male House Finch not only must secure the carotenoid pigments needed for ornamental display, but even more importantly, it has to ingest the nutrients and calories that are needed to grow strong feathers that will last for twelve months. A brightly colored male with structurally poor feathers will be dead before it ever gets a chance to display its colors to a female. The priority for males must be first to obtain the resources needed for the creation of healthy feathers and then to seek out the carotenoid pigments needed for ornamental display.

If the food sources that contain the best nutrition for feather growth are the same as those that hold carotenoid pigments, then there would be no conflict between foraging for nutrients and foraging for carotenoid pigments. The individuals that ingested the most food would necessarily also ingest the most carotenoid pigments. This may be how carotenoid acquisition works in the House Finch. However, I propose that for male House Finches the best sources of nutrition often are not the best sources of carotenoid pigments. This being the case, male House Finches will be faced with a time-budget problem. Only males that forage efficiently will be able to both ingest the basic resources needed for feather growth and find sufficient dietary carotenoids for maximum color display. The fact that House Finches feed together in flocks would have little to do with individual access to carotenoid pigments. The fact that some yearling and older males molt at different times, and that adult males must trade off late season breeding with early molt, however, could have a substantial effect on carotenoid access for these groups of birds.

There is some information available on the diet of House Finches, and House Finches appear to alter their diet during molt in a manner that is consistent with foraging for carotenoid pigments. In 1907, Beal published a study of the stomach contents of 1,206 House Finches collected in every month of the year in coastal California. Most of the food ingested by House Finches throughout the year was small seeds, but they also ate a fair amount of fruit. Beal (1907) summarized the results of his study in a table giving the composition of the diets of House Finches for each month of the year as the proportions that were weed seed, fruit, animal, or miscellaneous.

I plotted these proportions of food components by month (Hill 1995b) (Figure 5.8). Animal and miscellaneous foods combined never exceeded 3% of their diet; fruit and weed seeds made up virtually all of the diet. Consumption of fruit changed greatly over the course of the year (Hill 1995b), and it peaked during the period of molt. My first reaction to this pattern was that fruit must be primarily available from June to October and that finches simply shifted to fruit when it was available. However, according to Beal (1907), in central and southern California at the beginning of the twentieth century, fruit was available year-round and the shift in fruit consumption was not simply in response to a change in fruit availability. He wrote that "for some reason best known to itself [the House Finch] selected fruit" from June to October.

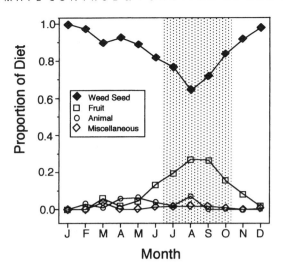

Figure 5.8. The proportion of the diets of House Finches composed of animal matter, weed seeds, fruits, or miscellaneous plant matter in each month of the year. Data from Beal (1907) based on analysis of 1206 stomachs of House Finches collected in central and southern California. The shaded region indicates the period of body molt in House Finches. From Hill (1995b).

A shift to a higher proportion of fruit in the diet during the molt period is predicted if, during molt, finches have to seek out sources of carotenoid pigments. Seeds are a good source of the calories, proteins, and minerals needed for body maintenance and growth of new feathers. In captivity, House Finches grow normal, healthy plumage at maximum rate when maintained exclusively on a diet of seeds. Most seeds, however, are a poor source of carotenoid pigments (Brockmann and Völker 1934, Brush and Power 1976, Goodwin 1980). In terms of providing nutrients versus carotenoids, fruit is virtually the opposite of seed. Fruit is mostly sugar, so it is a good source of calories but a poor source of the proteins and minerals needed for maintenance and feather growth. Most fruit, however, has high concentrations of carotenoid pigments (Gross 1987, Mangels et al. 1993). Thus, there is at least some evidence that finches use seed as their primary source of nutrition, but that they must also ingest fruit and perhaps other foods to obtain carotenoid pigments during molt.

One of the most neglected areas of research related to the function of color display, not just in the House Finch but in any bird species, is a detailed study of the sources of dietary carotenoid pigments. Knowledge of the sources of dietary carotenoids would lead to a better understanding of the potential challenges that birds face in obtaining the carotenoids that are needed for ornamental display.

## A Preference for Red Foods

If male reproductive success is determined largely by carotenoid coloration and carotenoid access plays a key role in carotenoid display, then we would expect male House Finches to seek out foods rich in carotenoids during molt. Carotenoids are colorful molecules, so their presence in foods is signaled by the coloration of food (Gross 1987). Thus, a basic prediction of the hypothesis that dietary access to carotenoid pigments is important to expression of ornamental coloration in House Finches is that males should seek out red foods during molt.

This hypothesis was tested in a series of experiments by a graduate student at Auburn, Cathy Stockton-Shields (1997). She presented male and female House Finches and male House Sparrows with a choice of peeled, diced apples that were colored either blue, yellow, or red. She then measured the quantity of each color of apple that they consumed. She conducted these food-preference experiments in both the spring, when neither species was molting, and during fall when both species were molting. House Sparrows have no carotenoid-based plumage coloration, and they were included in this study as a control for the possibility that all birds prefer red foods.

As predicted, in the fall during molt, male House Finches showed a preference for red over yellow and yellow over blue apples, while male House Sparrows showed no color preference (Stockton-Shields 1997) (Figure 5.9). Somewhat contrary to predictions, however, male House Finches also showed a preference for red apples during the spring, while female House Finches showed a preference for red apples in both spring and fall (Stockton-Shields 1997). The preference for red apple chunks by males in the spring may have been a carry-over effect of selection on male House Finches for red food in the fall. The preference for red foods shown by females, however, makes sense if carotenoid pigments are a critical resource to females just as they are a critical resource to males. The topic of female carotenoid coloration and the importance of carotenoid resources to females will be taken up in detail in chapter 9.

Overall, the observations of Stockton-Shields support the hypothesis that House Finches have evolved behaviors that help them ingest carotenoid pigments needed for color display. Few other studies of the food preferences of animals with carotenoid coloration have been published. A study of food color preference of Cedar Waxwings (*Bombycilla cedrorum*), a species with carotenoid-based plumage coloration, showed that they display a strong preference for red foods in the fall during molt, but no preference for red foods in the spring when plumage color is not affected by carotenoid intake (McPherson 1988).

## The Controversy Surrounding Carotenoid Access and Color Display

Based on my feeding experiments, I concluded that dietary access to carotenoid pigments had a large effect on the expression of plumage coloration in captive flocks of House Finches, and I speculated that access to carotenoid pigments may play an important role in determining plumage coloration of wild males (Hill 1990, 1992, 1993a; Hill and Montgomerie 1994; Hill et al. 1994a). My proposal that dietary access to carotenoid pigments affects expression of plumage coloration in wild birds evoked criticism from several researchers (Bortolotti et al. 1996, Hudon 1994, Thompson et al. 1997, Zahn and Rothstein 1999), and the topic has become

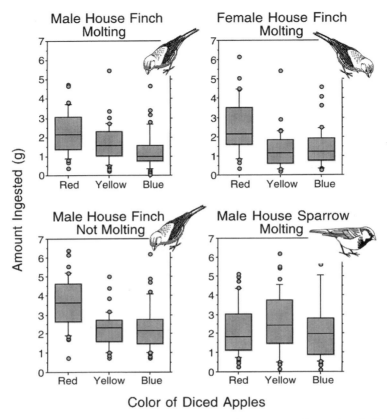

Figure 5.9. The food preferences of male and female House Finches and male House Sparrows presented with peeled, chopped apples that were colored either red, yellow, or blue with food coloring. In both the fall molt period and the spring, males showed a significant preference for apples dyed red (molt: $F_{2,114} = 8.62$, $P = 0.0003$; non-molt: $F_{2,108} = 16.86$, $P = 0.0001$). Female House Finches also showed a significant preference for apples dyed red ($F_{2,108} = 12.28$, $P = 0.0001$). In contrast, male House Sparrows showed no discrimination among apples dyed different colors ($F_{2,123} = 0.42$, $P = 0.67$). Adapted from Stockton-Shields (1997) with 1994 and 1995 data pooled.

a contentious issue in the field of condition-dependent indicator traits and the function and evolution of colorful plumage (Hill 1999b, Olson and Owens 1998). The debate centers largely around the question of whether carotenoids are a limiting resource in the wild and whether my feeding experiments appropriately tested the effect of dietary access to carotenoid pigments on expression of ornamental plumage.

Hudon (1994) first questioned the conclusions that I drew from my feeding experiments with House Finches in Michigan (Hill 1992). He proposed that House Finches in all of the treatment groups in my feeding experiments had access to sufficient carotenoid pigments to produce full expression of ornamental plumage, but that stress brought on by captivity inhibited the birds from producing red plumage. Hudon further contended that in supplementing male House Finches with β-carotene or canthaxanthin, I flooded them with carotenoids that would not naturally be part of their diet. The result, according to Hudon, was that all I had shown in my aviary experiments was (1) that in captivity House Finches lose their ornamental coloration (which was already an effect well known for many species of birds); and (2) that supplementing captive House Finches with red pigments can restore red coloration by flooding normal carotenoid control mechanisms. Hudon concluded by proposing that carotenoids were not limiting in the wild and that physiological condition is much more likely to affect plumage coloration than carotenoid access. Thompson et al. (1997) and Zahn and Rothstein (1999) largely echoed the criticisms of Hudon, but they emphasized even more the argument that a role for carotenoid access in expression of plumage coloration could be rejected if factors other than diet could be shown to affect expression of plumage coloration in the House Finch.

There are two key problems with these critiques of my feeding experiments: (1) The arguments by Thompson et al. (1997) and especially Zahn and Rothstein (1999) are logically flawed (Hill 1994b, 2001). The hypothesis that expression of plumage coloration is affected by dietary access to carotenoid pigments does not exclude the possibility that other factors such as disease or nutrition might affect expression of plumage coloration. And conversely, demonstrating that parasites have an effect on expression of plumage coloration does not exclude the possibility that dietary access to carotenoids might have an effect on plumage coloration. (2) The assertion that the diets of House Finches and other birds with carotenoid-based ornamental coloration contain excess carotenoid pigments relative to what is needed to pigment plumage is generally contradicted by field evidence or, at best, has no evidence in its support. No field study has shown that any species of wild bird with a carotenoid display has access to excess dietary carotenoids. As reviewed above, the few studies with wild birds and fish indicate that at least some species, including the House Finch, are limited in their color display by access to dietary carotenoids. In my feeding experiments, I found a dose-dependent response to supplementation with different levels of canthaxanthin (Figure 5.3), even though the lowest levels of supplementation in these experiments far exceeded what Hudon and others assume to be abundant dietary pigments. Thus, it appears that the carotenoid requirements of birds growing feathers with ornamental carotenoid coloration are very high.

The other point raised by Hudon (1994) was that in my feeding experiments there was likely a captivity effect that kept birds from metabolizing dietary carotenoids and achieving red plumage when fed a plain-seed or β-carotene-supplemented diet. Hudon was vague as to what exactly he meant by a "captivity effect," but generally he invoked poor physiological condition. The evidence presented by Hudon for a captivity effect focused on a study conducted by Weber (1961) on Common Redpolls, *Carduelis flammea*, a species of cardueline finch with ornamental red plumage coloration. In Weber's experiment, male Common Redpolls were housed in outdoor flight cages of three different sizes during molt. Weber contended that he carefully controlled access to carotenoid pigments and that the only difference among the cage treatments was the space in which the birds could move about. The author hypothesized that relatively long flight displays were necessary to maintain the physiological condition of these birds and to allow them to grow normal plumage. And, sure enough, Weber found that males held in the large cage grew redder plumage than males that were held in the smaller cages (Weber 1961). However, the flight cages in this experiment enclosed natural vegetation including, in the case of the larger cages, trees. Despite the author's attempts to standardize access to carotenoid pigments, one very reasonable explanation for the results of the study is that the large cages provided more diverse plant and insect resources and hence more or better carotenoid resources for the redpolls.

One effect of captivity, which may have altered the outcome of my feeding experiments with House Finches and which could possibly explain the observations that redpolls in different size cages grew differently colored plumage, is differential coccidial infection. (In the next section, I will review the evidence for coccidial infection impacting expression of plumage coloration in the House Finch.) It is possible that the more crowded conditions imposed by the smaller cages subjected birds to more exposure to fecal matter and higher stress, and led to higher coccidial infection. The higher coccidial infections then caused the males in the smaller cages to grow drabber plumage. I did not consider coccidia in my feeding experiments in Michigan, and it is possible that coccidia inhibited House Finches from using dietary pigments with full efficiency. In recent feeding experiments involving supplements with canthaxanthin and β-cryptoxanthin as well as plain seed treatments, I held birds free of coccidiosis, and all parasites for which I could screen birds. Controlling for parasites did not affect the outcome of these feeding experiments relative to feeding experiments conducted in Michigan, however.

Many of these points of debate related to dietary access to carotenoid pigments and plumage expression remain unresolved. As in many dichotomous arguments in biology, it is turning out that each side in this debate is partly correct. Dietary access to carotenoid pigments can have a large effect on expression of plumage coloration in at least some populations of wild vertebrates. But, other environmental factors such as parasite load and nutrition can also have a large impact on expression of carotenoid-based plumage coloration (see next section). The challenge for future research is to resolve the relative contribution of these various condition factors in determining expression of ornamental coloration in wild populations of birds.

Although in some of my papers I have stressed the role of dietary access to carotenoid pigments in determining the plumage coloration of males, it has always been clear that there is more to the story than just carotenoid content of diet dictating plumage coloration. There is now growing evidence that at least two factors related to carotenoid utilization—degree of parasite infection and access to food during molt—can have a significant effect on expression of carotenoid-based ornamental coloration independent of dietary access to carotenoids. I will discuss each of these factors in turn, relating observations back to what is known about the physiology of carotenoid pigmentation.

## Carotenoid Utilization: Effects of Parasites

### Background

With the exception of studies of dietary access to carotenoid pigments, relatively little research has been conducted on the environmental variables that might affect expression of ornamental carotenoid pigmentation. In recent years, however, a few studies have begun to look at the role of parasites in determining expression of carotenoid-based plumage coloration. The hypothesis that parasite load could affect expression of plumage coloration was first proposed by Hamilton and Zuk (1982) and has become known as the Hamilton–Zuk hypothesis. As originally proposed, the Hamilton–Zuk hypothesis was actually a model for host-parasite co-evolution that allowed for maintenance of genetic variation for disease resistance (Hamilton and Zuk 1982; see discussion of good genes benefits for female mate choice, chapter 7). Although it was not a focus of their paper, Hamilton and Zuk clearly predicted that, within a population, males with fewer parasites should have brighter plumage coloration than males that are more heavily parasitized and that more brightly colored, less parasitized males should be chosen as mates by females. Despite the focus on plumage coloration by Hamilton and Zuk, however, only a few studies, all conducted relatively recently, have addressed the Hamilton–Zuk hypothesis specifically as it relates to carotenoid-based coloration.

The first researchers to test the idea that ornamental carotenoid pigmentation was negatively affected by parasite load were Burley et al. (1991) who compared the bill coloration of wild Zebra Finches to the numbers of ectoparasitic lice and mites on their bodies. Contrary to predictions of the Hamilton–Zuk hypothesis, they found consistent positive correlations between bill coloration of both male and female Zebra Finches and ectoparasite loads (Burley et al. 1991). This study was followed by a study by Weatherhead et al. (1993) on the relationship of the size and coloration of the red epaulets in male Red-winged Blackbirds to infection by a wide range of ecto- and endo-parasites. Like Burley et al. (1991) they found no support for predictions of the Hamilton-Zuk hypothesis. In contrast, in studies of the brightness of carotenoid-based yellow plumage in Yellowhammers (*Emberiza citrinella*) (Sundberg 1995b) and Greenfinches (*Carduelis chloris*) (Merila et al. 1999), brighter males had lower infection by blood protozoa (hematozoa). In a similar study of plumage color and blood parasites in the Common Redpoll, however, there was no relationship between plumage brightness and level of parasitism (Seutin 1994).

Some of the best studies of the impact of parasites on expression of carotenoid-based plumage coloration have been conducted with fish. In experiments in which guppies were infected with a monogenean parasite, Houde and Torio (1992) showed that infection by the parasite significantly reduced the intensity of orange carotenoid coloration. Similarly, in the three-spined stickleback (*Gasterosteus aculeatus*), experimental infection with a protozoan parasite decreased the intensity of the carotenoid-based red coloration of males (Milinski and Bakker 1990).

Thompson et al. (1997) were the first to look at plumage coloration and parasites in the House Finch. Using a huge set of data amassed during three decades of banding by Elliot McClure, they divided male House Finches into three color categories—yellow, orange, and red—and compared individual plumage coloration to incidence of infection by pox (a viral disease) and feather mites. They found a consistent negative relationship between expression of plumage coloration and both pox and mites: more parasitized males grew drabber plumage (Figure 5.10). By themselves, these strong correlations based on large sets of data provided some of the best support for the idea that parasites are negatively associated with plumage redness. But Thompson et al. (1997) went further in their analyses. Because they had multiple measurements of plumage coloration for the same males before and after fall molt, they were able to test more directly for the effects of both pox and mites. For both parasites they found that heavily infected males decreased in plumage redness on average, while uninfected birds increased in plumage redness on average. Harper (1999) found a similar negative relationship between feather mite load and plumage coloration for nine species of European passerines. The studies by Thompson et al. are the most convincing field data yet published that parasites can reduce expression of carotenoid-based plumage coloration. These studies do not, however, establish a definite causal link between a specific parasite infection and reduced plumage coloration. The patterns reported by Thompson et al. (1997) are also consistent with the hypothesis that both parasite load and plumage coloration are actually affected by a third factor such as nutrition.

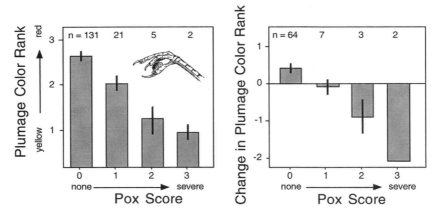

Figure 5.10. The relationship between pox infection and expression of plumage coloration in male House Finches. Among wild males in southern coastal California, individuals with higher pox infections had drabber plumage coloration than individuals with lesser pox infections ($H = 20.05$, d.f. $= 3$, $P < 0.001$). For those males captured before and after molt, individuals with pox infections tended to decrease in plumage color rank in proportion to severity of pox infection; males without pox infections tended to increase in plumage color rank ($H = 9.37$, d.f. $= 3$, $P = 0.03$). Adapted from Thompson et al. (1997). A similar loss of plumage redness was also associated with feather mites (Thompson et al. 1997).

*Interpretation of Parasite Data*

For a color display to evolve or be maintained by female choice, where the benefit to a female of choosing a brightly colored male is information about his health, the parasite implicated in the signaling system must have a demonstrable negative effect on the health of individuals (Clayton 1991, Endler and Lyles 1989, Read 1988). Moreover, there must be some mechanism by which infection by the parasite might depress expression of the ornament. Despite the strong negative relationships found between parasites and color display in some of the studies reviewed above, the detrimental effects of most of these parasites are questionable. There are also no known mechanisms by which any of these parasites could negatively impact the expression of carotenoid-based plumage coloration.

Various species of hematozoa have been the most popular parasites to compare to plumage coloration, primarily because they are relatively easy to count in blood smears. Several authors, however, have been unable to detect any measurable negative effect of various species of hematozoa on the condition of a variety of passerine birds (Bennett et al. 1988, Weatherhead 1990, Weatherhead and Bennett 1992). In two studies, only hematozoa in the genus *Trypanosoma* (malaria) negatively affected the health and condition of birds (Dufva 1996, Ratti et al. 1993). At best, it appears that only some hematozoa have a significant negative impact on some species of passerine birds. Thus, the lack of a relationship between hematozoa and plumage coloration does not falsify the hypothesis that plumage coloration signals parasite infection, and conversely, a negative relationship between hematozoa and plumage color is only weak support for the idea that the function of the plumage display is to signal infection by that parasite. Similarly, some authors have proposed that feather mites, which were a focus of the study by Thompson et al. (1997) described above, have little or no impact on the health or condition of their host (Blanco et al. 1997, OConner 1992). Feather mites feed on feather dander and oil, and it is not clear how they would affect the health of their host or especially how they would affect expression of colorful plumage. Lack of a basic understanding of the impact of parasites on hosts makes interpretation of correlative field data, either positive or negative, difficult. Of all the parasites that were compared to plumage coloration in the studies cited above, avian pox has the most serious impact on individual health and is the most likely to have a direct negative impact on plumage coloration, although the mechanism of such an effect on plumage coloration remains unknown.

Another factor that complicates interpretation of correlations from field studies between prevalence of parasites and plumage coloration is that one has to assume that the assessments of parasites are accurate and repeatable. This may not be the case for hematozoan counts for which there can be huge daily and seasonal fluctuations in prevalence (Weatherhead and Bennett 1992). Moreover, plumage coloration can only be affected by parasites during the period of feather generation, which occurs only once or twice a year for passerines and is temporally offset from nesting and usually from mate choice. Measurements of parasites have to be made during molt or negative relationships are impossible to interpret. In only a few studies, such as the study by Thompson et al. (1997), were parasites measured during molt.

## Coccidia and Poultry

Long before Hamilton and Zuk proposed their hypothesis that parasites depress expression of ornamental traits, it was well known by poultry scientists and poultry farmers that various diseases, particularly coccidia, suppress expression of carotenoid-based integumentary coloration in chickens (Bletner et al. 1966, Marusich et al. 1972). Moreover, poultry scientists have established clear mechanisms by which the coccidia inhibit expression of carotenoid pigmentation. The coccidia that have been studied extensively by poultry scientists are in the genus *Eimeria*. In contrast, passerine birds are more commonly infected with coccidia in the genus *Isospora*. In my description of the life history of coccidia, I will focus on coccidia in the genus *Isospora* because these are the parasites of House Finches. In contrast, coccidia in the genus *Eimeria* have been the focus of all research on the mechanism by which coccidia affect carotenoid pigmentation. A key assumption in this chapter is that isosporan coccidia affect expression of plumage coloration in a manner similar to that documented for eimerian coccidia (see Brawner 1997 for additional justification).

The life cycle of coccidia involves alternating sexual and asexual phases. Once they have been ingested by a suitable host, isosporan oocysts form haploid sporozoites that enter the epithelium of the intestine and undergo asexual reproduction through a variable number of cycles. This asexual phase is the damaging portion of the life cycle to the host as the sporozoites feed on host tissue to grow and divide. Eventually gametes are formed that fuse to form zygotes in the intestine. Zygotes produce a tough outer coating to become oocysts that can then pass out of the parent host, survive extreme environmental conditions, and then repeat the cycle if they are ingested by another suitable host (Brawner 1997).

Eimerian coccidians have a measurable detrimental effect on carotenoid pigmentation in chickens. Chickens with coccidial infection have consistently been shown to be less efficient in absorption of carotenoid pigments from food compared to uninfected or less infected chickens (Allen 1992, Ruff et al. 1974, Tyczkowski et al. 1991). The reason for decreased carotenoid absorption when coccidia are infecting the intestinal track is not fully understood, but Allen (1987b) showed that, associated with coccidial infection, there was a thickening of the intestinal epithelium (hyperplasia), even at sites distant from the area of coccidial infection. Although it has yet to be demonstrated definitively, Allen (1987a, 1987b) proposed that the hyperplasia induced by coccidiosis caused a decrease in the absorption of dietary carotenoid pigments.

Carotenoid transport may also be affected by coccidial infection. As discussed in chapter 4, high-density lipoproteins (HDLs) play a primary role in the transport of carotenoid pigments in birds. Some of these HDLs are manufactured in the small intestine, and it appears that epithelial hyperplasia not only interferes with carotenoid absorption, but also inhibits or depresses production of HDLs by as much as 30% (Allen 1987a). Reduced levels of HDLs associated with coccidiosis leads to the saturation of carotenoid carrier molecules and a reduction in the amount of carotenoids that can be transported. Thus, there are clear mechanisms by which coccidial infection can inhibit carotenoid absorption and transport and hence the display of carotenoid-based plumage coloration. Although the mechanisms of inhibition of carotenoid absorption and transport have been studied for eimerian coccidia, and not the isosporan coccidia that infect House Finches, both types of coccidia infect in the epithelium of the small intestine. It seems reasonable that isosporan coccidia might inhibit carotenoid pigmentation through a similar mechanism as eimerian coccidia. Thus, the effect of coccidiosis on expression of carotenoid pigmentation seemed like a logical system in which to test the Hamilton-Zuk hypothesis.

## Coccidia and Plumage Coloration

My students and I began studying the effects of coccidia on expression of plumage coloration in House Finches in the summer of 1996. As part of a class project, a master's student in my department, Blue Brawner, examined the feces of wild-caught House Finches for coccidial oocysts, the reproductive cells of coccidia that are shed into the intestine and pass from the host's body in feces. Brawner found that a high proportion of House Finches in the Auburn population were infected with coccidia of the genus *Isospora* (Brawner 1997). The presence of Isosporan coccidia in House Finches was an important discovery for my lab group in our investigation of the control of plumage coloration in male House Finches, but it wasn't a great surprise to my parasitological colleagues. Isosporan coccidia are relatively well-known parasites of many species of songbirds (Levine 1982). Isosporan coccidia are thought to form primarily chronic infections in songbirds, with a modest effect on the health and physiological condition of the host (Brawner 1997), but there were at least some reports of high mortality rates of cardueline finches associated with isosporan infection (Giacomo et al. 1997, Seroni 1994).

Given the clear results from numerous poultry studies showing that coccidia of the genus *Eimeria* inhibit expression of integumentary carotenoid pigmentation in chickens, we hypothesized that isosporan coccidia would also inhibit expression of carotenoid-based plumage coloration in House Finches. So, we conducted an infection experiment to experimentally test the effect of isosporan coccidia on the color of feathers grown by molting male House Finches. This was a conceptually simple but technically difficult proposition. Fortunately, my primary contribution was in conceptual development; Brawner was responsible for the numerous protocols to be worked out: what drugs could be used to treat coccidiosis in House Finches? Were there side effects of these drugs? How did we "harvest" the coccidia to use in the experiment? How would birds be inoculated? How would the dose be controlled?

With some trial and error, we found that coccidia were not particularly difficult parasites with which to work. Coccidia could be harvested by collecting the feces of infected House Finches and, once harvested, coccidia could be stored for years in potassium dichromate, a chemical that protected the coccidia from fungal and bacterial organisms—even parasites have parasites! From collected feces, coccidia could be purified and placed in a solution where a known density of viable coccidial oocysts could be established. We could then orally infect House Finches with a specific dose of coccidia. Moreover, Brawner found a drug, sulfadimethoxine, which proved to be very effective in the treatment of coccidiosis in House Finches. Within a couple of days of adding sulfadimexthoxine to the water of a flock of captive finches with isosporan infections, all birds stopped shedding oocysts, and clinical signs of the disease disappeared. However, even after weeks on sulfadimethoxine, if the treatment was stopped, birds would again begin passing oocysts. Thus, sulfadimethoxine suppressed but did not eliminate isosporan cocci-diosis in House Finches—it was coccidiostatic, but not coccidiocidal. Suppression of coccidiosis was fine for the infection experiment that we had planned, so this was an ideal drug for our purposes. Once these techniques were developed, we were ready to conduct infection experiments (see Brawner 1997, Brawner and Hill 1999, Brawner et al. 2000 for further details of coccidia sampling and inoculation tech-niques).

For the infection experiments, we used only juvenal birds, which had fledged about three to eight weeks before the start of our study and thus had, at most, a brief history of exposure to coccidia. Adult male House Finches, on the other hand, may have had years of exposure to coccidia and possibly could have acquired immunity to coccidia. It was clearly best to work with recently hatched House Finches, but by choosing to work with juvenal birds, we cut our sample size approximately in half. As detailed in chapter 2, House Finches cannot be sexed by morphology until they complete their first prebasic molt in their first fall. Consequently, going into the study we knew that on average half of our study birds would turn out to be females and not be of interest for the hypothesis that we were testing. We decided that the importance of using birds that were known to have had only a brief history of infection outweighed the decreased sample size.

The study was repeated over two years. In the first year, techniques had to be worked out and many mistakes were made, so we ended up with a small sample size of treated and control males and somewhat inconclusive results (Brawner et al. 2000). In the second year, learning from our mistakes, we conducted a much better experiment with an improved protocol and larger sample size. As a result, I will focus on the second year of the experiment. The basic protocol was to catch a large sample of recently fledged finches in mid-summer before they had started their prebasic molt. These birds were then randomly divided into two groups, a treat-ment group and a control group. Within two weeks of capture, we inoculated all birds with saline solution. Birds in the control group received untreated solution; birds in the treatment group received solution containing 2000 isosporan oocysts. The protocol called for us to use a bulbous syringe to inject the solution down the esophagus of birds, but the young House Finches were cooperative and drank the solution from the tip of the syringe. After inoculation, we added sulfadimethoxine to the drinking water of birds in the control group; birds in the treatment group received no sulfadimethoxine. In all other ways, the birds were treated the same. They had the same seed diet, and they were all had relatively small amounts of canthaxanthin added to their water (see Figure 5.3).

Throughout the experiment we examined the feces of birds in both treatment groups for the number of isosporan oocysts that were shed. Although both groups looked equally healthy and vigorous from visual inspection, the two groups differed significantly in the degree of coccidial infection. We found no or virtually no oocysts in the feces of birds in the control group, but we observed thousands of oocysts in the feces of birds in the experimental group (Brawner et al. 2000). The birds were held throughout the experiment in outdoor pens so they were exposed to natural photoperiod and they molted in synchrony with local wild House Finches. By October all birds had completed their prebasic molt, and we captured all birds in both treatment groups and scored their plumage coloration. The effects of the treatment were clear. Birds in the group infected with coccidia grew substantially less red and less saturated plumage than males in the control group (Brawner et al. 2000) (Figure 5.11). Thus, coccidiosis can have a large effect on expression of plumage coloration in the House Finch.

An unexpected result of our infection experiment was that coccidial infection had a significant effect on feather mite load: birds infected with coccidia were more likely to increase in mite numbers compared to birds that were not infected (Brawner 1997). Feather mites feed primarily on feather oils and dander, and their effect on the health of birds is thought by some parasitologists to be minimal (Blanco et al. 1997, OConner 1992). Feather mites thus appear to be more commensal organisms than parasites. This made the results of Thompson et al. (1997), which showed a consistent correlation between feather mites and drab plumage coloration in House Finches puzzling to me—how could small invertebrates that live on feather debris directly affect the plumage coloration of male House Finches? However, if feather mites are positively associated with coccidial infection in wild birds, as we found in our captive birds, then the pattern makes sense. Something about coccidial infection (perhaps increase build-up of feather oil or dander as a result of reduced preening by sick birds) causes mites to increase, and coccidia also

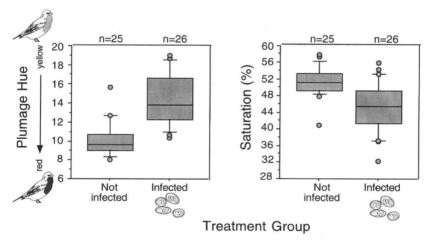

Figure 5.11. The effect of experimental coccidial infection on the expression of plumage hue and saturation in captive male House Finches. Control males were inoculated with saline solution and maintained on the coccidiostatic drug sulfadimethoxine; infected males were inoculated with 2000 isosporan oocysts and maintained without sulfadimethoxine. Males were maintained on their treatments throughout fall molt as they grew their ornamental plumage. Infected males grew significantly less red ($Z = -5.22$, $P < 0.0001$) and less saturated plumage ($Z = 3.92$, $P = 0.0001$) than control males. Adapted from Brawner et al. (2000). Note that, unlike previous figures, hue was measured with a Colortron in this study, and redder males had lower scores than yellower males.

directly inhibit expression of carotenoid pigmentation. Thus, mites are associated with but are not the cause of drab plumage. This would mean that the association between mite numbers and plumage coloration observed by Thompson et al. (1997) might actually have been a correlated response by mites to coccidial infection.

## Melanin Versus Carotenoid Ornaments

Another interesting observation in our study of plumage coloration and coccidial infection was that the same coccidial infection that suppressed carotenoid-based plumage coloration in male House Finches had no measurable effect on the melanin-based plumage coloration of tail feathers (Hill and Brawner 1998). This observation is consistent with the hypothesis that, while carotenoid pigmentation is a condition-dependent trait that is sensitive to environmental perturbations, melanin-based coloration is relatively unresponsive to environmental factors like parasites or nutrition (Badyaev and Hill 2000a, Gray 1996, Hill 1996a). The condition dependency of carotenoid-based color displays is the topic of this chapter and much of this book and is now supported by a large body of work. Melanin-based color displays, on the other hand, are fundamentally different than carotenoid displays because melanins are synthesized in birds from basic biological precursors (the amino acids tyrosine, tryptophan, and phenylalanine) that are not likely to be limiting for most birds (Gray 1996). Melanin-based plumage ornaments function most often in status signaling (see Senar 1999 for a review) in which signal honesty is maintained by social interactions, not by costs associated with producing the trait (see chapter 8). Rarely have melanin-based ornaments been found to function in female mate choice. Thus, melanin and carotenoid pigmentation appear to be fundamentally different types of ornamental traits.

Recently, with my student, Kevin McGraw, I performed a follow-up experiment to better test the idea that melanin-based and carotenoid-based ornaments respond differently to environmental stress. In the House Finch study described above, we measured the melanin-based color of tail feathers, but this coloration is not an ornamental trait—it is not showy or conspicuous and it does not differ between males and females. House Finches have no melanin-based ornamental traits. In contrast, male American Goldfinches, another species of cardueline finch, have both carotenoid-based ornamental coloration (brilliant yellow head and body feathers) and bold melanin-based ornaments (a black cap and black wings and tail). So, we repeated the coccidia infection experiment (see McGraw and Hill 2000b for details), but we used American Goldfinches as our study bird, and measured the effect of the treatment on expression of both melanin and carotenoid ornaments. We found that the carotenoid-based plumage coloration of male American Goldfinches was negatively affected by coccidial infection—infected birds grew pale yellow plumage compared to the relatively bright yellow plumage of controls. However, neither the blackness nor the size of black caps of male goldfinches was affected by coccidial infection (McGraw and Hill 2000b) (Figure 5.12). Even birds that became gravely ill from the infection grew black caps that were the same size and coloration as those of healthy birds. Thus, melanin- and carotenoid-based plumage coloration appear to be fundamentally different types of ornamental traits with different and probably largely independent information content.

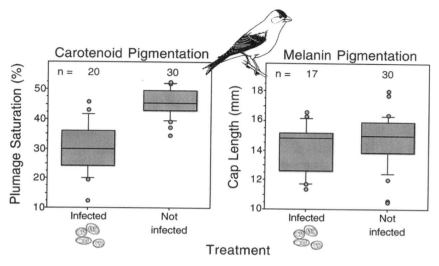

Figure 5.12. The effect of coccidial infection on expression of carotenoid-based yellow breast coloration (left box plot) and size of melanin-based black cap (right box plot) in male American Goldfinches (*Carduelis tristis*). Coccidia had a significant effect on the saturation of carotenoid-based yellow coloration ($t = 7.99$, $P = 0.0001$), but not on melanin-based plumage coloration including either cap size ($t = 1.24$, $P = 0.22$) or cap blackness (not illustrated; $t = 0.57$, $P = 0.57$). These observations support the hypothesis that carotenoid-based plumage coloration is affected more by parasitic infection than is melanin-based plumage coloration. Adapted from McGraw and Hill (2000b).

## Mycoplasmosis and Plumage Coloration

In the previous section, to keep my presentation of experiments on the effect of coccidiosis on expression of plumage coloration in male House Finches clear and comprehensible, I glossed over one detail: we picked the worst year in the history of House Finches to run these experiments. The start of our infection experiments coincided, almost to the day, with the start of the great House Finch mycoplasmal epidemic of the mid-1990s. This disease forced itself into our coccidia infection experiments in both 1996 and 1997, and it cannot be ignored in a discussion of parasites and plumage redness in the House Finch.

*Mycoplasma gallisepticum* (MG) is a bacterium that is a well-known parasite of the respiratory tract of chickens and turkeys. It was virtually unknown as a songbird parasite until 1994, when it was found infecting House Finches in Maryland (Ley et al. 1996, Luttrell et al. 1996). In finches, MG causes respiratory infection as well as severe infection and inflammation of the membranes surrounding the eyes. Among wild bird enthusiasts, mycoplasmosis in House Finches is frequently referred to as "conjunctivitis" because the disease causes severe swelling of the conjunctiva of the eye. MG is a devastating disease in finches. In 1996, mycoplasmosis killed about 60% of the House Finches in our Auburn study population, and by using a rough extrapolation to the entire eastern population, we estimated that it killed about one hundred million finches from 1994 to 1996 in eastern North America (Nolan et al. 1998; see also Hochachka and Dhont 2000).

MG spread rapidly from the site of first detection throughout the eastern United States (Hochachka and Dhont 2000). In 1996, just as we were assembling our captive flocks for the coccidia infection experiments, the worst phase of the mycoplasmal epidemic hit the local House Finch population in Auburn. We worked diligently to exclude the disease from our captive flocks, excluding from our experimental flocks all finches that we caught with symptoms of mycoplasmosis. About 60% of birds that we captured had symptoms of the disease, however, and more broke out with the disease once they were put in the aviary. It was very difficult to capture enough asymptomatic juvenal birds for our study. Although we caught dozens of House Finches, we ended up with small flocks for the infection experiment. And, despite courageous efforts to keep MG out of our aviary, MG still broke out in our captive flocks. As a result, in the infection experiments we ended up with four, rather than two, treatments: (1) control; (2) infected with coccidia; (3) infected with mycoplasmosis; and (4) infected with both mycoplasmosis and coccidia (Brawner et al. 2000).

Because we could not keep MG out of our captive flocks, we decided to study the impact of MG on expression of plumage coloration. In 1996, birds became very ill from MG while we were consulting with veterinarians and devising a treatment for the disease. We ended up periodically giving all birds tylosin tartrate, which lessens the symptoms of mycoplasmosis but does not cure birds of the disease. In 1997 we treated birds more aggressively with tylosin tartrate from the start of the experiment and, as a result, mycoplasmosis did not reach as severe an infection state in our captive birds in the second year of the experiment. Consequently, MG did not affect plumage coloration as much in 1997 as in 1996. Still, in both years of our study, we found that mycoplasmosis had a significant negative effect on expression of carotenoid-based plumage coloration. Birds that became infected with MG in 1996 grew plumage that was far drabber than control birds and substantially drabber even than birds with coccidiosis alone (Brawner et al. 2000). In 1997 the effect of the milder MG infections on expression of plumage coloration was less than the effect of severe MG of the previous years, but infected males were still significantly drabber than controls (Brawner et al. 2000) (Figure 5.13).

The observation that mycoplasmosis significantly depressed plumage brightness in male House Finches was interesting but obviously of less relevance than coccidiosis as an explanation for variation in expression of plumage coloration in wild male House Finches. MG is a new House Finch disease. It has conspicuous external symptoms (grossly swollen eyes), so females gain little information from drab plumage coloration that they cannot gain from the visible symptoms of the disease. At the intensity that it occurred in the captive flocks in our experiments, MG was a debilitating disease for finches. What we demonstrated in the MG portion of our disease research is that a bird that is very ill from a severe respiratory and eye infection during molt grows plumage that is less red than a bird that is not ill. This is hardly a surprising result. Nevertheless, it does further support the idea that environmental factors that negatively impact the health of a bird during molt will cause the bird to produce less elaborate carotenoid plumage ornamentation.

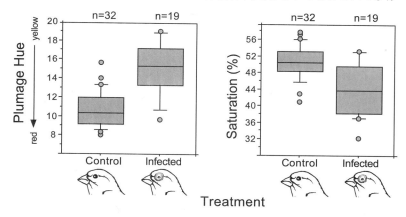

Figure 5.13. The effect of mycoplasmal infection on expression of carotenoid-based plumage hue and saturation in captive male House Finches. Control males developed no clinical symptoms of mycoplasmosis during captive molt; infected males had clinical symptoms of mycoplasmosis throughout the molt period. Infected males grew ornamental plumage coloration that was significantly less red ($Z = 4.58$, $P = 0.0001$) and less saturated ($Z = -3.39$, $P = 0.0001$) than control males. Adapted from Brawner et al. (2000).

## Carotenoid Utilization: Effects of Nutritional Condition

Another condition factor with the potential to have a significant impact on expression of plumage pigmentation that has received little consideration is the energetic cost of carotenoid utilization. As I mentioned earlier, males are expected to adopt a strategy that minimizes the energetic cost of pigmentation. At least in terms of the coloration of the plumage display, however, males are constrained by the mate preference of females. By choosing particular colors, females may force additional costs, in the form of energetically expensive oxidation of carotenoid molecules, onto the expression of bright plumage (Hill 1994c, 1996a) (see Figure 4.3). This would make carotenoid-based plumage pigmentation a more costly and therefore a more reliable signal. Moreover, as I discussed earlier, most or all carotenoid metabolism must be performed at the time of feather replacement when energy limitations are particularly severe (Lindstrom et al. 1993, Murphy and Taruscio 1995). Thus, the energetic costs of production of carotenoid displays may play an important role in determining the coloration displayed by males.

Before I focused on the topic in my studies with House Finches, the only test of the idea that energy constraints can limit expression of carotenoid-based plumage pigmentation was a study by Schereschewsky (1929) who treated one set of male Bullfinches, *Pyrrhula pyrrhula*, with thyroid tissue to accelerate their metabolic rates and left another set untreated. On the same diet, males treated with thyroid tissue grew feathers with drabber carotenoid coloration than birds that were not treated. Schereschewsky (1929) attributed the loss of coloration in birds treated with thyroid tissue to the loss of body fat induced by increased metabolic activity. Until my studies with House Finches, this study represented the only evidence that, independent of carotenoid access, available energy can affect expression of plumage coloration.

## Growth Bars and Plumage Coloration

As a first attempt to test the idea that nutritional condition at the time of molt affects expression of carotenoid pigmentation, I looked at feather growth rate in relation to plumage coloration. As birds grow feathers, material that is deposited at night is lighter than material that is deposited during the day. (Instead of the purposefully vague term "material" I would like to be able to state that "melanin pigments" or "keratin" are deposited more densely during the day. The physical cause of the banding pattern, however, is unknown.) The result is faint alternating light and dark bands on feathers that are called "growth bars" (Riddle 1908) (Figure 5.14). Long ago it was shown that in the House Finch one light plus one dark band equals 24 h of feather growth (Michener and Michener 1938). Moreover, it was shown by Grubb (Grubb and Cimprich 1990, Grubb 1989, 1991), through a series of field experiments, that the width of growth bars is an indicator of the nutritional condition of a bird during molt. Birds with access to more food grow feathers faster and hence have wider growth bars than birds that are food stressed. Thus, growth bars represent a convenient record of nutritional condition during feather growth.

In 1989 I started collecting an outer tail feather from male House Finches that I banded in the wild as well as from males that I held in captivity. By measuring the growth bars on these tail feathers, Bob Montgomerie and I were able first to compare the rate at which House Finches grew tail feathers in captivity, where they had

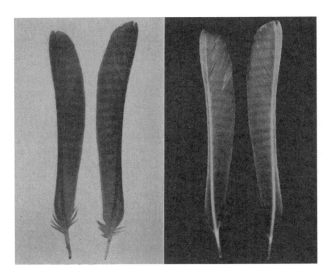

Figure 5.14. Growth bars on the tail feathers of House Finches. These faint alternating light and dark bands result from lighter material being deposited at night than during the day. One dark plus one light band equals 24 h of feather growth, so feather growth rate can be estimated by measuring growth bars. Figure from Michener and Michener (1938).

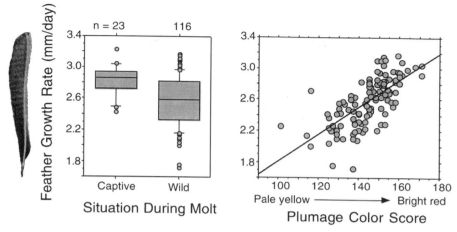

**Figure 5.15.** Feather growth rate, as measured by the mean width of growth bars, can be used as a measure of nutritional condition during molt. To test the hypothesis that feather growth rates are limited by food access in wild male House Finches, the feather growth rates of captive males held on *ad libitum* food were compared to the growth rates of wild males (left panel). Captive males grew feathers significantly faster ($t = 3.27$, $P = 0.001$, two-tailed $t$-test). There was a significant positive relationship between the growth rates of tail feathers and the plumage color scores of males from Ann Arbor, Michigan ($r = 0.68$, $n = 116$, $P = 0.0001$). These observations supported the hypothesis that nutrition at the time of molt is related to the expression of carotenoid-based plumage coloration. Adapted from Hill and Montgomerie (1994).

unlimited access to food, to the rate at which they grew tail feathers in the wild. We found that captive birds grew their feathers significantly faster than wild birds, indicating that in the wild male House Finches were growing their feathers at less than maximum speed and hence that food was limiting for wild birds (Hill and Montgomerie 1994) (Figure 5.15). We then tested the idea that expression of ornamental plumage coloration was related to nutritional condition during molt by comparing the mean width of growth bars on tail feathers to plumage brightness. We found a significant positive relationship; males that grew their feathers faster (i.e., birds in better nutritional condition) grew brighter plumage (Hill and Montgomerie 1994). These observations suggested that nutrition at the time of molt could affect expression of plumage coloration, but it was also possible that birds with access to better nutrition ingested more carotenoid pigments and hence grew redder plumage.

### Timing of Molt and Plumage Coloration

As another indirect test of the idea that plumage redness is related to nutritional condition at the time of molt, we compared the extent of molt of body plumage to plumage coloration for a set of males. This comparison was possible because, in August 1992, we captured and processed sixty-two yearling and sixty-nine adult male House Finches in a three-day period, giving us a snapshot of molt for a

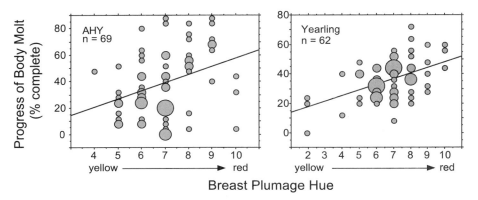

Figure 5.16. The relationship between timing of molt and plumage redness for AHY and HY (yearling) male House Finches in coastal California on 12–14 August 1992. Males that molted earlier tended to grow redder plumage (yearling: $r^2 = 0.26$, $P = 0.0001$; AHY: $r^2 = 0.11$, $P = 0.005$), suggesting that males that grow red plumage are in better condition during molt. Adapted from Hill and Montgomerie (1994).

population of males. For each of these males we scored the hue of ventral plumage, and we scored the extent of molt of body plumage. We compared the two scores, and we found a significant positive correlation; males that were growing redder plumage were also significantly farther advanced in their fall molt (Hill and Montgomerie 1994; Figure 5.16). In the previous section I presented experimental evidence that access to food affected the growth rate of feathers of male House Finches. Other studies have shown that nutritional condition can affect the timing of molt—birds in better condition begin and complete molt faster (Dunn and Cockburn 1999, Piersma and Jukema 1993, Robertson et al. 1998). Thus, the relationship between timing of molt and the redness of growing feathers is further support for the idea that nutritional condition affects the color display of males.

### Experimental Manipulation of Food Access

To directly test the idea that food access independent of carotenoid access could affect expression of plumage coloration, I established five flocks of juvenal House Finches in separate cages in Auburn during the summer of 1998. These cages had concrete floors with no vegetation growing within 3 m, so I could control the food access and the carotenoid access of these birds. All birds were provided with the same basic diet of sunflower seeds and millet. One group received no carotenoids in its water; two groups received canthaxanthin in their water; and two groups received tangerine juice as a source of β-cryptoxanthin in their water. In addition, one of the groups supplemented with canthaxanthin and one of the groups supplemented with β-cryptoxanthin were deprived of food for about 38% of daylight hours during molt (see Hill 2000 for details of experimental design). Food removal was designed to subject birds to periods of fasting like those experienced by wild birds. Thus, these treatments presented molting male House Finches with the following situations: (1) low carotenoid diet, unlimited food;

(2) precursor to red pigments, unlimited food; (3) precursor to red pigments, limited food; (4) red plumage pigments, unlimited food; and (5) red plumage pigments, limited food.

I found that there was a significant effect of food access, independent of carotenoid access, on color display (Hill 2000) (Figure 5.17). As predicted, males with access to β-cryptoxanthin, which is the primary precursor to the red pigment 3-hydroxy-echinenone (see chapter 4), grew significantly redder and more saturated plumage when they were not food stressed than when they were food stressed. I interpreted this observation as evidence that the metabolic conversion of pigment precursors to plumage pigments is an energetically costly process (see Figure 4.4). Contrary to what I predicted, however, males with access to canthaxanthin, a red pigment that can be deposited in feathers unaltered, grew redder and more saturated plumage when they were not food stressed than when they were food stressed. These observations suggests that costs beyond those related to carotenoid metabolism, such as costs associated with absorption, transport, and deposition, are also affected by access to nutrition during molt.

For those interested in the control of carotenoid pigmentation in birds, I suggest that the role of nutritional condition is a neglected topic of study.

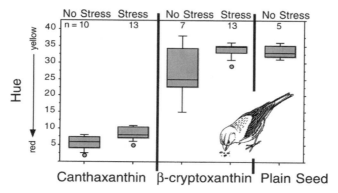

Figure 5.17. Effect of carotenoid supplementation and food stress on expression of carotenoid pigmentation. The label above each box plot indicates whether birds were maintained on *ad libitum* food (no stress) or if food was removed from the male's cages for 38% of daylight hours during molt (stress). Labels below the box plots indicate the type of carotenoid pigments that were provided to males in their drinking water. Note that the hue of the feathers of food-stressed males fed β-cryptoxanthin (a precursor to red plumage pigments) was similar to the hue of feathers of males fed plain seeds, but that males fed β-cryptoxanthin and not food stressed grew redder plumage. Males fed canthaxanthin (a red pigment that can be used directly to pigment plumage) and food stressed grew less red plumage than unstressed males fed canthaxanthin. These observations suggested that there are energetic costs to absorption, transport, or deposition of carotenoid pigments. For statistics see Hill (2000) from which the figure was adapted.

## Carotenoid and Free Radical Suppression

Some biologists have suggested further links between carotenoid display and individual condition by suggesting that carotenoid-based color displays evolved because of the antioxidant properties of carotenoids (Lozano 1994, Møller et al. 2000). This idea comes from the observation that many types of carotenoids are effective free-radical scavengers (Burton 1989, Machlin and Bendich 1987). A build-up of free radicals (charged molecules) is damaging to the bodies of organisms both because excess free radicals impair the immune system and because free radicals can cause tumor development (Bendich 1989, Chew 1993). In mammals, it has been shown in a number of studies that increased dietary intake of carotenoids, particularly β-carotene, can lead to lower rates of infection (Bendich 1989, Bendich and Shapiro 1986, Chew 1993) and reduced incidence of tumors (Krinsky 1989, Peto et al. 1981, Ziegler 1989). These studies with mammals have led some biologists to propose that animals with ornamental carotenoid-based coloration must trade off between using carotenoids as ornamental displays and using carotenoids to combat free-radical build-up and hence to enhance immune defense. Only individuals with good immune systems and access to large quantities of carotenoids would be able reach full ornament expression. Readers familiar with models of sexual selection (for a review of these models, see chapter 12) might note that such a display would be a true Zahavian handicap. Males would be "wasting" a critical resource just to show that they could survive even with such waste; only males in the best condition would be able to sacrifice sufficient carotenoids for a full ornament display.

The idea that carotenoid coloration signals immunocompetence because of the immune-enhancing properties of carotenoids is an intriguing and seductive idea because it potentially provides a direct link between male quality, ability to resist parasites, and expression of carotenoid-based color display. However, virtually all studies showing beneficial health effects of supplementation with carotenoids have been conducted on species of mammals that have diets with low levels of carotenoid pigments and typically that have very low levels of circulating carotenoid pigments. The levels of circulating carotenoid pigments in species with carotenoid-based color displays, such as the American Flamingo or Roseate Spoonbill, are typically two orders of magnitude greater than mammals lacking carotenoid display (e.g., human, rat, sheep) (Hill 1999a). It does not seem reasonable to assume that effects of carotenoid supplementation observed in these species lacking carotenoid display will be the same as in species with carotenoid display.

The hypothesis is also not consistent with what little is presently known about the carotenoid physiology of birds and fish with carotenoid-based ornamental displays (Hill 1999a). Most troubling are the observations that many species of birds with carotenoid-based ornamental display fail to absorb many types of carotenoid pigments. This selective absorption makes sense if, as was outlined in chapter 4, the physiology of birds has evolved to optimize use of carotenoids in ornamental display. "Wasting" carotenoids makes no sense if carotenoids are needed for critical immune function. Also at odds with the hypothesis that birds in general would experience immune benefits if they had access to more carotenoid pigments is the large difference in levels of circulating carotenoid pigments seen between males and females in many species of birds in which males have more extensive carotenoid ornamentation than females (e.g., see Figure 4.6). Similarly, among closely related species that differ in extent of carotenoid-based ornamentation, such as the Scarlet and White Ibis, the species with extensive carotenoid-based displays have much higher levels of circulating carotenoid pigments than the species with less ornamentation (Trams 1969). These sex-and-species differences make sense if birds have evolved mechanisms to use carotenoid pigments needed for ornament display (as reviewed in detail in chapter 4). If the immune systems of birds are in general carotenoid-limited, it is hard to explain why females have lower levels of circulating carotenoid pigments than males, and less ornamented species have lower levels than more ornamented species.

With my graduate student Kristen Navara, I tested the idea that birds with bright plumage coloration trade off use of carotenoids for plumage pigments and use of carotenoids as free-radical scavengers (Navara and Hill, in prep). We conducted this test with American Goldfinches, which have bright yellow body plumage, by manipulating the carotenoid access of males during the growth of ornamental feathers and then observing both the immune system function and disease resistance of males in different treatment groups. For this experiment, we fed male goldfinches quantities of dietary carotenoids that were (1) approximately the same that male goldfinches would ingest in the wild; (2) an order of magnitude lower than would be found in a wild diet; and (3) an order of magnitude greater than would be found in a wild diet. After we had maintained male goldfinches on these diet treatments for a month, we measured their immune responses. First, as a measure of acquired immunity, we quantified antibody production in response to a challenge with sheep red blood cells. Second, as a measure of cell-mediated immunity, we recorded the degree of swelling in the right wing tissue in response to challenge with a novel protein, phytohaemagglutinin (PHA). After these immune tests, we infected the males with *Mycoplasma gallisepticum* and recorded the resulting disease in males in the different treatments.

We found no evidence that the immune systems of male American Goldfinches were carotenoid-limited or that they traded off the use of carotenoids for immune function with the use of carotenoids for ornament display. There were no significant differences among males in the three groups in antibody production in response to sheep red blood cell challenge, nor did the degree of wing-web swelling differ among males with different access to carotenoids (Navara and Hill, in prep; Figure 5.18). Finally, there was no signifcant difference among males in the different treatment groups with respect to how ill they became in response to infection with MG (Navara and Hill, in prep). If the immune systems of birds with extensive carotenoid-based color displays are not carotenoid-limited, then there cannot be a direct trade-off between use of carotenoids for immune function and use of carotenoids for ornament display. Moreover, when we compared the plumage coloration of males at the time of capture to their immune responses when they were maintained on a low-carotenoid diet, we found no differences between any treatment groups in any of the measured immune responses (Navara and Hill, in prep; Figure 5.19). Again, this is not consistent with the idea that males trade off use of carotenoid pigments for ornamental display versus use of pigments for immune enhancement.

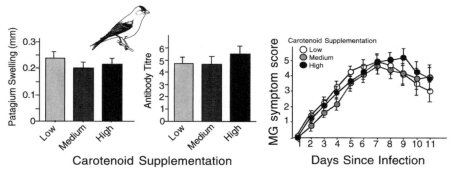

Carotenoid Supplementation                Days Since Infection

Figure 5.18. The immune responsiveness of captive male American Goldfinches in relation to dietary intake of carotenoid pigments. Males were fed lutein and zeaxanthin at a 70:30 ratio as follows: low treatment: 0.01 g/L water; medium treatment: 0.1 g/L water; high treatment: 1.0 g/L water. The left figure shows cell-mediated immune response (wing-web sweling) to injection of a plant protein, phytohaemagglutinin. The middle figure shows a humoral (antibody) immune response to injection of sheep red blood cells (SRBC), which mimics a parasitic infection. The right figure shows the progression of symptoms of conjunctivitis (1 = minor to 7 = severe) following infection with MG. There were no significant differences among treatment groups for any of the experiments (cell-mediated response: d.f. = 2, 36, $F = 0.91$, $p = 0.41$; humoral response: d.f. = 2, 56, $F = 0.38$, $P = 0.68$; severity of MG infection: d.f. = 2, 49, $F = 0.51$, $P = 0.60$, all ANOVA). Adapted from Navara and Hill (in prep).

This is only one test in one species, so these results cannot be used to completely reject the hypothesis that there is a trade-off between use of carotenoids for display versus use of carotenoids for free radical suppression. Similar tests with other species with carotenoid-based color displays will have to be conducted before we can confidentaly assess whether signaling the ability to resist free radical accumulation may be the primary signaling function of carotenoid-

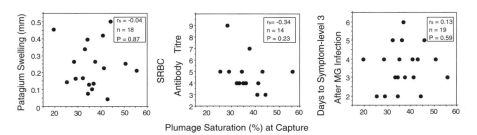

Plumage Saturation (%) at Capture

Figure 5.19. The relationship between saturation of yellow plumage coloration at the time of capture and immune response in captive male American Goldfinches maintained on low-carotenoid diets (see Figure 5.18 for types and concentrations of carotenoids and descriptions of immune challenges). The natural plumage coloration of males did not predict their immune performance, contrary to predictions of the hypothesis that birds trade off use of carotenoid pigments for immune defenses versus use of carotenoid pigments for ornamental display. Adapted from Navara and Hill (in prep).

based color display. If the trade-off that is proposed by the hypothesis proves correct, then an entirely new line of research into carotenoid-based display will begin. Given the results of the experimental test and the theoretical difficulties with the hypothesis, it seems unlikely to be a primary explanation for carotenoid displays in animals.

## The Relative Importance of Carotenoid Access, Nutritional Condition, and Parasites

In this chapter I have presented evidence that three environmental factors—carotenoid access, nutritional condition, and parasites—can affect expression of ornamental plumage coloration in male House Finches. However, I have skirted the key issue of the relative importance of these three environmental variables to plumage color of wild males. Some biologists have ascribed to me the viewpoint that all variation in expression of plumage coloration among male House Finches is due to differences in carotenoid access. In reality, I'm not that single-minded. The data now support the contention that carotenoid access plays an important role in color expression among wild male House Finches. But what is "an important role"? Does carotenoid access explain 90% of the variation in expression of plumage coloration for wild male House Finches? Or, is the effect of carotenoid access on expression of plumage coloration closer to 10%? At the present, no one can answer this question for any population of birds.

What do all these studies on environmental variation and expression of plumage coloration mean? There is now overwhelming evidence that carotenoid-based plumage coloration in the House Finch is a condition-dependent trait. It is becoming increasingly evident, however, that no one measure of condition predicts how an individual will express plumage coloration. In the laboratory, we can attempt to hold all but one environmental variable constant and individually test the effects of dietary access, parasites, and nutrition on plumage coloration. The results of such experiments indicate that each of these factors has the potential to have a significant impact on expression of plumage coloration. The real question is what do the results of these aviary experiments mean for variation in expression of plumage coloration among wild males? Specifically, what is the contribution of these various factors to color display in populations of wild House Finches?

A complete and confident answer to those questions is not yet possible. We need more data on the effects of condition factors outside of aviaries. Given what we do know about wild House Finches, I think that we can confidently state that carotenoid access, parasite infection, and nutrition each play at least some role in shaping the expression of plumage coloration in wild populations of House Finches (Figure 5.20). Moreover, it is not hard to imagine how these factors could interact in complex manners in the wild. For example, individuals that are parasitized might forage less efficiently thus eating less food and ingesting fewer carotenoid pigments. Conversely, males with access to less food due to poor foraging might ingest fewer carotenoids and at the same time be more susceptible to parasite infection. Working out the relative importance of these factors is the next great challenge to biologists

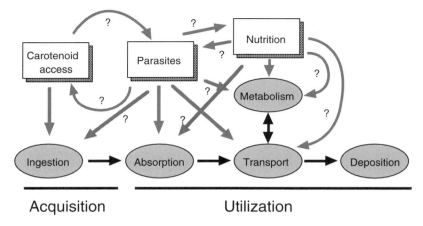

Figure 5.20. The effect of carotenoid access, parasites, and nutrition on the acquisition and utilization of carotenoid pigments needed for ornamental display. Dark arrows show the movement of carotenoids between utilizational processes. Light arrows show effects of environmental variables on both other environmental variables and on utilizational processes. These are the primary environmental factors that have been proposed to affect expression of carotenoid pigmentation. Laboratory and field observations and experiments have confirmed that parasites, nutritional condition, and access to carotenoid pigments can affect expression of plumage coloration. Moreover, these three environmental factors can interact with each other, creating a complex web of effects. Interactions or effects that are hypothesized to occur but that have yet to be demonstrated are labeled with a question mark. See Figure 4.2 for a discussion of the carotenoid acquisition versus utilization and of the physiological processes involved in carotenoid pigmentation.

who want to understand the proximate control of ornamental plumage coloration in wild birds.

To reach the next level of understanding of the proximate control and signal content of carotenoid pigmentation, we must turn the flowchart in Figure 5.20 into a path diagram. In other words, we must work toward assigning an effect (a number representing the importance of that link in the process) to each connection within the path. When that is achieved for the males in one population of birds, we will have reached a new level of understanding of carotenoid-based ornamentation.

To complicate things further, however, the effects of various environmental factors on plumage coloration is almost certainly not going to be the same in all years or for all populations of any given species, and it will certainly vary across species. So, we will have to work out the relative importance of environmental variables to carotenoid expression in several populations in a species like the House Finch and in several species of birds before we begin to grasp which, if any, of the environmental factors tend to exert a common effect and what factors tend to vary across time and space. Such an understanding could well lead to new insights regarding the evolution of species- and population-specific patterns of carotenoid display.

To quantify the relative importance of various environmental factors on the expression of carotenoid-based plumage coloration, investigators will need to be able to measure these variables accurately at appropriate time scales. First, to really start to understand the role of carotenoid acquisition in determining expression of plumage coloration, behavioral ecologists need to assess the carotenoid content of the foods of the birds under study during molt. Such information is currently not available for any species of bird. The only dataset on specific sources of dietary carotenoid pigments and the effects of food quality on color display is from a study on guppies (Grether et al. 1999). Colleagues are often surprised to hear that we have virtually no information on the specific dietary sources of carotenoids that male House Finches use to pigment their plumage. Obtaining such data will require careful observations of foraging birds during molt, perhaps incorporating the use of radio-telemetry, followed by quantification of the carotenoid content of the foods that are observed being eaten. I think that a combination of quantification of the gut contents of a set of birds (see Figure 5.5) and observation of the food consumption of males whose plumage coloration can be scored at the end of the molt period will begin to provide an accurate picture of how carotenoid acquisition affects expression of plumage coloration.

Assessing the role of parasites on expression of plumage coloration presents a different set of challenges. For instance, coccidia appear to be very important parasites with regard to expression of carotenoid-based plumage coloration. Unfortunately, coccidiosis is a relatively difficult disease to diagnose and quantify in wild birds. As yet, we have no thorough study of the effect of coccidiosis on expression of plumage coloration in a wild population of birds. Moreover, although coccidia appear to be particularly important parasites with regard to expression of carotenoid-based plumage coloration, they are only one of many types of parasites, from viruses to cestodes, that infect birds and that might impact plumage coloration. Studies that undertake a more comprehensive analysis of parasites will improve our understanding of the overall role of parasites in expression of carotenoid ornamentation.

A consideration of nutritional condition during molt adds further challenges. To date, nutritional condition has been measured by behavioral ecologists, including me, primarily by crude indices such as total mass corrected for body size or the mean width of feather growth bars. But more accurate measures of total body fat and body composition, which are more commonly used by physiologists, could be applied to give a better idea of the nutritional condition of an individual at the time it is growing ornamental feathers.

To me, the idea of simultaneously measuring nutritional condition, parasites, and carotenoid access in the same individual birds in a wild population is very exciting. Such a study holds great promise for taking the observations from aviary experiments that have been conducted to date and placing them in the context of the production of colorful plumage in real-world situations.

Finally, although experiments with House Finches have shown that expression of carotenoid coloration is largely phenotypically plastic and shaped by the environmental variables outlined above, individuals may also have heritable variation in their capacity to use ingested carotenoids. In virtually every complex system of any organism, there is heritable variation in how that system functions. It seems likely that such genetic variation exists in the metabolic and transport systems involved in the utilization of carotenoids. Partitioning variation in plumage coloration into environmental and genetic components by means of cross fostering experiments in the field and controlled feeding experiments in the lab would indicate the importance of genetic variation in the expression of ornamental coloration.

## Summary

The process by which carotenoid pigments go from potential food in the environment to pigments used to color the plumage of a male House Finch is complex and incompletely understood. It is useful to think of this process as involving two distinct phases—acquisition and utilization. Acquisition refers to the accruement of carotenoids from the environment. Aviary experiments in which I manipulated access to carotenoid pigments demonstrated clearly that carotenoid access could have a large effect on expression of carotenoid-based plumage coloration in male House Finches. A study of the carotenoid content of ingested food items relative to the color of growing feathers of wild male House Finches also supported the hypothesis that carotenoid access can affect expression of plumage coloration.

Carotenoid utilization begins after carotenoid pigments are ingested. It involves the absorption, transportation, metabolism, and deposition of pigments. Parasites have long been proposed to be a key environmental factor that can affect how carotenoid pigments are utilized. In a field study, the level of infection by avian pox and feather mites was negatively correlated with the redness of plumage grown by male House Finches. In infection experiments in an aviary, both isosporan coccidia and mycoplasmosis had a significant negative affect on the carotenoid-based plumage coloration grown by male House Finches.

A final environmental factor that may affect how ingested carotenoid pigments are utilized is nutritional condition. This hypothesis proposes that there are energetic costs to the utilization of carotenoid pigments. Observations from both field and lab studies indicate that access to good nutrition, independent of access to carotenoid pigments, can have a significant affect on expression of plumage coloration.

These studies provide a proximate basis for the relationship between plumage coloration and condition. All male House Finches have the potential to display a full range of color expression from drab yellow to bright red. The realized ornamentation of a given male is a function of its success at ingesting pigments, at avoiding infection by parasites, and at maintaining good nutritional condition. Despite recent interest in the condition factors that affect display of carotenoid pigmentation, the relative importance of these factors on the expression of males in wild population of birds remains to be determined.

# 6 Darwin Vindicated

*Female Choice and Sexual Selection in the House Finch*

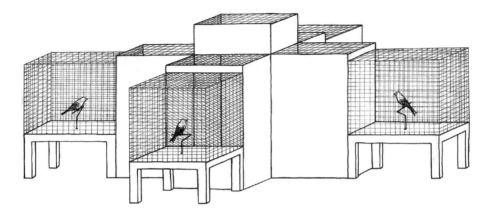

The truth will come out at last, and our difference may be the means of setting others to work who may set us both right.
> —A.R. Wallace (1868), as a postscript to a private letter to Charles Darwin referring to the running debate between the two scientists regarding the importance of sexual selection as an explanation for ornamental coloration; cited in Blaisdell (1992)

... the insufficient evidence that the final choice of the female is frequently determined by the gratification of this [aesthetic] sense, may, I think, be chiefly due to want of patient or discriminating observations upon wild animals in their natural conditions.
> —E. B. Poulton (1890), defending Darwin's idea for sexual selection through female mate choice

In nature . . . competing suitors are likely to be very much alike; this makes matters very difficult for the observer, who may easily pass over small differences which are plain enough to the eyes of the hen birds. This being so, experiment offers a better mode of solving the problem than ordinary observation, and is not difficult to carry out, provided a proper choice of subjects be made. What one needs is birds which are not domesticated, but display naturally sufficient difference in plumage of males to be readily appreciated by a human observer . . . to test female preference all one has to do is to confine them in such a way that, while the males cannot get at each other to fight, the hen may be able to declare her preference by associating with the suitor she favours.

—F. Finn (1907), in an outline of my dissertation research eighty years before I began my studies of House Finches

*We live in an age in which we are getting used to doing less and less for ourselves. We don't produce our own food, clothing, shelter, heat, or even our own entertainment. Most of us do one thing (like teach ornithology at a university) that has little relevance to daily necessities, and then we pay others to provide the essentials for life that we take for granted. The result of this life style is that most individuals seem to be losing the knack for devising solutions for simple problems or for inventing new things.*

*One of the things that I like most about field ornithology and behavioral ecology is that they are fields of study that demand innovation and creativity. For instance, in 1988 I decided that, to properly test the hypothesis that male plumage coloration is used as a criterion by female House Finches in mate choice, I had to manipulate the plumage brightness of wild males. But there was no protocol for what I intended to do. A few biologists had applied black dye to the feathers of wild birds, but I wanted to increase or decrease the redness of male plumage. I tried a number of products from the local "Arbor Drugs" in Ann Arbor. I found to my delight that hair lightener turned the red plumage of House Finches orangish—just the effect that I needed for my experiments.*

*Finding an appropriate red hair dye turned out to be more difficult. I ended up consulting the only true experts in the application of colorants to protein structures on living vertebrates—beauticians at a beauty supply store. Once I convinced them that I wasn't some crackpot and that I really was a biologist who needed a way to make the feathers of live birds red, the women at the beauty shop were an invaluable source of information. They helped me find the best human hair dyes for my experiments.*

*For some new graduate students, it is a bit unnerving to discover that they will have to invent the techniques for their research projects. I typically get questions like: "Could you explain the protocol to me? Is there a manual I can read?"*

*To which I reply "No, there is no protocol or manual. No one has done this before. We have to make up the techniques ourselves."*

*"But then how do we know that we are doing it the best way?"*

*"We don't, and we almost surely aren't. But all research programs have to begin with an idea and some guesswork as to the best way to proceed."*

*Once they've adjusted to the idea of inventing their own way of doing things, students have greatly enjoyed the challenge that field biology presents. Our lists of supplies from the local department store has led to more than a few inquires from the accounting office at Auburn: "What do duct tape and a plastic child-proof latch have to do with biological research?" My grad students Paul Nolan and Andrew Stoehr, who came to Auburn just as I was initiating the field study on the Auburn campus, spent many hours designing better traps, nest boxes, feeders, and banding techniques. Being a competitive lot, at the end of the first year, they wanted me to choose the best invention or innovation. In recognition of the best invention or innovation, we created the McGyver*

*Award, named after a TV character who created incredible devices out of whatever objects happened to be within reach.*

*It's funny to reflect now that what were once crazy new ideas have become standard techniques for my lab group. Lately, as we work within these established protocols, there have been fewer crazy new ideas. Maybe it's time to dust off the McGyver Award.*

As I discussed in chapter 1, when I started my graduate studies in the early 1980s, Darwin and Wallace's great debate over whether females of at least some species of birds showed a consistent mating preference for brightly colored males still had not been tested experimentally. The most extensive work on female mate choice relative to carotenoid display had been conducted on fish, and particularly guppies (Endler 1983, Houde 1987, Kodric-Brown 1985, Noble and Curtis 1939). There were also several avian studies conducted in the 1980s that convincingly demonstrated female choice based on leg band coloration (Burley 1981, 1986, Burley et al. 1982) and tail length (Andersson 1982a, Møller 1988). Most biologists assumed that there would also be female choice for male plumage coloration in many species of birds (e.g., Krebs 1979), but some attempts to demonstrate such choice relative to plumage coloration had failed (Searcy 1979, Slagsvold and Lifjeld 1988). In their 1989 review of plumage coloration, Butcher and Rohwer concluded that female choice for brightly colored males remained untested.

The first and primary goal of my research on the evolution and function of carotenoid-based ornamental coloration in the House Finch was to conduct definitive tests of the hypothesis that females used the quality of male color display as a criterion in choosing mates. I decided to test this idea with two different but complementary experimental approaches. First, I established captive flocks of House Finches and used a "mate choice box" (see Figure 6.1) to test the preferences of females in situations in which male–male interactions could be eliminated and male characteristics could be manipulated. This approach provided me with the opportunity to carefully control the environment in which mate choice occurred. To draw inferences from these trials, however, I had to assume that association preferences displayed by females in small white boxes reflected true mating preferences as would be observed in wild House Finches. I was not comfortable simply assuming that lab observations were representative of wild bird behaviors, so I conducted a companion field study. My approach to testing female choice relative to male color in the field was to change the coloration of wild male House Finches before they started breeding and to observe what effect the manipulation had on their pairing success. The goal was to use this combination of lab and field studies of mate choice to provide a definitive test of female mate preference relative to male plumage coloration in the House Finch.

## Laboratory Mate-Choice Trials

### Preparations

House Finches adapt to cages very well; after all, that's how they got to eastern North America. I started assembling captive flocks of finches for mate-choice trials in September 1987 by catching finches at backyard feeders around Ann Arbor. I continued adding House Finches to these captive flocks until February by which time I had twenty males and twenty-five females in captivity. In the fall and early winter of 1987 I had not yet developed a system for quantifying plumage coloration, so I simply banded the House Finches that I caught and put them in the large aviaries on the roof of the Museum of Zoology at the University of Michigan. I put males and females into separate cages, and I erected a cloth barrier between the male and female cages to visually isolate the sexes. In January 1988, after I had developed my protocol for quantifying plumage brightness, I recorded the plumage coloration of all males in my captive flock. Fortuitously, several of the males that I had captured in September had drab orange or yellow coloration. As a matter of fact, the mean plumage coloration of this group of captive males was much lower than mean coloration that I recorded for wild males in southeastern Michigan (Hill 1990). Although I will never know for sure, in hindsight it seems likely that I put these birds into cages before they had completed their prebasic molt, and without carotenoid supplementation (see chapter 4), they grew pale yellow or orange plumage coloration.

Because I had both drab and bright birds in my captive flocks, I had plenty of variation in plumage coloration with which to present females in mate-choice trials. For these experiments I wanted a set-up that would allow females to view and associate with males, but that would prevent males from competing with or even seeing each other. I adopted a mate-choice box design that had been published in a paper by Burley et al. (1982) and that had been used successfully for a study of mate choice in Zebra Finches (*Poephila guttata*) (Figure 6.1). The set-up consisted of a central chamber ($0.41 \times 0.41 \times 0.94$ m) with four side compartments ($0.31 \times 0.31 \times 0.74$ m). At the end of each side compartment was a Plexiglas window connected to an exterior cage, each of which held one stimulus male. Males were confined to these small ($0.5 \times 0.5 \times 0.5$ m) cages for the duration of a trial. The test female was introduced into the central compartment where there was food and water and from which she was visually isolated from all stimulus males. The female, however, could move about and visit males one at a time by moving under a 10-cm doorway at the base of a sidearm and up onto a perch adjacent to a Plexiglass viewing window. From such a perch the female could see and be seen by one of the four stimulus males. To visit a second male, the female had to move back down and under the doorway leading into the central chamber, through a doorway to a different sidearm, and up onto a perch adjacent to a different viewing window.

This seems like an elaborate set-up for a simple measure of the association preference of females relative to male coloration. The rather complex design of this mate-choice box, however, allowed me to eliminate any chance of male-male competition affecting the preference of the female, and it forced the focal female to actively seek out males if she wanted to associate with them (see Burley et al. 1982). In other words, I was not forcing the females to choose if they were not interested (Wang 1992). With this design, a pattern of association by a female relative to male plumage coloration could reasonably be interpreted as the result of active mate choice.

The one problem with this experimental set-up was that it left no way for an observer to watch the birds during a trial. In the methods section of the Zebra Finch

**Figure 6.1.** The experimental box used to test the mate preferences of captive female House Finches. During a trial, a stimulus male was confined within each of the four small cages at the ends of the arms of the chamber. The test female could move from the central part of the chamber, where she could view no males, to a side arm, where she could view one male by moving under a doorway, and up onto a perch facing the male. The movement pattern of the female was recorded automatically by switches connected to perches adjacent to the viewing windows.

mate-choice study that I had used as a model for my study of House Finches, Burley et al. (1982) gave no indication how they observed their birds. Captive-bred Zebra Finches are much more tolerant of human disturbance than wild-caught House Finches, so Burley and crew may have just sat near the cage and watched what the birds did. However, this approach did not work for my wild-caught House Finches. I tried perching on a ladder above the birds, but no matter how quiet or careful I was, having a human in the room caused the birds to remain silent and still. When I left the birds alone for a few minutes and then carefully approached the mate-choice box, the birds were noisy and active. Clearly, I had to find a way to record their behavior without disturbing them.

If humans could not be present, then an automated system to record the movement of birds had to be devised. In the basement storage of the museum I found an old Estaline-Angus event recorder. This device could not have been simpler to use or more appropriate for my study. Basically, this event recorder had twelve pens that would leave a track on chart paper as the paper slowly scrolled past at a constant rate. The pens each had two positions that would record the state of the electrical circuit to which they were connected: circuit open, pen up; circuit closed, pen down. I bought some simple switches, and I attached each of the perches adjacent to a viewing window to a switch so that, when the female landed on a perch, the switch connected to that perch would close a circuit connected to a pen in the event recorder. When she left the perch, the switch would again open the circuit. The movement of the female to and from the perch was recorded as a shift in the pen position on the paper. At the end of each trial, I had a record of the pattern of movement of the female, and I had the amount of time (length of chart record) that the female spent in each location.

Using a data logger, I could have potentially had all of this movement information recorded by a computer. Such direct collection of movement data would have required a computer committed to the experimental room and the development of software that properly organized the information being collected. I preferred the old fashioned ruler-on-paper approach. Transcribing data took me little time, and the system failed only a few times out of hundreds of mate choice trials. In the second and third years that I used this apparatus to measure female choice, I also connected the male perches to a chart recorder so I could monitor gross activity by males, and I placed sound-activated microphones, which worked just like the switches on the perches, in the male cages to monitor the amount of sound made by males. With this set-up, I could simultaneously monitor the movement pattern of females and the activity and sound production of males, and I didn't have to be present while all these data were being collected. I was free to set up a mate-choice trial and walk outside to make observations on the color-marked population of wild House Finches that surrounded the museum.

I ran mate-choice trials for four hours, and I arbitrarily designated the first hour of each trial as an adjustment period. For a trial to be counted as "successful" the female had to visit each male at least once. The behavior of females in the mate-choice box varied substantially. Some females moved rapidly among males for nearly the entire trial, while most females moved among the males and then sat in association with one particular male for an extended period. Because of the variation among trials in female behavior, I decided to summarize the movement pattern of females as association ranks (again following Burley et al. (1982)). The male with whom the female spent the most time received a rank of 4, and the male with whom she spent the least time was given a rank of 1. With rank scores summarizing each mate choice trial, I could use a simple statistical test to see if the observed ranks of males with different plumage coloration differed from that expected if the females were associating randomly with males.

## The Experiments

Mate-choice experiments consisted of twelve to twenty-one trials in which a different female was presented with four males. For each of the first two experiments that I conducted in 1988, I simply chose four males from the captive group that spanned the full range of plumage color variation—from drab yellow to bright red. In each trial within an experiment, I placed the stimulus males in their cages first and then, just before I left the room, I introduced the test female into the central chamber. I obtained successful trials for fourteen females in experiment 1 and seventeen females in experiment 2. In both experiments females showed a consistent preference for the reddest male presented to them, and the tendency for females to associate with the reddest male was significantly greater than expected by chance (Hill 1990).

These two experiments were certainly suggestive of female mate preference for bright red males, but because I used natural variation in plumage coloration, these initial experiments were not well controlled. To control for male characteristics that may have been correlated with plumage coloration and that could have confounded the results of mate-choice trials, in the third experiment I selected four males with approximately the same plumage scores (all started out orange-red). I then used human hair dye to make one male bright red, and human hair lightener to make a second male orange/yellow. I left two males unchanged. I obtained successful mate-choice trials for eighteen females, and again females displayed a strong association preference for the most colorful male (Hill 1990; Figure 6.2). In this experiment, females also showed a significant aversion to the least colorful male. Because I controlled for correlated effects, I think this third experiment provided more convincing evidence for mate choice relative to plumage coloration than the first two

**Figure 6.2.** Female mate choice relative to male plumage coloration in captive House Finches. Shown are the mean preference ranks for the four stimulus males, where the most preferred male in each trial was assigned a rank of 4 and the least preferred male was assigned a rank of 1. All males started with similar plumage coloration, and their plumage color was altered with hair dyes or lighteners. Females showed a significant preference for the most colorful male presented ($\chi^2 = 15.4$, d.f. $= 3$, $n = 18$, $P = 0.002$, where $n$ is the number of females tested). Adapted from Hill (1990).

experiments. However, in each of these first three experiments, I presented the same four males to each female, so all of these experiments suffered from pseudo-replication. Pseudoreplicaton is a concern in these sorts of mate-choice experiments because there is always the chance that the male designated as the most ornamented male is inherently attractive, independent of its ornamentation. By repeatedly testing females with the same red male, I could have created the illusion of multiple independent tests for a preference for red, when really the experiment was one test showing that a particular male was consistently preferred by many females. Thus, pseudoreplication leaves an experiment vulnerable to having patterns emerge from uninteresting events like accidentally making the best male the red male. If a different, randomly chosen male was used to represent the red male in each trial in an experiment, then such possibilities would be eliminated.

I had to wait until the spring of 1989 to resolve the pseudoreplication problem. The ideal way to have run mate-choice experiments would have been to use a different set of four stimulus males for each trial. However, for an experiment with twenty trials, this would have required eighty different males, and if plumage manipulations had been involved, it would have entailed eighty plumage manipulations. I did not have the aviary space to use that many males, so as a compromise I decided to use twelve different males with three males representing each color type. For each trial within an experiment, I selected a set of four males, one from each color type. I assembled these sets of stimulus males so that no females saw the same set of four males and so that all males were used in approximately the same number of trials. Thus, each female was presented with a unique combination of four males, but all saw approximately the same plumage variation among males (Hill 1990).

In this experiment I was also able to randomize for correlated effects by assigning males to a plumage type. I did this by manipulating the diet of the males at the time of molt (as described in chapter 5). I created yellow and orange males by capturing bright red males prior to their fall molt and feeding them plain-seed diets (on which they grew yellow plumage) or β-carotene-supplemented diets (on which they grew orange plumage). I wasn't able to create red-orange males through diet manipulation, so I dyed some orange males to make them red-orange. Finally, I obtained red males for the experiment by capturing colorful males after they had completed their fall molt. The result was a set of adult males, all with about the same natural coloration, now representing the full range of color variation in House Finches (Hill 1990).

I completed twenty-one trials in this experiment and found again that females displayed a very strong preference for the most colorful male (Hill 1990) (Figure 6.3). The most colorful male ranked first in association preference in eighteen out of the twenty-one trials, and second in the remaining three trials. The mean ranks of the other three males were ordered as expected, but the differences were small and not significant. For these experiments, I also monitored the sound production and movement rate of males as well as their dominance position in their captive flocks. In contrast to the response to coloration, there was no association preference for the most active male. In this comparison I simply ordered males according to their relative activity, just as I ordered males according to their relative coloration in the previous comparison. I also found no association preference related to vocal activity. Finally, there was no significant association preference related to domi-

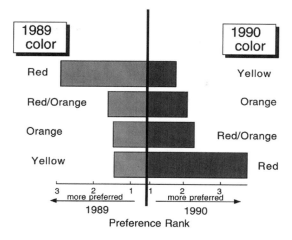

Figure 6.3. The mean mate preference ranks of female House Finches relative to the twelve males used as stimulus males in 1989 and 1990 mate-choice trials. Three males represented each color category, and I assigned males to plumage color categories by controlling access to carotenoids during molt. In both years, females showed a strong and significant association preference for the most colorful male presented (1989: $\chi^2 = 31.34$, d.f. $= 3$, $n = 21$, $P = 0.0001$; 1990: $\chi^2 = 17.10$, d.f. $= 3$, $n = 12$, $P = 0.001$, where $n$ is the number of females tested). The same males that represented the reddest color category in 1989 were used to represent the drabbest category in 1990, and the same males used to represent the drabbest category in 1989 were the brightest males in 1990. Females altered their choice of males accordingly, demonstrating that they were attracted by the coloration of males, not by some other feature of the males. Adapted from Hill (1990) and (1994a).

nance rank. From this experiment I concluded that size, sound production, movement rate, and dominance had no significant effect on female association preference, but that plumage coloration had a highly significant effect (Hill 1990).

As a final definitive test of female mate choice relative to male coloration, I conducted a female choice trial in 1990 using the same twelve stimulus males used in the experiment described above. However, in the summer of 1989, prior to the prebasic molt of these birds, I placed them on diets that were the reverse of their diets in the previous year. The males that initially had yellow plumage were put on a diet supplemented with the red pigment canthaxanthin, and they grew bright red plumage. The males that initially had red plumage were put on a plain-seed diet, and they grew pale yellow plumage. The six birds representing the intermediate color groups were again fed β-carotene, and they grew orange plumage. For these latter groups, however, the three birds that had been dyed to look red/orange in 1989 were not dyed in 1990, and the other three males were dyed to look red/orange. Thus, the most highly ornamented and most preferred males of the 1989 trials were now the least ornamented males, and the least ornamented and least preferred males of the 1989 trials were now the most ornamented males. In this experiment I tested the mate preferences of females captured in California (the rational for using California birds will be explained in chapter 11), but these

California females were the same subspecies as House Finches from Michigan and there was no reason to think that they should behave differently.

I obtained successful trials for twelve females, and these females showed a strong preference for bright red males. The red male ranked first in association preference in ten of these twelve trials, and female association with the reddest male was significantly greater than expected by chance (Hill 1994a) (Figure 6.3). This last experiment is particularly convincing evidence of female mate preference for red plumage coloration because the males for which the females showed strong preferences were the same males for which a different group of females had shown the least preference in the previous year. The only difference was that the plumage color of the stimulus males had been switched between years. The same males were highly attractive with bright red plumage, but not attractive with dull yellow plumage (Hill 1994a).

In a final series of experiments using this mate-choice box, I tested the mate preferences of female House Finches relative to patch size as well as plumage coloration. These studies were conducted in the context of a comparative study of the evolution of plumage pattern in the House Finch, and I will present these results in detail in chapter 11. Here I will simply state that female House Finches showed a significant preference for males with large patches of color over males with smaller patches of color (see Figure 11.4), but when they were forced to choose among traits, they showed a significant preference for small bright patches over large drab patches (see Figure 11.5). Thus, females assess male patch size in mate choice, but they use plumage brightness as their primary criterion in choice.

## Field Studies of Mate Choice

### Correlational Evidence: Plumage Color and Pairing Success

Female House Finches begin breeding in the first spring following the year in which they hatch. All males are also sexually mature and attempt to breed in the spring after they hatch in most populations (including the Michigan and Alabama populations that were the focus of my studies). In most years, however, a substantial proportion of males is unable to attract a mate and breed. The number of males failing to breed is hard to determine exactly because non-breeding males form a floater population, but I estimated that up to 50% of males in some populations may fail to obtain mates (Hill et al. 1994b). In the Ann Arbor population and in the Auburn population before 1997, the percent of unpaired males was closer to 10–20%. The reason that many males go unpaired is primarily due to the highly male-biased sex ratio observed in virtually every population of House Finches (Hill et al. 1994b). Asynchrony in female breeding may also allow some males to monopolize more than one female per season even within the constraints of social monogamy, and this may further increase the proportion of males that go without a mate (Hill et al. 1994b). A mycoplasmal epidemic that began in the summer of 1996, described in chapter 5, disproportionately killed males and caused the sex ratio in Auburn and probably throughout eastern North America to change from male biased to female biased (Nolan et al. 1998). This event substantially decreased the number of unpaired males in the Auburn population in subsequent years. For now, however, I want to focus on "typical" years in which the sex ratio was male biased and a substantial number of males went unpaired.

As described in chapter 2, House Finches form strong pair bonds beginning about six weeks before the first clutch of eggs appears in a population. Prior to the onset of incubation by females in a population, if a male is observed alone for any length of time it can reliably be assumed to be unpaired. Conversely, a male can be categorized as paired if it is seen in association with a female and the relationship appears cooperative (Hill et al. 1999). Using these criteria, my students and I recorded the pairing status of males through the early spring in four season in Ann Arbor and four seasons in Auburn. Once females in a population began incubating, categorizing males as paired or unpaired became much more difficult because a lone male could have been either unpaired or paired but with its mate incubating eggs. Thus, we could categorize a male as being paired with certainty, but we could never be sure that a lone male was really unpaired. Our assumption was that, over a long breeding season, most paired males would be seen at some time in association with a female.

As a first field test of the hypothesis that female House Finches prefer to mate with brightly colored males, I compared the plumage redness of paired males to that of males that were not observed to be paired. This is a relatively conservative test because the unpaired-male category included males that did not pair and males that paired but were not seen associating with their female. Despite the conservative nature of the test, I found that paired males were consistently redder than unpaired males. This pattern held for four years of observation in Ann Arbor (Hill et al. 1999) (Figure 6.4), and it has held for four years of observation in Auburn (1996 and 1997, published in Hill et al. 1999, and 1998 and 1999 in Figure 6.5). The significant relationship between redness and pairing success is especially interesting for the 1997–99 Auburn data because, due to the *Mycoplasma gallisepticum* epidemic, there was little variation among males in plumage coloration in these years, but plumage coloration still predicted pairing success. The observation that paired males are redder on average than males not observed to be paired is not a definitive test of the hypothesis that female House Finches choose redder males as mates, but it does corroborate a primary prediction of the hypothesis.

One important variable that is not ruled out in these simple comparisons is the effect of male–male competition. If brighter males are dominant to drabber males, then bright males could monopolize females and generate the relationship between redness and pairing success without female choice. As I outlined in chapter 2, House Finches are not territorial, and it appears that a male House Finch can only pair with a female who accepts it as a mate. Thus, it seems unlikely that a correlated effect of dominance and coloration could have created the pattern. Moreover, as I will show in chapter 8, drab males—not bright males—tend to be socially dominant in House Finches.

Another obvious variable to consider in a comparison of plumage redness and pairing success is age. As described in chapter 2, yearling males are drabber on average than older males. Therefore, it is possible that female House Finches preferentially mate with older males, irrespective of their coloration, and that the

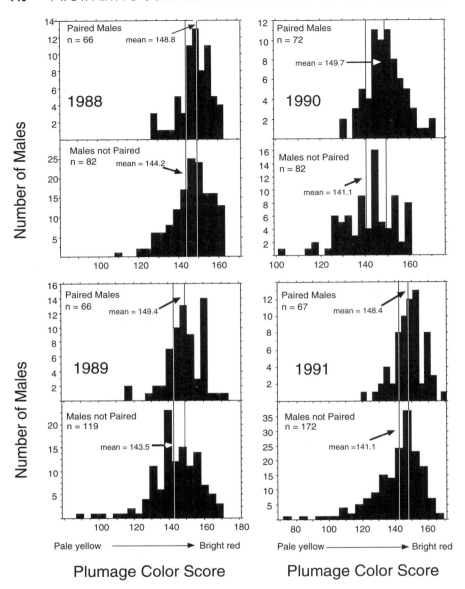

Figure 6.4. The distribution of plumage scores of males known to have paired versus males that were not observed to be paired in the Ann Arbor, Michigan, population in the years 1988 to 1991. Means are indicated with vertical lines. Paired males consistently had higher mean plumage color scores than males that were not paired (one-tailed $t$-test comparisons: 1988: $t = 2.80$, $P = 0.006$; 1989: $t = 2.97$, $P = 0.002$; 1990: $t = 5.09$, $P = 0.0001$; 1991: $t = 3.69$, $P = 0.0002$). The effect was at least partly independent of age. When I compared only males known to be in their second or subsequent breeding season, I also found that paired males were brighter than males not observed to be paired ($n = 84$, 40 respectively, $t = 2.13$, $P = 0.017$). Adapted from Hill (1990) and Hill et al. (1999).

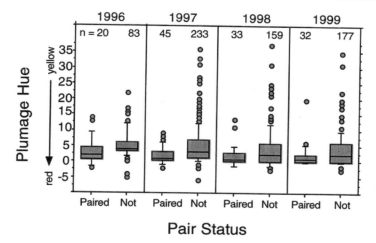

Figure 6.5. Box plots comparing the plumage hues of males known to have paired versus males that were not observed to be paired in the Auburn, Alabama, population in the years 1996 to 1999. Paired males consistently were redder than males that were not paired (one-tailed Mann-Whitney $U$-test comparisons: 1996: $U = 1177$, $P = 0.002$; 1997: $U = 7016$, $P = 0.0002$; 1998: $U = 3383$, $P = 0.004$; 1999: $U = 2614$, $P = 0.008$). Color measurements were made with a Colortron, so redder males received lower hue scores. Adapted from Hill et al. (1999) with 1998 and 1999 data added.

apparent choice for redder males is only a correlated effect of choice for older males. However, when I compared the plumage redness of paired versus unpaired males known to be in at least their second breeding season, I still found that paired males were significantly redder (Figure 6.4). Thus, the effect of redness on male pairing success was not simply a correlated effect of male age.

*Patch Size and Pigment Symmetry*

My published studies of mate choice relative to carotenoid ornamentation have focused almost entirely on the hue and saturation of plumage. Carotenoid ornaments of male House Finches vary not only in the degree of pigment elaboration (redness and saturation of color); they also vary in the size of the ventral patch of colored feathers and in the left/right symmetry of pigment deposition (Hill 1992, 1998b; see chapter 2). In field observations, I found that, not only did paired males have significantly brighter plumage, they also had significantly larger patches of ornamental coloration (Figure 6.6) and more symmetrical crown pigmentation (Hill et al. 1999; Figure 6.7) than unpaired males. Unfortunately, I stopped measuring the patch sizes of males in Auburn at the same time that I started measuring pigment symmetry. When my students and I switched to recording plumage data with a Colortron and started recording pigment symmetry, the processing time for birds was getting excessive, and we decided to reduce the variables that we recorded. So, I have data for pairing success and patch size only for Michigan. In all four years in which I monitored the Michigan population, paired males had

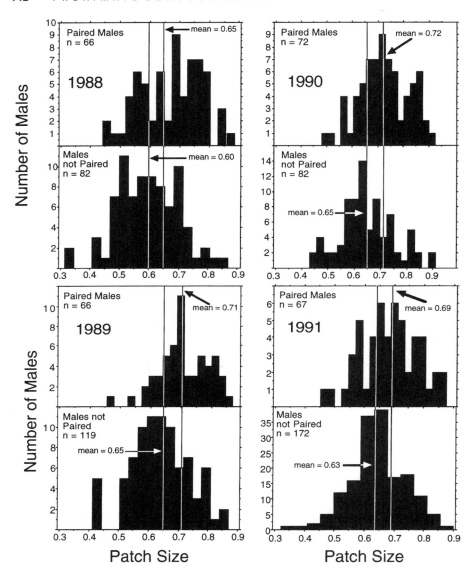

Figure 6.6. The distribution of the proportion of ventral area with carotenoid pigmentation (patch size) of male House Finches known to have paired versus the patch sizes of males that were not observed to be paired in Ann Arbor, Michigan, population in 1988 to 1991. Means are indicated with vertical lines. Paired males consistently had larger patches, even when I controlled for the correlated effect of color by using the residual values from the correlations between plumage color score and patch size (one-tailed $t$-test comparisons of data adjusted for plumage coloration: 1988: $t = 1.63$, $P = 0.05$; 1989: $t = 2.97$, $P = 0.002$; 1990: $t = 5.09$, $P = 0.0001$; 1991: $t = 3.69$, $P = 0.0002$). Not previously published.

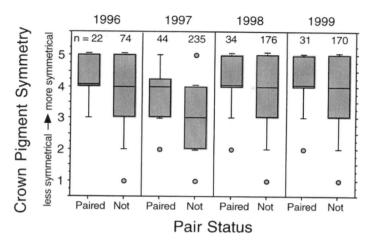

Figure 6.7. Box plots comparing the degree of pigment symmetry of crown feathers of male House Finches known to have paired versus males that were not observed to be paired in the Auburn, Alabama, population in the years 1996 to 1999. Paired males consistently had more symmetrically pigmented plumage than males that were not paired (one-tailed Mann-Whitney $U$-test comparisons: 1996: $U = 1054$, $P = 0.02$; 1997: $U = 7017$, $P = 0.0001$; 1998: $U = 3549$, $P = 0.03$; 1999: $U = 2999$, $P = 0.09$). Adapted from Hill et al. (1999) with 1998 and 1999 data added.

significantly larger patches of color than unpaired males (Figure 6.6). This relationship between patch size and pairing success held even when I removed the effects of plumage coloration by taking the residual values from a regression analysis of plumage coloration (independent variable) versus patch size (dependent variable).

My students and I recorded pigment symmetry scores only for the males in the Auburn population. For all years of comparison, we found that paired males had significantly more symmetrical crown pigmentation than males that were not paired (Hill et al. 1999) (Figure 6.8). As with patch size, this relationship held even when we removed the effects of plumage coloration by taking the residual values from a regression analysis of plumage coloration (independent variable) versus pigment symmetry (dependent variable) (Hill et al. 1999) (Figure 6.7).

Thus, all three components of carotenoid ornamentation that I measured—coloration, patch size, and pigment symmetry—had a significant effect on male pairing success. Laboratory mate-choice experiments showed that female House Finches used male plumage coloration as a primary criterion in mate choice, and male patch size as a secondary criterion. No experiments have been conducted to test the relative effects of pigment symmetry versus plumage coloration. It is intriguing to speculate, however, that in choosing a mate, female House Finches use several different components of carotenoid ornaments, perhaps obtaining different information about male condition from an assessment of each component (see also Badyaev et al. 2001).

## A Field Experiment

To eliminate all confounding variables in a test of female mate choice relative to male plumage coloration in wild House Finches, I conducted a plumage-manipulation experiment. The breeding biology of the House Finch made it an ideal subject for an experiment in which the plumage coloration of males was changed before or during pair formation. Unlike migratory species that have been the subject of most field studies of sexual selection, the breeding population of House Finches that I studied was present on the study site for weeks before nesting was initiated. This meant that I had weeks in which I could capture males and alter their plumage coloration before pairs became stable.

Working with the Ann Arbor population in the springs of 1989 and 1990, I changed the plumage coloration of a set of wild males and observed the effect of the manipulation on pairing success. In 1989 I altered the appearance of forty males, making twenty more colorful and twenty less colorful. In 1990 I altered the appearance of an additional sixty males, making twenty more colorful, twenty less colorful, and using twenty males as sham controls. This experiment included all unbanded males that I captured on the study site between 10 February and 21 March, which is before the start of nest building. Most unbanded males captured in late winter and early spring are yearling males that have been newly recruited into the breeding population. The males that had bred previously on the study site were virtually all banded, and hence excluded from manipulation. So I manipulated a population of young males that had not bred previously and that were trying to attract a female for the first time.

As in my mate-choice experiments, I performed the plumage manipulations using human hair dyes and lighteners. Sham controls had color buffer applied in place of dye or lightener, but they were otherwise treated the same. In all cases, the plumage scores of manipulated males fell within the range of natural plumage coloration in the Ann Arbor population. The manipulation simply moved individuals toward the color extremes—drab yellow or bright red (Hill 1991). These manipulations removed all variation in both patch size and pigment symmetry among males, so I was testing the effect of plumage coloration independent of variation in the other components of carotenoid pigmentation. The results of the experiments were very similar in both years, so I pooled observations from 1989 and 1990 (Hill 1991).

There did not seem to be any mortality or differential dispersal associated with any treatment. Of the forty males that were brightened, twenty-three were resighted after being released; of the twenty sham controls, ten were resighted after being released; and, of the forty males that were lightened, twenty-six were resighted after being released. The manipulation had a rather dramatic affect on pairing success, however. Twenty-two of the twenty-three brightened males that were resighted paired. In contrast, only seven of the twenty-six lightened males and only five of ten sham controls paired (Figure 6.8). Moreover, among those males in the experiment that paired, bright males paired on average faster than sham controls which paired on average faster then lightened males. Unlike the correlational data on which I had focused up to this point, this experimental manipulation

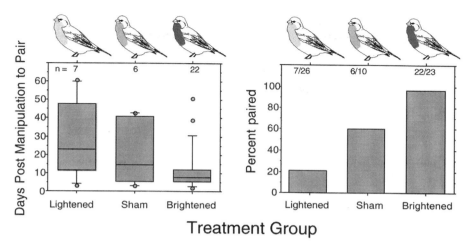

## Treatment Group

**Figure 6.8.** The effect of plumage color manipulation on pairing success of wild House Finches in Ann Arbor, Michigan. The left panel shows the time from plumage manipulation until pairing for males in each of the three treatments. Brightened males tended to pair more quickly and lightened males less quickly than controls ($H = 3.99$, d.f. $= 2$, $P = 0.07$). Note that this comparison excludes males that failed to pair. The right panel shows the percentage of the males in each treatment group that remained on the study site that paired. A higher percentage of brightened males and a lower percentage of lightened males paired relative to controls ($\chi^2 = 24.75$, $P = 0.0001$). Adapted from Hill (1991).

enabled me to rule out mating success related to any effect except plumage coloration.

Thus, both laboratory and field experiments indicated that male plumage coloration is an important criterion for female mate choice in the House Finch, and that females prefer to mate with the brightest male available. One hundred and thirty-one years after Darwin had proposed the idea that female birds choose males based on expression of plumage coloration, there was definitive experimental support for the hypothesis.

## The Payoff to Males for Having Bright Plumage Coloration

One fundamental question related to my studies of female mate choice and male plumage coloration is: What benefits do males gain by investing in colorful plumage? The obvious benefit, which was the focus of most of the field studies outlined above, is improved pairing success. Brightly colored males have a significantly better chance of attracting a social mate than do drably colored males. Some 10–20% of male House Finches go without mates in most populations. The failure of some males to breed generates variance in male reproductive success, and is undoubtedly important in maintaining bright color display in male House Finches. However, this means that in most populations 80–90% of males end up pairing with a female. This begs the question: among males that pair, do redder males gain benefits from their plumage display? To answer this question, I looked at the success of paired males relative to their plumage coloration.

## Nest Initiation Date and Offspring Production

Whether there are benefits of ornamentation in monogamous mating systems in which most or all males pair is a key issue for how sexual selection operates in nature. Birds are among the most ornamented of all animals and yet, like the House Finch, 90% of birds are socially monogamous (Ligon 1999). Darwin understood the significance of this problem for his theory of sexual selection (Darwin 1871:261). His solution was to point out that the game of reproduction was not won simply by securing a mate. Darwin proposed that within any population, females would vary in condition and experience. As a result, contests among males would extend well beyond simply being successful in attracting a social mate. Among males that paired, Darwin proposed that some males would attract the highest quality females, while other would attract lower quality females. Darwin further speculated that there would be a fitness payoff for males that attracted the highest quality females in terms of more or higher quality young produced. Specifically Darwin, and later Møller (1988), proposed that, for seasonally breeding birds in temperate regions that have the potential to raise more than one brood, a key determinant to breeding success, beyond simply finding a mate, would be nest initiation date. Pairs that started nesting earlier generally would produce more young in a season than pairs that began nesting later (e.g., Møller 1988, Wolfenbarger 1999c). Females that initiated nesting earliest in the year should be the most attractive as mates, and should have choice of all males. The prediction, therefore, was that there would be a negative relationship between nest initiation date and male coloration—brighter males should be associated with earlier nests.

With my students, I looked at nest initiation date and plumage coloration in four seasons in Michigan and four seasons in Alabama. We found a consistent relationship between male plumage redness and nest initiation date, with redder males nesting earlier (Hill et al. 1994b, 1999) (Figure 6.9). Yearling males tended to be less red on average than older males (see Figure 2.2), and yearling males are, by definition, less experienced breeders than older males. Thus, it seemed possible that the pattern of plumage redness and nest initiation was simply a correlated effect of age. However, when we controlled for age in the analysis, redder males still initiated nesting sooner (McGraw et al. 2001).

Darwin (1871) proposed that not only would the most highly ornamented males pair with females that initiated breeding early, but that they would also pair with females who were more experienced and likely to produce more young. In a number of passerine species, female nesting success has been shown to increase with age as females gain experience in choosing nest sites, constructing nests, and provisioning young (Curio 1983, Harvey et al. 1979, Part 1995, Perrins and McCleery 1985). I tested the hypothesis that more brightly colored males would pair with more experienced females by comparing the plumage color of males to the age of their mates (Figure 6.10). I found that redder males tend to pair with older, more experienced females (Hill 1993c).

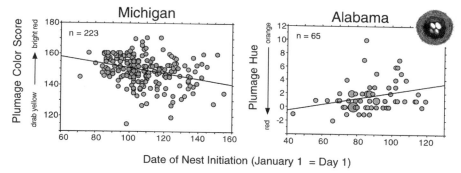

Figure 6.9. Male plumage coloration in relation to the date of nest initiation, measured as date on which the first egg was laid in a male's nest. In both Michigan ($r = -0.31$, $P = 0.0001$) and Alabama ($r = 0.29$, $P = 0.02$), pairs with redder males tended to initiate nesting earlier than pairs with drabber males. For each location, data are pooled across four years (Michigan: 1988–91; Alabama: 1997–2000). Note that the slope of the two regressions is reversed because brighter males received a higher score in Michigan (book scoring) while brighter males received lower scores in Alabama (Colortron scoring). Adapted from Hill et al. (1994b, 1999).

These observations are logical extensions of the observation that redder males are preferred as mates. Males that are most attractive will have their choice of females (see chapter 9 for a detailed discussion of male mate choice), and these males would be expected to pair with females with the most experience, that are in the best condition, and that can initiate nesting early. The question then becomes: Is

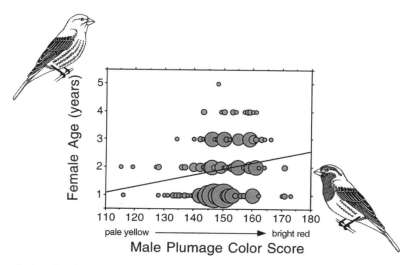

Figure 6.10. The plumage color score of male House Finches in relation to the age of their mates in Ann Arbor, Michigan. Redder males tended to pair with older, more experienced females ($n = 270$, $r_s = 0.20$, $P = 0.001$). This pattern supports the hypothesis that males with redder and more saturated plumage are preferred as mates by females. Adapted from Hill (1993c).

there a fitness benefit to nesting early with experienced females as predicted by Darwin (1871)?

My students and I looked at annual reproductive success of paired males relative to plumage coloration and nest initiation date in the Auburn population of House Finches. I was unable to conduct this study in the Ann Arbor population because too many nests were inaccessible, and hence I had only incomplete data on reproductive success. In the Auburn population, however, we found and had access to virtually every nest in the population, which allowed us to compute the annual reproductive success of males. As I mentioned above, in all years of observation in both Michigan and Auburn, there was a correlation between plumage redness and nest initiation date—on average, redder males started to breed earlier. With complete reproductive data from the Auburn population, we were able to show that this earlier nest initiation date led to more broods per year on average and more chicks fledged (McGraw et al. 2001; Figure 6.11). So, plumage redness did translate into greater reproductive success, even among those males that attracted a social mate. Moreover, this effect was independent of age because for this group of males neither male coloration nor male reproductive success was related to age (McGraw et al. 2001). These observations show that, at two distinct points in the reproductive process, plumage redness enhances male reproductive success. First, plumage redness increases a male's chance of attracting a social mate; and second, brightly colored plumage allows males to attract higher quality females, which leads to earlier nest initiation and to more offspring produced.

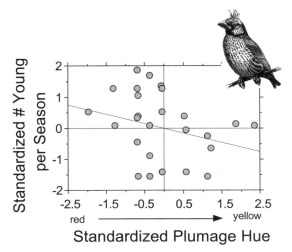

Figure 6.11. The relationship between male plumage redness and reproductive success measured as chicks fledged per season. Redder males tended to produce more chicks than drabber males ($r_s = -0.38$, $n = 29$, $P = 0.046$). In this dataset, male age was unrelated to either male plumage coloration ($r_s = -0.16$, $n = 29$, $P = 0.39$) or the number of young fledged ($r_s = 0.02$, $n = 0.35$, $P = 0.90$), so this effect of early breeding was independent of age effects. Data are for male House Finches breeding on the Auburn, Alabama, study site in 1997 and 1998. Before being pooled between seasons, plumage hue and reproductive success were standardized to a mean of zero. Adapted from McGraw et al. (2001).

*Extra-Pair Paternity*

The final reproductive process through which males may gain an advantage by having brightly colored plumage is mating that occurs outside of pair bonds. Extra-pair mating and extra-pair paternity generally go unrecorded in field studies. Only a few years ago, virtually all field ornithologists assumed that, in monogamous species with biparental care, all the chicks in a nest were fathered by the male attending the nest. The discovery of high levels of extra-pair paternity in many species of monogamous birds with biparental care (e.g., Westneat et al. 1990) has revealed that fitness benefits of ornamental traits may accrue outside of the context of attracting social mates. Thus, I wanted to check for the effect of extra-pair mating in my study populations.

Two components of extra-pair mating might be affected by male plumage coloration: (1) bright plumage might aid a male in avoiding being cuckolded; and (2) brightly plumaged males may gain a disproportionate share of extra-pair matings. Extra-pair mating has the potential to reinforce the selective advantage conferred to brightly colored males during pairing. It could also counter the sexual selection resulting from female mate choice if brightly colored males are cuckolded more than drab males or if they do poorly in gaining extra-pair copulations. By working with Bob Montgomerie and in the molecular genetics laboratory of Peter Boag at Queen's University, I was able to test whether plumage coloration affected a male's likelihood of being cuckolded.

In my last field season in Michigan, I collected blood samples from the chicks and the attending male and female at thirty-five nests. We then used DNA fingerprinting to determine the genetic mother and father of the chicks. We found that the attending female was the genetic mother of all chicks that we tested. In contrast, ten of the 119 chicks (8.4%) were fathered by a male other than the attending male. These extra-pair young were found in five different nests, so extra pair young were detected in the nests of 14.3% of males (Hill et al. 1994b). Unfortunately, determining the actual paternity of the extra-pair young was beyond the scope of our study. Using DNA fingerprinting we could confirm or rule out the paternity of the attending male, but we could not search this large House Finch population for the genetic father of extra-pair young. DNA fingerprinting simply did not produce sufficiently detailed banding pattern to compare the banding patterns of a chick and dozens of potential fathers. Newer paternity techniques using microsatellites apparently have such discriminating power (Double et al. 1997, Hanotte et al. 1994, Primmer et al. 1995), and I hope one day to repeat this study with assignment of paternity.

We could and did answer the question: are redder males cuckolded less than drabber males? The answer was a pretty definitive "no." We had substantial variation in plumage coloration among the males that we included in the paternity study. We found that the mean coloration of cuckolded males was virtually identical to the mean coloration of males that were not cuckolded (Hill et al. 1994b) (Figure 6.12). Male age also did not predict who was cuckolded. As a matter of fact, the only variable that we measured that seemed to predict who would be cuckolded was nest crowding. Nests that were in close proximity to the nests of other House Finches were significantly more likely to have extra-pair young compared to nests placed

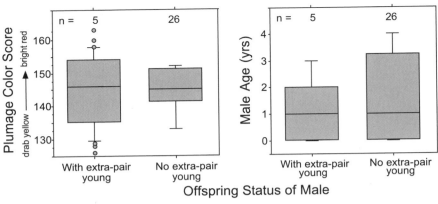

Figure 6.12. The effect of male plumage coloration and age on cuckoldry. DNA fingerprinting was used to determine whether the male attending a nest was the genetic father of the chicks in that nest. Males that were cuckolded did not differ significantly in mean plumage coloration from males that were not cuckolded (Mann-Whitney $U = 62$, one-tailed $P = 0.44$) (left panel). Age also did not predict whether or not a male House Finch would be cuckolded (Mann-Whitney $U = 56$, one-tailed $P = 0.31$) (right panel). The only variable that differed significantly between males that were cuckolded and males that were not cuckolded was crowding; nests closer than 10 m were significantly more likely to have extra-pair young than nests farther than 10 m apart ($G = 4.39$, $P = 0.04$). Adapted from Hill et al. (1994b).

more than 10 m from the nearest House Finch nest. From these studies we concluded that there is no evidence that redder males either gain or lose extra-pair benefits from their bright plumage. Thus, extra-pair mating appears to neither oppose nor reinforce sexual selection on plumage coloration generated by mate choice in the House Finch. A study of plumage coloration in relation to success in obtaining extra-pair copulations remains to be conducted.

## *Mate Choice for Carotenoid Pigmentation Beyond House Finches*

Since I published my results that female House Finches prefer to mate with males with the reddest and brightest carotenoid-based plumage coloration, similar female preference for carotenoid-based plumage coloration has been documented in a few other species of birds. In experimental studies of male choice relative to plumage coloration in the American Goldfinch (Johnson et al. 1993) and the Yellowhammer (Sundberg 1995a), females were found to prefer as mates those males with the most saturated, carotenoid-based plumage coloration. Moreover, as I found in House Finches, Wolfenbarger (1999c) found a significant negative relationship between plumage redness and first egg date and a significant positive relationship between redness and reproductive success in the Northern Cardinal, suggesting that redder males are preferred as social mates. However, in laboratory mate-choice experiments, she found no evidence that female cardinals prefer redder males as mates (Wolfenbarger 1999a). In my experience, cardinals are difficult to keep in captivity—they adjust poorly to the confined space of cages—and the inability of male cardinals to adjust to captivity may have affected the outcome of these mate-choice trials. In a field study of the Common Rosefinch, *Carpodacus erythrinus*, a congener of the House Finch, no relationship was found between pairing success and redness (Björklund 1990). In this study, however, expression of carotenoid-based plumage coloration was scored only as "bright" or "dull," and only five bright and five dull males were used in the comparison (Björklund 1990).

One form of female choice that is particularly difficult to assess is choice of extra-pair sexual partners. This form of mate choice can be studied indirectly by determining the paternity of offspring (Birkhead and Møller 1992). Studies to date provide mixed support for the idea that carotenoid-based plumage coloration is used by females in choosing extra-pair mates. In Yellowhammers, drably colored males were no more likely to be cuckolded than brightly colored males, but males that gained paternity through cuckoldry were brightly colored (Sundberg and Dixon 1996). In contrast, S. U. Linville and R. Breitswich (pers. commun.) found that male Northern Cardinals that gained paternity through extra-pair copulations were no brighter on average than males that had no such paternity gain and that bright males were cuckolded more than drab males. In Zebra Finches, males with red leg bands (simulating bright carotenoid pigmentation) gained more paternity through extra-pair fertilizations and were cuckolded less than males banded with green bands (Burley et al. 1994, 1996).

Although it is tempting to extrapolate from my studies of House Finches and from the studies of goldfinches and Yellowhammers to the idea that mate attraction is a general function of carotenoid-based plumage coloration in birds, too few species have been tested and almost half the studies published to date refute the hypothesis. Interpretation of the generality of the patterns of mate choice and carotenoid pigmentation that I have found in House Finches will have to await further study.

## Mechanisms of Female Choice

It is ironic that, although female choice is acknowledged as the primary mechanism driving sexual selection for ornamental traits and has been a focus of so much of my research, it is a process that I have barely paused to consider in my experiments and observations. I have generally approached my studies of plumage coloration with the implicit assumption that female House Finches have a single mate preference, and I've used experiments to reveal this single preference. Clearly, this is an overly simplistic view of how females choose mates.

Mate preferences must necessarily be conditional. As Darwin pointed out, females are not going to forsake breeding just because they do not have access to a maximally ornamented male. Presumably, females mate with the most attractive male available to them. But what is the search strategy used by females? Do they establish some threshold and then accept the first male they encounter that exceeds this minimum? Do they sample for a given period of time or until a given number of males is viewed and then return to the most preferred male in the sampled pool? (See Real 1990 for a summary of hypothesized search tactics.) Such questions about female mate sampling have been addressed primarily in studies with fish and insects. Few studies of the process of mate selection have been conducted on birds, and none has been conducted with House Finches or any bird with carotenoid-based signals. Moreover, at least some females choose both a social and extra-pair mate. Do females use the same choice algorithms for both types of mates?

The optimal mate-choice strategy for a female to follow is likely to be dependent on context, because the signal content of male traits is likely to be context-dependent (Qvarnström 2001). Viewing female mate choice as a condition- and context-specific character will make the study of the evolution of female mating preferences more complex, but it should provide a framework that will lead to a better understanding of why females make the choices that they make.

## Summary

Both laboratory and field experiments with House Finches show that females prefer to mate with males with the reddest and most saturated carotenoid-based plumage coloration. Females also show a preference for males with larger patches of coloration and more symmetrical crown pigmentation. In the populations of House Finches that have been most intensively studied, about 10–20% of males fail to pair, so success in attracting or failing to attract a social mate is a strong selective pressure maintaining expression of male plumage coloration. The benefits to males of having bright red plumage coloration, however, extend beyond pairing success. Among those male House Finches that pair, redder males nest earlier, and they pair with older, more experienced females. As a result, among males that pair, redder males have higher annual reproductive success than drabber males. The effect of plumage coloration on extra-pair paternity in the House Finch has not been adequately studied. One study to date indicates that male plumage coloration has no effect on the likelihood that a male will be cuckolded, but it is unknown whether plumage coloration affects the success of males in gaining extra-pair paternity.

These observations confirm Darwin's hypothesis that female mate preference provides the selective advantage that maintains bright plumage coloration in populations of birds. It also sets the stage for further studies of plumage coloration in the House Finch, including comparative studies that test hypotheses for the evolution ornamental plumage coloration. The evolution of plumage coloration will be taken up in chapter 11. For now, having addressed the question, Why are male House Finches red?, I will turn to the corollary question, What do female House Finches gain by choosing red males?

Fine Fathers and Good Genes

## The Direct and Indirect Benefits of Female Choice

Consider the case of an ancestral woman who is trying to decide between two men, one of whom shows great generosity with his resources to her and one of whom is stingy. Other things being equal, the generous man is more valuable to her than the stingy man. The generous man may share his meat from the hunt, aiding her survival. He may sacrifice his time, energy, and resources for the benefit of the children, furthering the woman's reproductive success. In these respects, the generous man has higher value as a mate than the stingy man. If, over evolutionary time, generosity in men provided these benefits repeatedly and the cues to a man's generosity were observable and reliable, then selection would favor the evolution of a preference for generosity in a mate..
—D. Buss (1994), in *The Evolution of Desire*

*Since I began conducting experiments with aviary birds and traveling to different parts of North America to sample House Finches, I've had to rely on the generosity of the bird-feeding public to provide me with locations at which to trap. I never cease to be amazed at the devotion of these folks to the care of their wild birds. I'm also constantly amazed at how misinformed many of the most avid bird feeders are about the feathered denizens that visit their yards.*

*On at least a half-dozen occasions I've contacted a homeowner with a bird feeder about catching birds in his or her yard. After they've heard me present my justification for disturbing the birds, they agree to have me trap at their feeder, but they add, "I only have ten birds that use the feeder. You are welcome to band those, but that's the most I ever see." At the end of the day the bird enthusiast is astounded to hear that I've banded thirty or fifty or even, in one case, sixty-five finches at the feeder. As I well know going into the yard, at any one moment there may be a maximum of ten birds using the feeder, but in the course of the day, dozens and dozens of House Finches might cycle past.*

*Then there was the very devoted bird enthusiast on Long Island who kindly agreed to let me catch and band finches in his backyard. He had fed House Finches for years and considered them part of his family. As a matter of fact, he had replaced one of his bedroom windows with a glass box that projected into his bedroom, bringing feeding birds literally into his house. He told me that he spent countless hours watching the birds from a few feet away.*

*When I described my project and my interest in recording variation in male plumage coloration, he shook his head. "There's no plumage variation in the finches that come to my feeder. I'm afraid that you'll find that they are all red."*

*I told him that I'd like to catch the birds anyway. If they were all uniformly colored I needed to record that as well. After I had been banding at a picnic table in the backyard for a while, the homeowner came out and politely asked if he could watch me work. I was more than happy to oblige. I happened to be banding a drab orange male that I had just caught within three feet of his bedroom window. When I showed the drab male finch to homeowner, he was absolutely astounded.*

*"I've been watching my finches for over ten years," he said. "I see lots of yellow goldfinches at my feeders, but I've never seen a male House Finch that wasn't red before." I pulled another orange male out of a bag and then a drab yellow male. I explained to him that I see this sort of variation in male plumage coloration to a greater or lesser degree in all the populations of House Finches that I sample. When I left he said that he would start to watch more closely for color variants. (I didn't say anything, of course, but I couldn't help but think, "How could you watch more closely than having birds feeding a few feet away from you in your bedroom?") I got a note from the man a few weeks later saying that, indeed, he was now regularly seeing males with quite a range of plumage color from yellow to red.*

*This man is not the only bird enthusiast I've met who was quite familiar with the House Finch, but was adamant that all males are red. I can only guess that because field guides depict House Finches as being red (or a least they did until recently), that is what people expect to see. And, of course, most people see only what they expect to see.*

*It would be easy to scoff at such a bird enthusiast as ignorant or unobservant. However, we all go through life with our own set of blinders. The challenge for a scientist is to remember to look around once in a while and not take for granted the things that we "know" to be true.*

What do females gain by choosing to mate with males that have bright red plumage? In chapter 4, I summarized current understanding of the biochemical basis for red/orange/yellow plumage coloration in House Finches and other birds. In chapter 5, I reviewed the evidence indicating that environmental factors such as nutrition, parasites, and access to carotenoids can affect expression of carotenoid-based displays. These studies provide evidence that carotenoid-based plumage coloration is a condition-dependent trait in the House Finch. The implications of this conclusion are that carotenoid-based ornaments encode information that is useful to females and that females benefit by choosing to mate with brightly colored males. Such benefits for females cannot be taken for granted, however, and the studies that have been undertaken to test the idea that females benefit through their choice of a well-pigmented male will be presented in this chapter.

As has been pointed out repeatedly in the sexual selection literature (Andersson 1994, Møller 1994), females stand to benefit in two primary manners if they choose to mate with well-ornamented males. First, they can receive direct benefits such as food for themselves or food for their offspring if the ornament reliably indicates the potential of the male to provide resources. Second, if ornament expression is related to the genetic quality of the male, they can receive genetic benefits for their offspring in the form of good genes.

Measuring the resource benefits gained by females has proven to be relatively straightforward in a variety of bird species and, as I will present below, it was a relatively simple exercise in my studies of House Finches. Good genes benefits, on the other hand, have proven universally difficult to quantify. The best place to test for good genes effects is in species with lek mating systems in which males contribute no resource benefits to the female (Andersson 1994, Kirkpatrick and Ryan 1991). In such circumstances, the only potential benefit to the female is good genes. However, in a species like the House Finch, which is monogamous and in which males contribute substantially to the care of young, disentangling resource benefits from indirect genetic benefits is extremely difficult.

## Direct Benefits

### Male Provisioning

In the House Finch, there is certainly the potential for females to gain material benefits through mate selection. As I outlined in chapter 2, females are dependent on their mates for food during incubation. Males also feed nestlings and fledglings. To see if male coloration might be a predictor of paternal contribution, I conducted focal observations of nests that had females incubating a complete clutch of eggs. At the time I was designing my study of male provisioning in 1989, the standard means to measure parental effort was to count the number of feeding visits made by birds during a set observation period (typically one hour). This approach worked well for small insectivorous birds that made frequent trips to the nest. For instance, in a study of the Pied Flycatcher, during one-hour nest watches males made between thirteen and thirty-one feeding visits to nests (Saetre et al. 1995). Similarly, male Barn Swallows made an average of about ten feeding trips per hour (Møller 1994). For these species, there is adequate variation in the number of visits per hour for provisioning to be estimated. Unlike insectivorous passerines, House Finches eat plant material and regurgitate to offspring relatively large boluses of food that they store in their crops. As a result, males make rather infrequent trips to nests. In addition, like males of most cardueline finch species, male House Finches feed their mates throughout the incubation period and even after the chicks have hatched. Thus, I was faced with several problems in measuring male provisioning in the House Finch: males fed infrequently and I was worried that, instead of directly feeding young themselves, some males would pass food to the female who would then feed the young. Subsequent video analysis of male provisioning showed this concern to be unfounded, but it affected how I conducted nest watches at the time. In addition, it seemed likely that brood size and chick age would affect male provisioning and that these variables would somehow have to be standardized between nests—a concern that proved well founded (Nolan et al. 2001, Stoehr et al. 2001).

For all these reasons, I decided to focus on the rate of male provisioning during incubation feeding. The advantage of observing male attentiveness during that period was that the food requirements of the female through the incubation period were much more standardized than were the food requirements of chicks of different ages or for broods of different size. I still had the problem of infrequent visits to the nest, and I dealt with this problem by measuring the interval between provisioning visits rather than the number of visits for a set observation period. This gave me a more accurate index of the rate at which males were bringing food to the incubating females. Thus, I did not conduct nest watches for a fixed time period; rather, I watched until I recorded two provisioning visits by a male. If I happened to start a nest watch just after a visit by a poorly provisioning male, it was sometimes an hour before I recorded the first visit by the male and then another hour until its second visit. (Nobody said that behavioral ecology had to be exciting all the time.)

In my Michigan study population, I monitored the amount of time between feeding visits by males at nests with incubating females. I then assessed the relationship between the plumage coloration of the males under observation and their attendance rate. I found a significant positive correlation between male feeding rate and plumage brightness: redder males tended to provide more food than drabber males (Hill 1991) (Figure 7.1). In chapter 2, I showed that there is a significant tendency for yearling males to be less colorful than older males, so it was important to show that this relationship between feeding rate and male coloration was not simply an artifact of drabber plumaged younger males feeding less. Thirteen of the thirty males that I included in this analysis were males that I had banded in a previous spring, so they were known to be at least two years old. When I repeated the analysis using only these two-years-old-or-older males, and hence removing age effects, the relationship persisted (Hill 1991). Thus, it appears that, by choosing a colorful male, females tend to choose a male that will provide resource benefits during nesting.

In the Auburn population, we filmed every nest for eight-hour periods on the seventh day after eggs hatched and transcribed the rate at which males fed chicks. The long recording periods provided by these cameras overcame the problem of accurately measuring male provisioning rate when feedings were infrequent—feeding per eight hours provided an accurate measure of the provisioning rates of males. So, I compared the provisioning rates of males to their plumage coloration. As in the Michigan study, redder males tended to provide more food than less colorful males, but the pattern was not quite significant (Figure 7.2). These provisioning data were collected in years following the mycoplasmal epidemic when most drably plumaged males had been eliminated from the population (see Figure 7.6) and there was relatively little variation in coloration among males. Despite the lack of statistical significance, the effect of plumage coloration on provisioning rate was similar to what I observed in Michigan for incubation feeding.

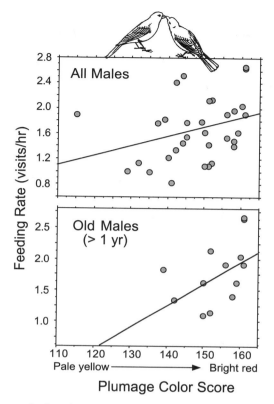

Figure 7.1. Rates at which male House Finches provisioned incubating females relative to the males' ornamental plumage coloration in Ann Arbor, Michigan. The top panel shows the relationship for all males including both yearling and older males. Brighter males tended to provision at a higher rate than drabber males ($n = 32$, $r_s = 0.42$, $P = 0.02$). The bottom panel shows the relationship when only males known to be in at least their second breeding season are included. There was still a significant positive relationship between feeding rate and plumage coloration ($n = 13$, $r_s = 0.63$, $P = 0.03$). Adapted from Hill (1991).

Interestingly, when I repeated the analysis of incubation feeding for Ann Arbor males by comparing the patch size of the attending male to its provisioning rate, I found no significant relationship (Figure 7.3). Males with larger patches of carotenoid pigmentation did not feed females more than males with smaller patches of carotenoid pigmentation. This result is especially interesting and becomes easier to interpret when one considers that feeding experiments indicate that dietary access to carotenoid pigments, which is likely a function of foraging ability in wild birds, had an over-riding effect on expression of plumage coloration. After the same dietary treatments, however, substantial variation remained in patch size (see Figure 5.4). So, plumage coloration is the primary criterion in female mate choice, and plumage coloration provides reliable information about the likely resource investment by mates. Patch size is a secondary criterion in mate choice, and this plumage trait provides no information about provisioning.

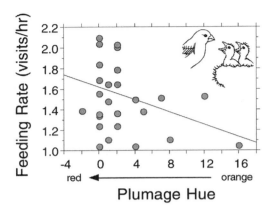

Figure 7.2. The relationship between male plumage hue and the rate at which males provisioned six-day-old chicks in Auburn, Alabama. Redder males tended to provide more food to young than drabber males, but the relationship was not statistically significant ($r = -0.35$, $n = 28$, $P = 0.068$). Despite the fact that the relationship was not significant, the effect of plumage color on provisioning rate was similar to the effect of plumage color on the rate at which males fed incubating females in the Ann Arbor, Michigan, population (see Figure 7.1).

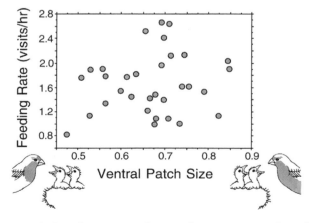

Figure 7.3. Rate at which males in Ann Arbor, Michigan, provisioned incubating females relative to the proportion of their ventral plumage with carotenoid pigmentation (patch size). In contrast to the significant relationship between plumage coloration and male feeding rate (see Figures 7.1 and 7.2), there was no significant relationship between patch size and male feeding rate ($n = 32$, $r_s = 0.19$, $P = 0.30$).

*Divorce and Renesting*

Females are dependent on their mates for food during nesting, and the food provided by males during incubation and chick feeding appears to play a critical role both in the quality of young produced by a female and her ability to maintain her body condition during the stressful period of nesting. A male also gains by providing resources to its mate. Food provided by a male leads to the production of more and healthier offspring, and by feeding its mate a male may increase the number and size of clutches laid in a season. However, the interests of males and females are not identical. Male House Finches can benefit not only by providing food to their social mate and chicks, but also by pursuing extra-pair matings (Trivers 1972). Although little is known about what determines which male House Finches are successful in obtaining extra-pair matings, one would imagine that males that are available and engaged in courtship activity such as singing would be more likely to gain extra-pair matings than males that stay close to their social mate and provide food. Thus, there is a fundamental conflict of interest between males and females regarding how much a male should invest in its nest and social mate versus how much it should invest in attracting additional sexual partners (Trivers 1972).

In the House Finch, it appears that females enforce their demand for high resource investment by males through the threat of divorce. At the Ann Arbor study site, about 11% of females abandoned their nests before eggs hatched (Hill 1991). In all of these cases of abandonment, the nest was undisturbed. No eggs were missing, and there was no sign of predation, human disturbance, or anything that might have changed a female's perception of the value of a nest. Such abandonment always occurred at the egg stage, never after eggs had hatched. A clutch of eggs represents a substantial investment in time and energy by a female. Why would a female abandon such an investment?

The answer appears to be that the females are really abandoning their social mates, and that abandoning a social mate requires that the entire nesting effort be abandoned. Evidence for nest abandonment being tightly linked to divorce comes from an analysis of the rate of renesting by pairs associated with abandoned nests. After nest abandonment, only 11% of pairs remained together for a subsequent nesting attempt; 89% of females that abandoned a nest renested with a different male. In contrast, following a successful nesting attempt, if the female renested in the same season it was always (42/42) with the same social mate. After a nest failure, which in most cases meant loss of eggs or chicks to predators and which was unrelated to male provisioning, 93% of females that renested did so with the same social mate (Figure 7.4).

There is potentially much to be gained for a female if a more desirable (higher quality) male becomes available as a mate (Ens et al. 1993, Otter and Ratcliffe 1996), and that alone could explain nest abandonment. However, rather than a strategy to make a good situation better, I think that the abandonment of eggs by female House Finches is usually a response to a bad situation. As I just discussed, during incubation males provide their mates with food, and the amount of food brought by males varies considerably. I hypothesized that females would initiate divorce and abandon their clutch of eggs if the male-provisioning rate dropped below a minimum level (Hill 1991). If male feeding rate was low enough, the

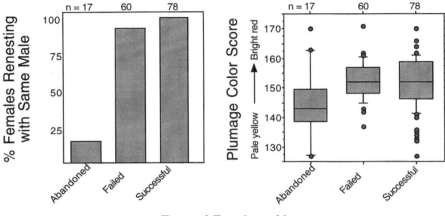

**Fate of Previous Nest**

Figure 7.4. Mate switching in relation to nesting success and male plumage coloration. About 11% of female House Finches in Ann Arbor, Michigan, switched mates during a breeding season. The left panel indicates that a disproportionate number of these divorces are associated with the abandonment of a nest at the egg stage ($\chi^2 = 55.87$, $P = 0.001$); if nests were successful, or even if nests failed due to predation, there was a high probability that the male and female would renest together. Divorce was also related to male coloration. Males associated with nests that were abandoned were significantly less colorful than males associated with either successful nests or nests lost to predators ($F = 6.02$, $P = 0.003$). Adapted from Hill (1991).

female could lose weight during incubation which could compromise her current nesting effort, her potential for additional nests that year, and possibly her potential for nests in future years. Such carry-over costs of a large investment in a single clutch have been documented in other species of birds (Dijkstra et al. 1990, Nur 1984, Ots and Horak 1996).

If the female continued to invest in a nesting effort with a poor mate simply because she had already invested resources, she would be committing what behavioral ecologists call "the Concorde fallacy." This is a term coined by Dawkins and Carlisle (1976), who used the analogy of the continued investment in the supersonic transport, the Concorde, by the French and British governments even when it was clear that the benefit of the program would never repay the investment. Politicians argued that they could not stop the project because they had already invested so much. Because House Finches lack rational thought, they are spared such irrational, emotionally based decisions. Probably based on some hunger-driven trigger mechanism, females appear to cut their losses when they end up with a mate who provisions poorly.

I have no direct support for the hypothesis that females abandon males who provision poorly, but indirect evidence does support the hypothesis. In the previous section I showed that redder males provided more resources to nesting females than did drabber males (Hill 1991; Figures 7.1 and 7.2). If females abandoned their social mates because of insufficient provisioning by males, one would predict that drabber males would be abandoned more often than bright males. Indeed, when I compared the plumage coloration of males that attended nests that were eventually abandoned versus males that attended nests that were not abandoned (either successful or lost to predators), I found that males associated with abandoned nests were significantly less colorful (Hill 1991; Figure 7.4). So, it appears that by choosing to mate with brightly colored males, females not only gain the greater food resources provided by redder males, but they also benefit by reducing their chances of being forced to switch mates, which is costly in terms of both time and energy.

## Carotenoid Displays and Direct Benefits Beyond House Finches

Although I was able to show that male plumage coloration is a reliable signal of resource investment in the House Finch, it is not clear that this is the typical component of male quality that is signaled by songbirds with carotenoid-based color displays. In the Northern Cardinal, Linville et al. (1998) observed that males with brighter or redder carotenoid pigmentation tended to provide more food to mates or offspring than did drabber males. However, in the Linnet (Drachmann 1998), Yellowhammer (Sundberg and Larsson 1994), and Zebra Finch (Burley 1988), males with drabber carotenoid-based color displays fed at nests more than males with brighter carotenoid displays. Moreover, both positive and negative associations have also been found between feeding rate and expression of non-carotenoid plumage coloration (Grant and Grant 1987, Palokangas et al. 1994, Studd and Robertson 1985b).

The allocation of resources by a male to its mate or offspring represents a reproductive strategy on the part of the male (Trivers 1972), and such strategies are likely to be context dependent (Höglund and Sheldon 1998, Qvarnström 2001). When the primary means by which a male may enhance its reproductive output is through investment in its social mate and offspring, as seems to be the case with House Finches in eastern North America, then we would expect a large parental investment and a positive association between ornamentation and parental care (Kokko 1998). When males can benefit by mating outside the pair bond, then parental investment should decrease and the association between ornamentation and parental care should decline (Kokko 1998). Testing hypotheses related to the evolution of investment strategies and carotenoid ornamentation will require comparative analyses with more observations from more species with carotenoid pigmentation.

## Good Genes

### The Hamilton–Zuk Hypothesis

Resources in the form of food in their own mouths and in the mouths of their offspring are the most tangible benefits that female House Finches gain by choosing to mate with brightly colored males. It may be that resource benefits are the main driving force in the evolution of female preferences for colorful males. There are, however, other not mutually exclusive benefits that females might gain through their choice of well-ornamented males. By choosing to father their offspring with a male with bright carotenoid-based coloration, females may gain genetic benefits, if the genes that are contributed to her offspring are better than genes that would have been contributed by a less colorful male. The concept of genetic benefits for offspring has intrigued biologists since before the formal concept of genes was even developed—both Darwin and Wallace suggested the idea that females would benefit by mating with well-ornamented males because these tended to be vigorous males that would produce vigorous sons (Darwin 1871, Wallace 1889). With the rebirth of interest in sexual selection in the age of molecular biology, interest in the "good-genes hypothesis" has intensified.

The basic concept of good genes benefits is that there exists variation in genes that have an effect on individual fitness but that cannot be assessed directly by a female choosing a mate. These good genes, however, affect or are genetically correlated with the expression of a trait that females *can* assess directly (Andersson 1986a, 1994). Thus, females benefit by basing their choice of mates on expression of the display trait because the display trait is reliably linked to specific genes that have a direct effect on the fitness of offspring. The fitness benefits of the good genes drive the evolution of female choice for the trait, and female choice drives the evolution and maintenance of the trait in males.

The heritable traits that affect fitness could be anything—genes for digestive physiology, blood-clotting agents, the pattern of cone cells in the retina—but most models and studies of good genes have focused on traits related to disease resistance. The Hamilton–Zuk hypothesis (Hamilton and Zuk 1982), which proposes that ornamental traits signal genetic resistance to parasites (as discussed in the parasite section of chapter 5), is the most famous specific hypothesis of good genes selection. The reason that Hamilton and Zuk and many subsequent researchers focused specifically on genes for disease resistance, among all the potential genes that might vary among individual and that might affect individual fitness of an organism, is that resistance to disease is known to vary substantially within any population, to have a large effect on fitness, and, at least in a few model systems, to be heritable (Hamilton and Zuk 1982). Moreover, it has long been appreciated that parasites can affect expression of ornamental traits. But perhaps the most important reason that parasite resistance has been the focus of good genes research is that parasites present a constantly changing environmental challenge to a population of organisms.

One important theoretical problem with good-genes models is that female choice for traits associated with good genes soon leads to the fixation of the more fit alleles and loss of any advantage to females for choices based on the display trait. Maintaining additive genetic variance for fitness traits in the face of strong selection for those traits has been viewed as a major obstacle for good-genes models (Dominey 1983, Maynard Smith 1978a, Williams 1978). A focus on parasites generally solves the problem (Hamilton and Zuk 1982). Parasites evolve in response to the evolutionary changes in their host, and because most parasites have a very short generation time, they generally evolve at rates much faster than their hosts. So, the best genes for resisting a parasite today are potentially useless tomorrow. New means for dealing with the parasite must constantly evolve, so there is continually renewed genetic variation that affects fitness and that can be tied to expression of ornamental traits (Hamilton and Zuk 1982).

## Previous Tests of the Hamilton–Zuk Hypothesis

Surprisingly, there have been very few studies of any organisms that have demonstrated definitively the four necessary conditions of the Hamilton–Zuk hypothesis: (1) parasites affect the fitness of their host; (2) there is heritable resistance to parasites; (3) there is a correlation between or a direct causal link between parasitic infection and ornament expression; and (4) females use expression of the ornamental trait in choosing mates. Numerous studies have verified two or three of these predictions, but virtually no studies of a single population have tested and verified all four key assumptions. No studies of plumage coloration have demonstrated these four conditions within a single species. (I reviewed evidence that parasites affect expression of plumage coloration in chapter 5.) The most complete and convincing support for the Hamilton–Zuk hypothesis comes from studies of tail length in the Barn Swallow conducted by Møller (see Møller 1994 for an overview).

Møller's studies focused on elongated tails and blood-sucking mites. First, using a cut-and-paste experiment that eliminated potential confounding effects like male age, he showed that female Barn Swallows prefer to mate with males with the longest tails (Møller 1988, 1990c). Møller then showed that longer-tailed males had fewer mites than shorter-tailed males and that males with fewer mites returned the following year with relatively longer tails while males with more mites returned with relatively shorter tails (Møller 1990a). To test the heritability of resistance to mite infestation, Møller conducted a partial cross-fostering experiment in which he compared the mite load of a breeding male with the mite load of (1) its own offspring in its own nest; (2) its own offspring in a foreign nest; (3) another male's offspring in its own nest; and (4) another male's offspring in that male's nest. Møller found a high degree of similarity between the mite load of a father and the mite load of its offspring whether that offspring was raised in its own nest or a foreign nest (Møller 1990a). Moreover, longer-tailed males produced offspring with fewer mites, regardless of the environment in which they were raised (Møller 1990a). This set of studies stands as the most complete and convincing test of an ornamental trait functioning as an honest signal of the genetic quality of a male.

## Good Genes in the House Finch

At present, there is no convincing evidence that female House Finches gain genetic benefits through their choice of mates. As I presented in chapter 5, coccidiosis, mycoplasmosis, and pox infections can negatively affect expression of carotenoid-based plumage coloration. However, there is currently no evidence that there is genetically based resistance to these diseases or that females produce more fit offspring by pairing with brightly colored males. A valid alternative hypothesis to the idea that females gain good genes by choosing to mate with brightly colored males with lower parasite loads is that brightly colored, less parasitized males are simply in better physical condition, and that this allows them to provide more resources to females. Another alternative to the good genes hypothesis, given the link between disease, plumage coloration, and mate choice, is that females are choosing mates so as to avoid direct transmission of the parasites to themselves and their offspring (Lombardo 1998). We cannot assume that female House Finches gain genetic benefits for offspring just because it has been established that expression of plumage coloration is related to parasite load. The key missing observation in our attempts to test the Hamilton–Zuk hypothesis in the House Finch remains a test of whether or not disease resistance is heritable.

The most direct test for good-genes benefits to females for choosing brightly colored males in my studies of House Finches was the paternity analysis that I conducted at Queen's University (Hill et al. 1994b; see chapter 6). I've already presented this paternity study as a test of possible extra-pair benefits to males having bright red coloration (i.e., do red males get cuckolded less or gain more extra-pair copulations than drab males?—discussed in chapter 6). Our primary intent in this study, however, was to test the good-genes hypothesis. When I conducted the paternity analysis, I had already shown that female House Finches prefer brightly colored males as social mates, and I had established that females gained resource benefits through their choice of a brightly colored male. Paternity analysis and an assessment of extra-pair mate choice provided a means of testing for good-genes benefits of mate choice.

By looking at choice of extra-pair partners, we were able to disentangle choice for resource benefits from choice for good genes. Extra-pair partners provide no resource benefits to female House Finches that we could observe. Therefore, if females were choosing among extra-pair partners, we concluded that the choice had to be related to genetic benefits and not to resource benefits. We hypothesized that if females could gain genetic benefits for offspring by mating with a colorful male, then females paired to drab males would engage in more extra-pair copulations than females paired with bright males. Moreover, extra-pair matings should be with colorful rather than drab males, if red males hold the good genes. As I mentioned in chapter 6, we couldn't test the prediction regarding who would father extra-pair young, but we were able to look at cuckoldry relative to male plumage coloration.

We found that overall there were low levels of extra-pair paternity in House Finches. Only about 8% of chicks were the result of extra-pair matings. Many species of passerine birds that have been studied have a higher proportion of extra-pair young than we found in House Finches (Møller and Ninni 1998). Moreover, and most critically, drab males were no more likely to be cuckolded than bright males (see Figure 6.12). This paternity study is far from a conclusive test of the hypothesis that females gain genetic benefits through their choice of mates, but it does suggest that having drab males father offspring is not enough of a fitness detriment to increase the rate of cuckoldry.

A few other observations are consistent with the idea that plumage redness is related to male genetic quality. First, plumage redness predicts over-winter survival in males. I looked at over-winter survival across four seasons in Michigan. Like many passerine birds, male House Finches show high site fidelity to their breeding area. So, working under the assumption that males that failed to return to the study site between years had died, I compared the plumage coloration of males that returned (survived the winter) to the plumage coloration of males that did not return (died). I found that returning males were significantly more colorful in all three years of comparison (Hill 1991; Figure 7.5). One problem with this approach to testing traits relative to survival is that if the trait under investigation is associated with reproductive success (as I have shown plumage coloration to be) and if reproductive success affects the site fidelity of males (as has been shown in some birds), then a false pattern of differential survival could be created if unsuccessful, poorly ornamented males tended to leave the area and were counted as dead. To get around this problem, I repeated the analysis of plumage coloration of surviving and non-surviving males, but this time I considered only those males that attracted a mate and bred in the previous year. By considering only successful males, I was confident that I had eliminated any dispersal bias related to reproductive success from my analysis. Once again, surviving males were significantly more colorful than males that did not survive (Hill 1991). If over-winter survival is an indicator of genetic quality as proposed by some authors (Kempenaers et al. 1992, Svensson and Nilsson 1996), then this could be taken as at least indirect evidence that male plumage coloration signals genetic quality.

The survival analysis presented above concerned typical over-winter survival episodes for eastern House Finches. A particularly severe selection episode for eastern House Finches occurred when a mycoplasmal epidemic swept through the population. This epidemic killed an estimated 60% of finches in its first year

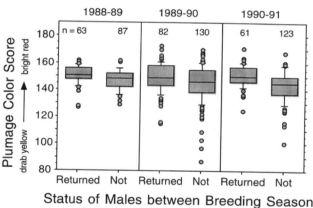

Figure 7.5. Over-winter survival in relation to plumage coloration for male House Finches in Ann Arbor, Michigan. In each year, the mean plumage coloration of males returning between breeding seasons was higher than the mean plumage coloration of males that did not return (1988–89: $t = 1.66$, $P = 0.05$; 1989–90: $t = 2.45$, $P = 0.0007$; 1990–91: $t = 4.74$, $P = 0.0001$). When the comparison was repeated considering only males that paired in the previous year, returning males were still significantly brighter than males that did not return (1988–89: $t = 1.70$, $P = 0.05$; 1989–90: $t = 1.89$, $P = 0.03$; 1990–91: $t = 2.05$, $P = 0.02$). Adapted from Hill (1991). Patch size did not differ significantly between males that returned and males that did not return. (All males: 1988–89: $t = 1.30$, $P = 0.10$; 1989–90: $t = 1.25$, $P = 0.11$; 1990–91: $t = 1.04$, $P = 0.15$. Paired males only: 1988–89: $t = 0.84$, $P = 0.20$; 1989–90: $t = 0.74$, $P = 0.23$; 1990–91: $t = 0.79$, $P = 0.21$; unpublished data not illustrated.) All comparisons are one-tailed $t$-tests.

(Nolan et al. 1998; see also chapter 5). It had a major impact on the population, reducing the number of breeding birds by about half (Hochachka and Dhont 2000). Fortuitously, we were in place monitoring the breeding population of House Finches on the Auburn University campus before and after the epidemic, so we could look for an effect of plumage coloration on survival. In this analysis we simply compared the mean male plumage score of the population before the epidemic to the mean plumage score after the epidemic. We found that males surviving the epidemic were significantly redder than males that did not survive the epidemic (Figure 7.6; Nolan et al. 1998). Put another way, drably plumaged male House Finches appeared to be more susceptible to a novel pathogen than were brightly colored males. As in the previous survival analysis this is not a direct test of the good-genes hypothesis, but it does suggest that redder males have better genes for disease resistance than drab males. Alternatively, it could be that red males were simply in better physical condition than drab males and, independent of any genetic advantage, their better physical condition allowed them to better survive the epidemic.

One final observation that has important implications for the good-genes hypothesis relative to male plumage coloration in the House Finch is that I found a significant correlation between the plumage coloration of fathers and their sons. Parent–offspring analyses that require measurements from offspring more than a few weeks after they leave the nest are difficult to conduct with House Finches because, like many species of passerine birds, nearly all individuals disperse away from their natal area to breed. Of the 345 chicks that I banded on my study site in Michigan only seventeen (4.9%) returned to the study area as adults (Hill 1993b), eleven of which were males. I had the father's plumage score for eleven of these

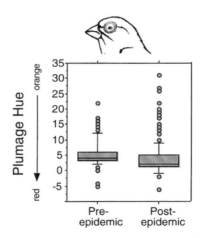

Figure 7.6. The mean plumage coloration of male House Finches in Auburn, Alabama, before and after an epidemic of *Mycoplasma gallisepticum* that killed about half of the individuals in the population. The mean plumage coloration of the population was significantly redder following the epidemic ($t = 2.94$, d.f. $= 302$, $P = 0.004$), suggesting that drably colored males died at a higher rate than brightly colored males. Adapted from Nolan et al. (1998).

returning males, and I compared the plumage coloration of these returning male finches to the plumage scores of their fathers. I found a relatively strong correlation: redder males tended to have redder sons (Hill 1991) (Figure 7.7). However, in a comparison of thirteen fathers and sons from the population in Auburn, I found no relationship between the coloration of fathers and the coloration of their male offspring. In this latter comparison, however, there was little variation in male plumage coloration and this may have made detection of the sort of pattern that I found in the more variable Michigan population difficult. Interestingly, when I used the same set of fathers and sons in Michigan to compare similarity of patch size, I found no significant relationship. The patch size of a male did not predict the patch size of its male offspring.

The good-genes hypothesis proposes that well-ornamented males possess better than average genes that are passed on to the offspring that they sire. It follows from the good-genes hypothesis that there should be a positive correlation between the plumage brightness of fathers and sons. Thus, the relationship that I found between the plumage coloration of males and their male offspring in Michigan supports a basic prediction of the good-genes model. However, the positive correlation between the coloration of fathers and the coloration of their male offspring should not be viewed as a measure of the heritability of plumage coloration in the House Finch. There are many alternative and more likely explanations for the observed relationship between plumage coloration of fathers and sons than direct inheritance of colorful plumage. In chapter 6, I showed that brightly colored males tended to nest earlier with higher quality females than drabber males. In many

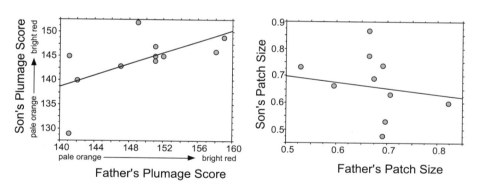

Figure 7.7. The similarity of carotenoid-based plumage coloration and the extent of ventral carotenoid pigmentation between fathers and sons. There was a significant positive relationship between the plumage coloration of fathers and sons for House Finches in Ann Arbor, Michigan ($n = 11$, $r^2 = 0.37$, $P = 0.048$). Adapted from Hill (1991) with 1990 chicks added. There was no such relationship between fathers and sons for patch size ($n = 10$, $r^2 = 0.13$, $P = 0.36$). All offspring scores were taken from birds banded in the nest that returned to the study site in first basic plumage. Genetic paternity of offspring used in this study was not tested, but DNA fingerprinting of other males and nestlings in this population indicated that only about 8% of chicks in this population were illegitimate (Hill et al. 1994b). In Auburn, Alabama, there was no significant relationship between the plumage scores of fathers and the plumage scores of the thirteen male offspring that returned in basic plumage ($n = 13$, $r^2 = 0.01$, $P = 0.96$; unpublished data not illustrated).

species of birds, it has been shown that young that fledge earlier are at a competitive advantage compared to young that fledge later in the season (Visser and Verboven 1999). So, the young of redder males may simply fledge earlier, gain an advantage in foraging during their first prebasic molt, and hence grow redder plumage. In addition, earlier in this chapter I presented data that males with more colorful plumage provide more food to nesting females than more drably colored males. Thus, chicks in the nests of brightly colored males may have access to better nutrition, putting them in better condition and enabling them to grow redder plumage during the fall molt. While the relationship between the plumage coloration of fathers and sons is consistent with the good-genes hypothesis, it is not conclusive evidence for it. Given the large effect of environment on expression of plumage coloration, as documented in chapter 5, genetic effects on expression of plumage brightness are likely to be small and indirect, but this remains to be tested with cross-fostering experiments.

## Summary

By choosing to pair with a brightly colored male, female House Finches gain direct resource benefits. Male plumage redness is positively correlated to the rate at which males provision incubating females and the rate at which males feed chicks in the nest. Moreover, females paired to males with redder plumage are less likely to divorce their mate and abandon their first nest of the season, which is a costly behavior.

What remains to be tested definitively in the House Finch—and in virtually every other species of animal with an ornamental display—is the hypothesis that females gain good genes through their choice of a well-ornamented male. Presently, there is only weak and indirect evidence that female House Finches gain genetic benefits for offspring by choosing bright red males as mates. Male House Finches that survive between years were redder than males that did not survive, and males that survived an epidemic caused by a novel parasite were also redder than males that did not survive. In Michigan, but not Alabama, I also observed that there was a significant positive relationship between the plumage coloration of fathers and sons. These observations are consistent with the idea that plumage coloration indicates the genetic quality of males, but all of these observations can also be explained simply by brightly colored males being in better condition than drab males, as presented in chapter 5. Thus, a definitive test of the idea that plumage redness is a signal not only of male resource quality but also of male genetic quality will have to wait for future studies.

# 8 Studs, Duds, and Studly Duds

## Plumage Coloration, Hormones, and Dominance

> Color, then, seems to be of small importance, if it plays any role, in determining social status.
>
> —W. L. Thompson (1960b), commenting on the role of plumage redness of male House Finches

> The establishment of territorial rights involves frequent disputes, but these are by no means all mortal combats; the most numerous, and from our point of view, therefore, the most important cases are those in which there is no fight at all, and in which the intruding male is so strongly impressed or intimidated by the appearance of his antagonist as not to risk the damage of conflict. As a propagandist the cock behaves as though he knew that it was as advantageous to impress the males as the females of his species, and a sprightly bearing with fine feathers and triumphant song are quite as well adapted for war-propaganda as for courtship.
>
> —R.A. Fisher (1958:155)

*As my mother tells it, when I was about four years old, I was in the backyard of our house playing in a sandbox with a new metal pail and shovel. I had barely gotten started with the new toy when a five-year-old from across the street, Dicky McCasky, came over and tried to take the shovel away from me. Despite the fact that Dicky was substantially bigger than me, I wouldn't surrender the shovel and the tussle escalated. As we struggled for possession of the toy, Dicky momentarily released his grip (probably to go for a more direct attack on me), and I took the opportunity to strike him hard in the face with the shovel. Dicky ran home crying, ended up with a half-dozen stitches in his cheek, and I was sentenced to two days in my room. But, an amazing thing happened. Dicky stopped harassing me and stealing my toys.*

*On his way through primary school Dicky failed a grade or two and ended up in my grade in high school. By this time, his childhood misbehavior had turned into serious juvenile delinquency. He was a cruel and relentless bully at my school, terrorizing all who were smaller or weaker, which was virtually everyone. By all rights, I should have been the easiest of pickings for Dicky, but Dicky avoided me completely. It was as if I moved invisibly through the hallways and classrooms, right past Dicky as he extorted lunch money and traumatized the other boys in the class.*

*It was probably that I was too small and pathetic to pick on. Even a delinquent like Dicky had to draw the line somewhere. I can't help but think, however, that by standing my ground in the sandbox and with an impulsive swing of a shovel I sent a message to Dicky that he still carried a decade later. Although clearly he was never consciously apprehensive around me, his lack of action against me suggests that my willingness to escalate the conflict in the sandbox established a boundary between Dicky and me. For Dicky, there were many other victims, and it was easier to simply leave me alone.*

*For social animals with a long memory, contests over resources and dominance interactions play out in a complex manner. Sometimes they make sense only if one is familiar with the history of association between the contestants.*

As first conceptualized by Darwin and Wallace—and now taken for granted by modern evolutionary biologists—sexual selection can act through two primary processes. Males can potentially increase their mating success either by charming females with ornamental traits and displays, or by winning access to them through direct competition with rival males. Selection for enhanced ability to attract mates is hypothesized to lead to the evolution of ornamental traits. In the previous chapters of this book, I presented studies that clearly demonstrate a mate-choice function for carotenoid-based plumage coloration. As a matter of fact, a focus on mate choice dominated my research on House Finch plumage coloration for the first dozen years or so. Until recently, I simply ignored male–male competition as a factor in the evolution or maintenance of ornamental plumage traits in this species.

The reason that I felt justified in ignoring male–male competition for so long is that male aggression seemed to play such a minor role in breeding biology of House Finches. In most of the best-studied songbirds in the world, like the Red-winged Blackbird, Great Tit, or Pied Flycatcher, the onset of breeding is marked by male song and display as males stake out territories centered around food or nest sites. In these species, aggressive male–male interactions dominate the period of pair formation and nesting, and there is no doubt that success in these contests is essential to male reproductive success. Males that win battles with other males end up with more or better resources, attract more or better females, and produce more or better offspring. As a matter of fact, in such typical songbirds, male–male aggression is so overt that it is difficult to determine whether female mate choice plays any role in the mating success of males.

The mating system of House Finches is very different. Males do not defend food or nesting resources. To the contrary, breeding onset is marked by increasingly stronger associations between males and females, and male dominance seems to play no role in these interactions. Females drive away males whose company they do not want, and once a male and female establish a pair bond, they work together to drive away rival males and females. Behavioral observations gave no indication that male aggressive behavior or dominance provided any benefit in the form of access to social mates (Hill 1990, 1993b).

In a number of other cardueline finches, however, the dominance status of males is linked to melanin-based plumage coloration (Senar 1999) and likely to improved reproductive success. Even though male aggression and dominance interactions did not seem to play a central role in the reproductive biology of House Finches, to reach any sort of a comprehensive understanding of the function of plumage coloration, I had to formally test the role of plumage coloration in dominance interactions. Moreover, because so much is known about the proximate control of carotenoid-based plumage coloration and its role in female mate choice, plumage coloration in the House Finch presented an ideal system in which to integrate a study of ornamentation relative to male–male competition and female mate choice.

Most of the research that I have presented in this book was published after 1989. However, dominance in the House Finch was the focus of a detailed study by William Thompson in 1960, and the basic patterns of dominance and ornamentation that I will present in this chapter were established in Thompson's work (1960a, b). More recent research by Jim Belthoff and his colleagues and by my lab group has extended the observations of Thompson by experimentally testing the role of carotenoid coloration in dominance interactions and by looking at the hormonal control of aggressive behavior, behavioral display, and plumage coloration.

## Carotenoid Coloration as a Signal of Social Status

### Redness and Social Status

As documented in chapter 5, carotenoid-based plumage coloration in male House Finches is a condition-dependent trait. Research has shown that only males with access to large quantities of suitable carotenoid pigments, which avoid disease and are in good nutritional condition, can produce the reddest and most saturated plumage pigmentation. A logical prediction, then, is that these high-condition, brightly colored males would also be socially dominant to drabber males. This is not, however, the pattern that we see in wild or captive House Finches. No study has found that brighter males are dominant to drabber males.

Like virtually all birds, a group of House Finches held in an aviary quickly establishes a stable hierarchy of dominance—a pecking order. These hierarchies are linear, meaning that the most dominant bird defeats all other birds in the cage, the second most dominant bird defeats all but the top bird, and so forth. Because House Finches form linear hierarchies in cages, several researchers have studied the relationship between male plumage coloration and dominance rank in captive flocks of finches.

In the first published description of dominance and plumage coloration in the House Finch, Thompson (1960b) found no relationship between plumage redness and dominance in three captive flocks with four males in each flock (Figure 8.1). These early observations of dominance in the House Finch were followed by studies by Belthoff and coworkers (Belthoff et al. 1994, Belthoff and Gauthreaux 1991a,

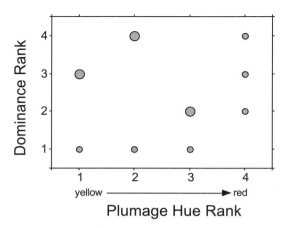

**Figure 8.1.** Plumage rank versus dominance rank of captive male House Finches captured in Berkeley, California. Data are pooled across three cages each with four males. In each cage, males were ranked for social dominance and ornamental color. The relationship between color and dominance is not significant ($r_s = 0.07$, $n = 12$, $P = 0.83$). Drawn from data in Thompson (1960b).

Belthoff and Gowaty 1996), who reported negative correlations between plumage coloration and dominance. In flocks of captive birds, they observed a tendency for drabber males to be socially dominant to brighter males, but the relationships between plumage color and dominance were not statistically significant within any particular flocks. Kevin McGraw and I recorded the same tendency for drab males to be dominant to brighter males in our observations of captive male House Finches (McGraw and Hill 2000a, c). Much like Belthoff et al. found in their studies, our observations indicated a significant negative relationship between color and dominance in some of our captive flocks, but not in others (Figure 8.2). We observed this tendency for drab males to be dominant to bright males in both the breeding and non-breeding seasons (McGraw and Hill 2000a, c). If we combine observations of dominance in relation to coloration from Belthoff's study and the two studies by my lab group into a single meta-analysis, then the number of flocks in which plumage coloration was negatively correlated with dominance is greater than we would expect by chance. (Thompson's studies are excluded from this meta-analysis because he ranked rather than measured color and he had only four birds per group.)

To further test whether plumage color predicts the resource-holding potential of males, Belthoff et al. (1994) staged paired contests in the winter between captive males of the same age and size but with different natural plumage brightness. They conducted twenty-nine such trials, and they observed that the drabber male won eighteen (62%) of the trials. This result suggested that drab plumage coloration was a predictor of social dominance, but the trend was not statistically significant. Kevin McGraw and I conducted similar paired-dominance trials in aviaries during the breeding season by pitting naturally bright males against naturally drab males. The paired males had no prior experience with each other, and we matched

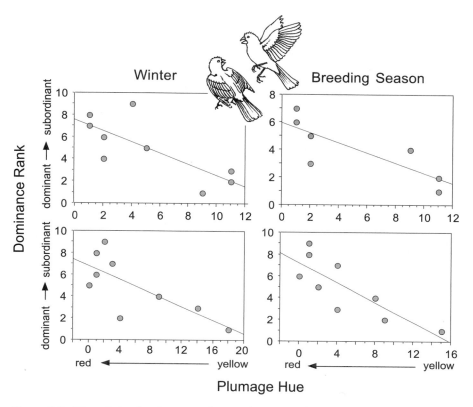

Figure 8.2. The relationship between plumage hue (scored with a Colortron) and dominance status for four flocks of captive male House Finches. These were the four flocks, out of ten flocks observed, in which there was a significant negative relationship between color and dominance (clockwise from upper left: $r_s = -0.73$, $n = 9$, $P = 0.04$; $r_s = -0.92$, $n = 7$, $P = 0.004$; $r_s = -0.82$, $n = 9$, $P = 0.007$; $r_s = -0.68$, $n = 9$, $P = 0.05$) (from McGraw and Hill (2000a, c). Combining observations from these two studies with those of Belthoff et al. (1994), there was a negative relationship between plumage coloration and dominance in twelve of sixteen flocks observed ($\chi^2 = 4.0$, $P = 0.05$).

these males for dominance rank in their captive flocks (Figure 8.3). Both the drab and bright males in these trials were held without food for about 12 h before the trial, so we assumed that both birds would be motivated to gain access to the small bowl of food that we provided in the cages into which the males were released. We observed that drab males were dominant in thirteen out of the seventeen trials, which is more wins by drab males than one would expect by chance (McGraw and Hill 2000c). If we once again pool the observations of Belthoff et al. (1994) with our observations, there is a strong negative relationship between the plumage brightness of a male and its resource-holding potential. Thus, both the consistent negative correlation between male coloration and male dominance and the dominance of drab males over bright males in paired trials support the idea that, in the House Finch, drab ornamental plumage is associated with social dominance.

Rather surprisingly, there are no published observations of the dominance patterns of drab versus bright male House Finches in the wild to corroborate the patterns documented in captive flocks. Based on observations at feeders, Brown and Brown (1988) and Shedd (1990) reported that drab House Finches tended to dominate brightly colored House Finches, but in both studies females and males in drab plumage were lumped together. This makes it impossible to deduce the dominance interactions of males of different plumage coloration. Until observations suggest otherwise, we have to assume that the patterns that we observe in aviaries are the same that we would see in the wild.

## Status Signaling

Given the observation that males with drabber plumage tend to be socially dominant to males with brighter plumage, the question then becomes: is plumage coloration serving as a signal of status or is drab coloration simply correlated with high social status? To get at this question we first have to define status signaling and then review previous work that has been conducted on status signaling and plumage coloration.

A status signal is a morphological trait that reliably indicates the resource-holding potential (fighting ability) of the individual displaying the trait. The concept of status signaling was originally developed by Rohwer (1975) who proposed that, in situations such as those that occur in winter flocks of seed-eating birds, it would be advantageous for birds of both high and low dominance rank to develop easily assessable markers of status, such as patches of color. Such status signals would reduce the number of aggressive encounters among flock members because dominant birds would be less inclined to challenge subordinate birds that were already advertising their subordinance. Likewise, subordinate birds would be less likely to challenge dominant males that were signaling their high status. A potential problem with this sort of signaling system is the invasion of cheaters—males with low resource-holding potential who dishonestly signal high status (Maynard-Smith and Harper 1988). However, as Rohwer (1975, 1977) and others (Jarvi et al. 1987, Owens and Hartley 1991, Watt 1986) pointed out and as has been verified through experimental tests (Rohwer 1977, Rohwer and Rohwer 1978), such cheating strategies do not work because cheaters cannot afford the high cost of aggression when they are tested by dominant males. Thus, in this sort of status-signaling system, signal honesty is maintained by social mediation. (For a challenge to this explanation see Slotow et al. 1993.)

Readers should note that this is a fundamentally different form of honest signaling than I have proposed for carotenoid-based plumage coloration throughout this book. I have been presenting plumage coloration in the House Finch not as a signal of status, but as an honest indicator of male health and condition. The status signaling and honest advertisement hypotheses were proposed at about the same time (Kodric-Brown and Brown 1984, Nur and Hasson 1984, Rohwer 1975, 1982, Zahavi 1975, 1977), but the literature related to these topics developed largely independently. By the status-signaling hypothesis, plumage traits are arbitrary markers of high status that are cheap to produce but costly to maintain. Carotenoid-based plumage coloration, on the other hand, is a trait whose expression is intricately tied to various components of individual condition. In contrast to what is proposed in the status-signaling hypothesis, cheating is not possible in the carotenoid-pigment system of House Finches because the production costs of bright coloration are too high. (See chapter 12 for a detailed discussion of models for the evolution of ornamental traits.) Only males in the best condition with access to enough of the right types of carotenoid pigments, that resist parasites, and that have access to good nutrition produce bright red plumage coloration. Independent of any sort of maintenance cost such as social mediation, carotenoid-based plumage coloration is a signal of male quality.

All of the early studies and tests of the status-signaling hypothesis involved melanin-based coloration. Rohwer conducted extensive work with black breast coloration in Harris' Sparrows (*Zonotrichia querula*) (Rohwer 1975, 1977, 1985, Rohwer and Ewald 1981, Rohwer et al. 1981, Rohwer and Rohwer 1978), various groups in Europe studied the black breast stripe of the Great Tit and the black bibs of other European titmice (Brotons 1998, Högstad 1989, Högstad and Kroglund 1993, Jarvi and Bakken 1984, Jarvi et al. 1987, Lemel and Wallin 1993), whilst, most recently, Senar looked at black badges in European Cardueline finches (Senar 1999, Senar et al. 1990, 1993, Senar and Camerino 1998). The results of these studies have presented consistent support for the idea that melanin-based black patches of color serve as signals of social status. Perhaps it is not coincidental that support for plumage coloration functioning as a status signal has come exclusively from studies of melanin rather than carotenoid pigmentation. As outlined in chapter 5, carotenoid-based and melanin-based plumage ornaments appear to be under very different proximate control (Badyaev and Hill 2000a, Gray 1996) with melanin pigmentation less sensitive to environmental stress than carotenoid pigmentation (Hill and Brawner 1998, McGraw and Hill 2000b). For melanin-based ornaments, signal honesty appears to rely on social mediation.

Although the emphasis in the status-signaling literature has been exclusively on social mediation, I can see no reason why status signaling could not also work with ornaments whose honesty is maintained through production costs. Condition-dependent traits, such as carotenoid-based color displays, are costly to produce so that only high-quality males display the most elaborate form of the trait. It seems logical that high-quality males would be dominant males and hence that carotenoid-based plumage coloration could serve as a reliable signal of social status. As I showed in the preceding section, however, correlations between plumage coloration and status in captive flocks of male House Finches falsify this prediction; drabber males tend to dominate brighter males. This brings us to a much less intuitive possibility: could *drab* plumage coloration be a reliable signal of high status? To test this idea, we had to demonstrate that, rather than simply being correlated to dominance status, plumage coloration was used by one male to assess the likely dominance status of another male.

## Experimental Tests of Status Signaling

Balph et al. (1979) and Whitfield (1987) pointed out that true status signaling should allow the assessment of resource-holding potential not simply between individuals in different age cohorts but also between individuals within an age cohort. In the House Finch, I showed that, although older males tend to be brighter than yearling males, there is no age-specific plumage ornamentation (see Figure 2.2). Carotenoid-based plumage coloration in male House Finches, therefore, fulfills a basic requirement for within-age-group status signaling. The key assumption of the status-signaling hypothesis, however, is that conspecifics alter their treatment of an unfamiliar conspecific in response to the expression of its status signal. Thus, to convincingly demonstrate that a trait is used as a signal of status, one must manipulate expression of the trait and measure how the change affects the responses of conspecific males of the same age (Senar 1999, Whitfield 1987).

With Kevin McGraw, I conducted aviary experiments to determine if carotenoid-based plumage coloration served as a signal of social status. Our goal was to pit males of the same age and dominance status, but with different plumage coloration, in contests over food. In these experiments we wanted plumage to be randomized with respect to health, vigor, and true resource-holding potential, so we randomly assigned males to a color type and colored their feathers with felt-tip pens. (See McGraw and Hill 2000a for details on the plumage manipulations, including the reflectance spectra of the artificial color showing that it closely mimicked carotenoid-based coloration, with no hidden UV coloration.) To run this study, we captured juvenal male House Finches in July at two different locations and held them in cages that were visually isolated. Catching birds at different sites and holding them apart ensured that males in these flocks had no prior experience with each other. Males in all four flocks were maintained through their fall molt on diets of millet and sunflower seed, supplemented with a small amount of the red carotenoid pigment canthaxanthin. On this diet, all males grew a nearly identical prebasic plumage with pale red/orange coloration (see chapter 5). For each cage of males, we then determined a dominance hierarchy in early winter.

After determining the dominance rank of each male in these cages, we conducted paired trials in which we pitted males from different cages with the same dominance position (Figure 8.3). As determined by the flip of a coin, one of the males in each of these paired contests had its plumage treated with a red marker, giving it a bright red appearance; the other male had its plumage treated with a yellow marker, giving it a pale yellow/orange appearance. The males were held from late afternoon until the next morning with no access to food to ensure that they would be hungry at the start of the trials and motivated to gain access to food. We then simultaneously released the two males into an empty cage with one small food bowl. We were most interested in initial assessment of one male by the other male, so trials lasted for only 20 min or until one male had won seven encounters.

We were able to establish male dominance clearly in seventeen of the twenty trials that we conducted, and there was no obvious effect of manipulated plumage coloration on dominance. Bright males displaced the drab males in eleven trials, and drab males out-competed the bright males in six trials. This rate of wins by bright males was within the range expected by chance if there was no effect of plumage coloration on dominance (Figure 8.3). Moreover, the trend for bright males to win the encounters is in the opposite direction from that predicted if drab plumage coloration signals dominance. Thus, while naturally drab males tend to be dominant to naturally bright males (see previous section), carotenoid-based plumage coloration in the House Finch appears not to serve as a signal of social status.

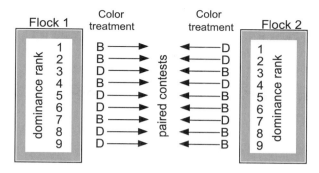

Figure 8.3. The experimental design used to test whether carotenoid-based plumage coloration serves as a status signal in male House Finches. Males of the same dominance rank from two different flocks were pitted in contests over food. (Nine birds are shown for this example, but there were seventeen males in each of the flocks that were tested.) Within each pair of males, one was assigned to a color treatment that made its plumage bright red (B), the other to a treatment that made its plumage drab yellow/orange (D). Males with bright plumage won eleven out of seventeen contests, which was not statistically different than random ($P = 0.33$, two-tailed binomial test). Based on McGraw and Hill (2000a).

The only other published test of status signaling in relation to expression of carotenoid-based plumage ornamentation was conducted by Wolfenbarger (1999b) on male Northern Cardinals. In captive flocks of cardinals, Wolfenbarger found that redder males tended to be dominant toward less red males. When she manipulated the plumage redness of males, however, she found no effect of artificial coloration on success in contests over food, and hence no support for status signaling. Thus, while most studies of melanin-based ornaments have found that such ornaments serve as signals of status, the two studies of carotenoid-based ornamental coloration have found no evidence that they function as status signals.

## Testosterone, Plumage Coloration, Display, and Aggression

Given the detailed case that I have made for plumage redness in the House Finch being a condition-dependent signal of male quality, how do we explain the tendency for drab males to be dominant to bright males? This seems to present a major challenge to the idea that red males are the highest quality males and that they are in the best condition. To begin to resolve this question, we have to consider the hormonal basis for the elevated aggressiveness of males and how this might link back to plumage coloration.

It has long been known that, in many vertebrates, aggressiveness is controlled largely by the androgen hormone testosterone (Wingfield et al. 1987). Moreover, in birds, testosterone controls not only the propensity for an individual to be aggressive but also a range of associated behaviors from song to parental care (Wingfield and Moore 1987). Testosterone is secreted primarily by the testes, and hence for many species it provides a controlling link between aggressive behavior, parental behavior, and reproduction (Eisner 1960, Murton and Westwood 1977). Another hormone linked to dominance and aggressiveness in birds is the glucocorticoid hormone, corticosterone. Corticosterone mobilizes energy reserves and is often associated with stress (Siegel 1980). Birds subjected to a variety of stresses from extreme temperatures to parasites to aggressive conspecifics elevate corticosterone, presumably as an adaptive response to mobilize energy and deal with the situation (Wingfield et al. 1982). Thus, corticosterone is typically inversely related to testosterone and the behaviors associated with testosterone (Wingfield et al. 1982).

In a number of studies of songbirds, levels of circulating testosterone have been linked to dominance (Hegner and Wingfield 1987, Ramenofsky et al. 1992, Wingfield 1985, Wingfield and Moore 1987). However, the effect of testosterone on social status is generally restricted to the period of the initial establishment of the dominance hierarchy; during this brief period, males with higher circulating testosterone levels gain higher status than males with lower circulating testosterone levels (Wingfield et al. 1990). Thereafter, the levels of circulating testosterone are often virtually the same for males regardless of their dominance status (Wingfield et al. 1990). The idea that male birds spike testosterone over short periods of time in response to specific behavioral challenges is called the "challenge hypothesis" (Wingfield et al. 1990). Corticosterone also has been shown to be correlated with dominance status in some birds (Nunez-De La Mora et al. 1996), including the House Finch (Belthoff et al. 1994). Subordinate birds generally have significantly higher corticosterone levels than dominant birds, and unlike testosterone, the differences in corticosterone levels among birds of different dominance rank tend to persist (Nunez-De La Mora et al. 1996).

Testosterone has also been linked to expression of plumage coloration in some birds, but the link appears to be indirect (Owens and Short 1995). In general, testosterone must be present for males of some species of birds to express male-typical plumage coloration (Owens and Short 1995). In some sex-reversed species in which females have showier plumage than males, females also have higher levels of circulating testosterone (Witschi 1961). The role of testosterone in controlling expression of carotenoid-based plumage coloration was, until my students and I began an investigation, unstudied.

With my graduate students, Andrew Stoehr and Renee Duckworth, I studied the roles of testosterone and corticosterone in determining patterns of male dominance, courtship display, parental behavior, and plumage coloration in the House Finch. In the field, we looked at condition, parasite load, and coloration in relation to levels of circulating testosterone and corticosterone. We also experimentally elevated the circulating testosterone of wild males to observe the effects on song rate, parental care, and plumage coloration. In aviaries, we implanted males with testosterone and looked at the effect of the elevated testosterone on disease resistance and expression of plumage coloration.

## Testosterone and Dominance

Belthoff et al. (1994) were the first to look at circulating testosterone in relation to the dominance status of male House Finches. In a study conducted in the fall, during formation of winter flocks, they found no significant relationship between levels of circulating testosterone and the dominance status or plumage coloration of males in captive flocks. Overall, the levels of circulating testosterone that they detected in males were low, and the power of their tests for detecting an effect of testosterone on dominance status and plumage coloration was low (Belthoff et al. 1994). Thus, they could not definitively rule out a potential role for testosterone in mediating aggression in finches. In contrast to the lack of a relationship between testosterone and dominance, they observed that subordinate males consistently had higher levels of corticosterone than did dominant males (Belthoff et al. 1994).

Given the difficulty of correlating the testosterone level of captive males with their dominance rank at the critical period of flock formation, Renee Duckworth and I decided to conduct an experiment to test the effect of testosterone on dominance status (Duckworth et al., in prep). During the winter, we divided thirty-two captive males into five flocks and recorded the dominance rank of males in each flock. We then randomly assigned males to one of three treatment groups. Eight males, including one or two in each flock, were implanted with two Silastic tubes packed with testosterone, which released testosterone slowly into the bloodstream and caused these implanted males to maintain high levels of circulating testosterone. Eight males, including one or two in each flock, had their testes surgically removed (were gonadectomized) so they had little or no circulating testosterone. The remaining sixteen males were not manipulated. We then looked for changes in dominance status over the next few weeks. The results were rather striking. Testosterone-implanted males rose in dominance rank by an average of 2.4 positions; gonadectomized males dropped in dominance status by an average of 2.5 positions; and unmanipulated males retained virtually the same average dominance position (Duckworth et al., in prep; Figure 8.4). I should note that, without hormone manipulations, it is rare to see any change in dominance rank among male House Finches in a captive flock. Thus, testosterone clearly had a direct effect on male aggressiveness and led to higher dominance status among males.

## Testosterone, Song, and Parental Behavior

Increased dominance and aggressiveness must necessarily come at the expense of other behaviors, particularly parental behaviors. As a general rule, animals must trade off investment in a particular mate and the offspring produced with that mate and investment in attracting additional mates (Trivers 1972). It appears that for many species of birds the degree of investment in parental care versus mate attraction is mediated by levels of circulating testosterone. In several studies in which the levels of circulating testosterone of wild males has been experimentally elevated, it has been shown that testosterone stimulates increased territorial defense (Hegner and Wingfield 1987, Wingfield 1984) and increased song output (Enstrom et al. 1997, Ketterson et al. 1992, Silverin 1980) but decreased parental care (Enstrom et al. 1997, Hegner and Wingfield 1987, Ketterson et al. 1992, Moreno et al. 1999, Saino and Moller 1995).

With Andrew Stoehr, I looked at the effect of elevated testosterone on the song rate and parental investment of breeding male House Finches. To conduct this research, we captured males on our Auburn study site from January through

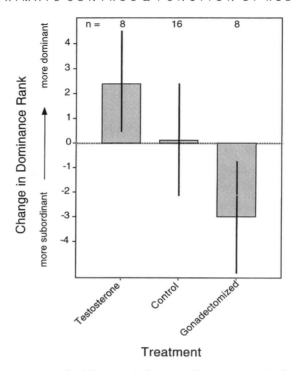

**Figure 8.4.** Mean (± standard deviation) change in dominance rank of captive male House Finches after treatments to alter their circulating levels of testosterone. Males in the testosterone-treatment group were implanted with two Silastic tubes packed with crystalline testosterone that raised their circulating testosterone level to the maximum observed in wild male House Finches. Males in the gonadectomized group had their testes surgically removed, dropping their level of circulating testosterone below detectable levels. Control males were not manipulated and had natural levels of testosterone. Treatment had a significant effect on change in dominance rank ($F = 9.53$, $P < 0.001$). Males implanted with testosterone rose significantly in dominance rank; gonadectomized males fell in dominance rank; and control males retained the same average dominance (from Duckworth et al., in prep).

April—just before they had begun nesting. We implanted these males either with two Silastic tubes packed with testosterone (treatment males) or with empty tubes (sham control males) and released them back into the study population. These testosterone implants elevated the circulating levels of testosterone to the maximum levels observed in wild males (Stoehr and Hill 2000).

Over two breeding seasons, we found that testosterone implants had a significant effect on both singing behavior and parental care. Males with testosterone implants sang at significantly higher rates than males that were not implanted with testosterone (Stoehr and Hill 2000) (Figure 8.5). In addition, males with testosterone implants made significantly fewer provisioning trips to nests than males that were not implanted (Stoehr and Hill 2000). Thus, just as has been observed in other species, testosterone appears to control the time and energy

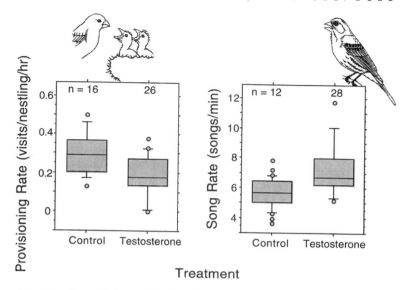

Figure 8.5. The effect of elevated levels of circulating testosterone on the singing and parental behaviors of wild male House Finches. At the start of the breeding season, males were implanted either with Silastic tubes filled with testosterone, or with empty tubes. Testosterone implants significantly decreased the rate at which males visited their nests to feed nestlings ($t = -3.32$, $P = 0.001$) and significantly increased the rate at which males sang ($t = -3.49$, $P = 0.001$). Adapted from Stoehr and Hill (2000).

devoted to the care of offspring versus that spent in pursuit of additional mates in the House Finch. Previously I showed that brighter males tend to provide more parental care than drabber males (see Figures 7.1 and 7.2), and that drabber males tend to be socially dominant to brighter males. Thus, it appeared that males were using testosterone to determine whether they invested in parental care or display, with brighter males following the former strategy and drabber males the latter. Before we could tie the whole story together, however, we had to consider the effect of testosterone on plumage coloration.

### Direct Effects of Testosterone on Plumage Coloration

In the testosterone-manipulation experiment of wild males that I described above, we had no way to recapture males dependably at the end of the breeding season to remove their hormone implants. Thus, inadvertently, this manipulation also became a test of the effect of testosterone on the expression of plumage coloration in male House Finches. Hormonal control of individual variation in expression of ornamental plumage coloration is virtually unstudied (Kimball and Ligon 1999, Owens and Short 1995). Hormonal control of molt is also little studied, but it appears generally that testosterone inhibits molt in songbirds (Nolan et al. 1992, Payne 1972). Thus, for most temperate songbirds that undergo one complete molt of feathers per year just after breeding, the onset of molt appears to be stimulated, at least in part, by a drop in circulating testosterone associated with the curtailment of breeding.

Fortuitously, the Silastic capsules that we used to implant male House Finches with testosterone in this study lasted about six months (give or take a month). In other words, after about six months most tubes had discharged all of their testosterone and had stopped having any direct effect on the hormone levels of the implanted males. This meant that for males implanted early in the season (January to early March) their tubes ran out by August, in time for them to initiate a normal molt sequence in August and September. In contrast, the tubes of males implanted later in the season (April and May) did not run out until past the time when these males would normally have begun molting (Stoehr and Hill 2001). For these late-implant males, elevated testosterone levels seemed to delay the onset of molt. We captured three of these late-implant males in October or November, after all unmanipulated wild males had completed molt, and they were still in the process of growing feathers.

The effect of testosterone treatment on plumage coloration was dramatic. Unmanipulated males showed minimal change in plumage coloration from one year to the next. Likewise, males that were implanted with testosterone early in the season, before about March 10, showed little change in plumage coloration between the year in which they were implanted and the following year. The group of males implanted with testosterone late in the spring, however, grew plumage that was much drabber than their pre-treatment plumage (Stoehr and Hill 2001) (Figure 8.6). Virtually all of the males in this latter group started out bright red or reddish orange in coloration, but after molt they were pale yellow. Elevated levels of testosterone during the period when males should have been molting had a large effect on the color of the carotenoid-based plumage that they grew.

To further test the effect of testosterone on plumage coloration, Andrew Stoehr and I also conducted an experiment with captive males. For this study we held thirteen hatch-year males on a seed diet supplemented with canthaxanthin. We implanted three of these males with testosterone prior to the molt period, and left ten males without implants to serve as controls. Because testosterone suppresses molt, it is not possible to look at the effect of testosterone on growing feathers—feathers won't grow in the presence of testosterone. By early October all control males had completed their molt, but the testosterone-implanted males had not started or had barely started molt. At that time, we removed the testosterone implants from males, and by early December these previously implanted males had molted much of their plumage. We observed that the males that had been implanted with testosterone grew significantly drabber plumage than males that were not implanted (Stoehr and Hill 2001) (Figure 8.6). Thus, even when they had access to the same carotenoid pigments during molt, males with elevated testosterone grew drabber plumage than males with normal low levels of testosterone.

The obvious question from the above observations is: how did elevated testosterone affect plumage coloration? We proposed that there are two likely mechanisms that are not mutually exclusive (Stoehr and Hill 2001). First, testosterone may have affected plumage coloration of the free-living males indirectly by delaying the onset of their molt. This explanation assumes that molt in House Finches coincides

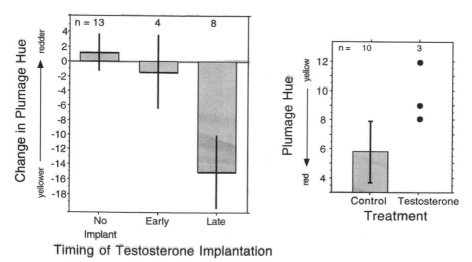

**Timing of Testosterone Implantation**

**Figure 8.6.** The effect of testosterone on expression of carotenoid-based plumage coloration in male House Finches. The left figure shows the mean (± SD) change in plumage hue of wild male House Finches subjected to different testosterone manipulations. Males were either not implanted, were implanted with testosterone early in the year so that their implants ran out before fall molt, or were implanted late in the spring so that their implants did not run out until after molt should have begun. Males that were implanted late showed a significantly greater decrease in plumage redness than did males that were implanted early or that had no implants ($F = 47.13$, $P = 0.0001$, one-way ANOVA). The right figure shows the mean (± SD) plumage coloration of captive male House Finches that were either implanted with testosterone or maintained with natural hormone levels. Males implanted with testosterone grew significantly drabber plumage than males that were not implanted ($U = 1.5$, $P = 0.02$). Adapted from Stoehr and Hill (2001).

with a period of maximum availability of the dietary pigments that are needed to color plumage. It further proposes that by causing males to delay their fall molt by a month or two, testosterone implants pushed the molt period for these males into a time of year when dietary carotenoids were relatively scarce, thus preventing them from acquiring the carotenoids that were needed to produce brightly colored plumage. This idea remains purely speculative, however, because there are no data to show that carotenoids become less available over the time periods involved.

Another possibility is that testosterone may have affected the ability of males to utilize dietary carotenoid pigments. Such utilizational disruption could have been a result of elevated parasite levels associated with elevated testosterone. Unfortunately, we were not able to measure parasite loads of manipulated males. Utilizational effects could also have been more direct if elevated testosterone affected the process of absorption, transport, metabolism, or deposition of carotenoid pigments, for instance, by reducing the production of carrier proteins or enzymes needed for carotenoid metabolism (see chapter 4). While observations from the field-manipulation experiment suggested that timing of molt and change in carotenoid access contributed to the decrease of coloration in testosterone-implanted males, the aviary experiment, in which access to carotenoids was controlled, indicates that testosterone may also have a direct effect on expression of plumage coloration (Stoehr and Hill 2001).

In summary, although the mechanism by which the effect occurs is still not entirely clear, elevated testosterone in male House Finches is consistently associated with drab plumage coloration. This is a very interesting observation given the previous observation that drab males are commonly dominant to bright males and that high testosterone is linked to social dominance. These links suggest that male House Finches must trade off high testosterone, dominance, and aggression with plumage coloration and paternal behaviors. What remains to tie this line of explanation together is establishing a cost of elevated testosterone, and that is the topic of the next section.

## Testosterone and Immunocompetence

In vertebrates, expression of display traits and behaviors that function in sexual selection are commonly controlled by levels of circulating testosterone (Andersson 1994, Ligon 1999). Earlier in this chapter we saw that testosterone can have a large direct effect on display behaviors such as song in male House Finches, and can have at least an indirect effect on expression of plumage coloration. At the same time, in several species of vertebrates, testosterone has been shown to have a suppressive effect on the immune system (reviewed in Folstad and Karter 1992). These observations prompted Folstad and Karter (1992) to propose the Immunocompetence Handicap Hypothesis, which hypothesizes that high levels of circulating testosterone in vertebrates act as a double-edged sword. Males must have elevated levels of testosterone to express sexually selected display traits, but elevated testosterone may also compromise their immune systems. By this idea, only males in the best condition and with the most efficient immune systems can afford the cost of elevated testosterone and maximum ornament display. Thus, it is the cost of testosterone that makes testosterone-dependent sexual traits honest signals of condition (Folstad and Karter 1992).

From its inception, the Immunocompetence Handicap Hypothesis evoked the question: "Why don't males simply evolve mechanisms to express ornamental traits independent of costly circulating testosterone?" This line of questioning led to the proposition that it is not circulating testosterone per se that is costly to males; rather, it is both ornament production and immune defense that are costly. Testosterone simply serves as the mechanism for adaptively distributing energy between body maintenance, including immune defense, and ornament display (Wedekind 1994, Westneat and Birkhead 1998). Consequently, as suggested by Folstad and Karter, only males with the best immune systems and in the best condition will have high testosterone levels and hence elaborate ornament displays. While the Immunocompetence Handicap Hypothesis is widely cited as an explanation for how ornamental traits serve as honest signals of male condition, it has been tested on few species of songbirds (see Evans et al. 2000 for a review).

## Is Testosterone Immunosuppressive?

With Renee Duckworth and my faculty colleague Mary Mendonça, I tested a basic premise of the Immunocompetence Handicap Hypothesis—that elevated testosterone levels reduce immune function in male House Finches (Duckworth et al. 2001). We established a group of captive males and then randomly assigned them to one of three treatment groups: testosterone-implanted, gonadectomized (testes removed), and control. As in studies described previously, gonadectomized males had essentially no circulating testosterone, so these groups represented high, low, and no testosterone treatment groups. A month after these males had been manipulated, we inoculated them with coccidia (see chapter 5 for details of coccidia as House Finch parasites). We then assessed the effects of testosterone treatment on responses to the parasite. We found that treatment had no effect on whether or not birds became infected, but it did have a significant effect on the time that it took coccidia to become established in the finches' digestive tracts (Duckworth et al. 2001). Males with testosterone implants began passing coccidial oocysts (a sign of established coccidial infection) significantly faster than either gonadectomized or control males (Duckworth et al. 2001) (Figure 8.7). Thus, elevated levels of circulating testosterone do appear to reduce the immunocompetence of males. This link between elevated testosterone and reduced immune system function supports the contention that males that elevate testosterone to increase their aggressiveness and dominance status also increase their risk of disease.

## Testosterone, Corticosterone, Condition, and Disease

Another prediction of the Immunocompetence Handicap Hypothesis is that only males in the best condition will be able to maintain high levels of circulating testosterone (Folstad and Karter 1992). We tested this prediction in the Auburn population of House Finches. As described above, on our Auburn study site we measured the circulating testosterone and corticosterone levels of males captured in 1999 and 2000. We also recorded the condition of these males, measured both by hematocrit level (proportion of blood volume made up of red blood cells) and mass adjusted for body size. We found that males with higher levels of circulating testosterone were in better condition than males with low levels of circulating testosterone; testosterone was significantly positively correlated with both hematocrit and with mass corrected for body size (Duckworth et al. 2001; Figure 8.8). At the same time, corticosterone—which you will recall is associated with stress in birds—was negatively related to both hematocrit and mass corrected for body size. So, males in better condition do maintain higher levels of circulating testosterone, as predicted by the Immunocompetence Handicap Hypothesis.

The experiments outlined above indicated that elevated testosterone put males at risk of disease and that in the wild only males in good condition maintained elevated levels of testosterone. Once a male became infected with a parasite, however, it would be expected to reduce its level of circulating testosterone to channel resources away from sexual display and into immune defense. Likewise, parasitized males are subjected to stress and should have higher levels of circulating corticosterone. Thus, we would expect males with parasitic infections to have lower levels

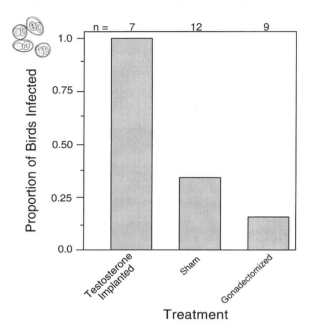

Figure 8.7. The effect of circulating testosterone on disease resistance in captive male House Finches. Males were either implanted with testosterone so they maintained high levels of circulating testosterone, gonadectomized so they had little circulating testosterone, or implanted with empty tubes so they had natural levels of circulating testosterone (sham). All males were then inoculated with isosporan coccidia. After one week, a significantly higher proportion of testosterone-implanted males were passing isosporan oocysts (a sign of infection) than control or gonadectomized males ($\chi^2 = 12.41$, $P = 0.001$), indicating that the testosterone compromised the ability of males to resist infection. Adapted from Duckworth et al. (2001).

of circulating testosterone and higher levels of circulating corticosterone than males that are not parasitized. To test this idea we determined whether or not the wild males that we captured were infected with coccidia or *Mycoplasma gallisepticum*. We then compared the levels of circulating testosterone and corticosterone of parasitized and unparasitized males. We found that males with either coccidiosis or mycoplasmosis had significantly lower levels of circulating testosterone and significantly higher levels of circulating corticosterone than males that were not infected (Duckworth et al. 2001) (Figure 8.9). Again, this is consistent with predictions of the Immunocompetence Handicap Hypothesis.

Taken together, these observations support the idea that testosterone serves to regulate energy investment in ornament display versus energy investment in body maintenance, including the immune system. In experiments in which the testosterone levels of males were elevated artificially, there were negative effects on immune function. In this case, by assigning hormone levels to males that were higher than they would have had naturally, we induced males to allocate energy away from immune defense to sexual display, and the result was increased vulnerability to

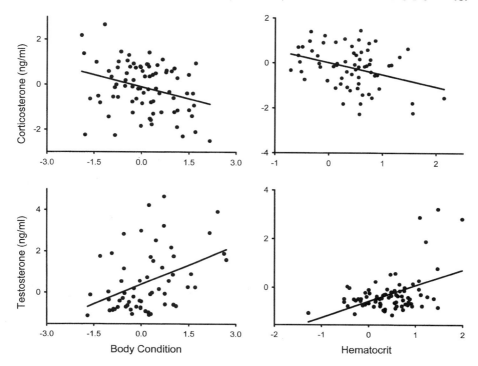

**Figure 8.8.** The relationship between male condition and circulating testosterone and corticosterone in House Finches in Auburn, Alabama. Condition was measured either as hematocrit level (proportion of blood volume that was red blood cells) or mass corrected for size (body condition). There was a significant positive relationship between testosterone and both body condition ($r = 0.42$, $t = 3.56$, $P = 0.0007$) and hematocrit ($r = 0.46$, $t = 4.50$, $P < 0.0001$). Conversely, there was a significant negative relationship between corticosterone and both body condition ($r = -0.24$, $t = 2.37$, $P = 0.01$) and hematocrit ($r = -0.21$, $t = 2.88$, $P = 0.005$). Adapted from Duckworth et al. (2001).

parasitic infection. In the correlational study in the field, where hormone levels were monitored but not manipulated, we found further evidence that males regulate hormone levels based on their physical condition. Males in better body condition had higher levels of circulating testosterone than males in poor condition, and males infected with parasites had lower levels of circulating testosterone than males that were not parasitized.

### Testosterone and Plumage Coloration in Wild Males

To complete this line of investigation of male dominance, hormone levels, disease resistance, and plumage coloration we looked at the relationship among males in our wild population between circulating levels of testosterone and plumage coloration. Our predictions going into this analysis were somewhat ambiguous. We had established that drab male House Finches display behaviors that are consistent with high levels of testosterone—they are dominant to bright males and they provision

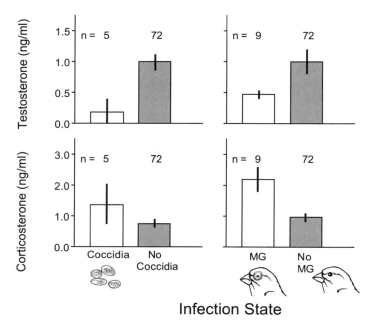

Infection State

Figure 8.9. Levels of circulating testosterone and corticosterone in relation to coccidiosis and mycoplasmosis in wild male House Finches in Auburn, Alabama. Males with coccidiosis and mycoplasmosis had significantly lower levels of circulating testosterone ($P = 0.03$, $P = 0.02$ respectively, sign test) and significantly higher levels of circulating corticosterone ($P = 0.003$, $P = 0.002$ respectively, sign test) than males without these diseases. Adapted from Duckworth et al. (2001).

offspring poorly. Among wild males, however, we had observed that higher levels of circulating testosterone were associated with good condition and avoidance of parasites. On the one hand, we expected a positive relationship between testosterone and plumage coloration, because more brightly colored males should be in better condition. On the other hand, to explain the aggressiveness and poor parental behaviors of drab males we expected a negative relationship between circulating testosterone and plumage color.

We measured the levels of circulating testosterone (see Mendonça et al. 1996 for details of the hormone assays used) of wild male House Finches during the prebreeding period (December through February), when pairs are forming and males are not being affected by nesting cycles. As might be expected given the ephemeral nature of hormone levels in an individual, the patterns were messy. Somewhat paradoxically, there was a trend for males with redder plumage to have higher levels of circulating testosterone (Figure 8.10).

In interpreting the pattern of association between plumage coloration and circulating levels of testosterone it is important to note that almost all the measured concentrations of circulating testosterone in male House Finches were low. We rarely measured levels of circulating testosterone above 5 ng/mL. A few males, however, had testosterone levels above 14 ng/mL (unpublished data). Our inter-

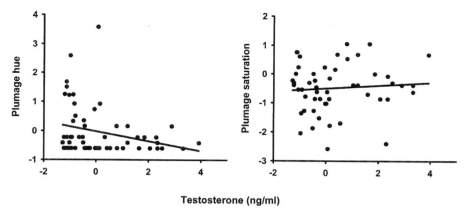

**Testosterone (ng/ml)**

Figure 8.10. The relationship between circulating testosterone and the hue and saturation of carotenoid-based plumage coloration of male House Finches captured in Auburn, Alabama. Samples were taken from December through February during the period of pair formation when males should not be subject to the cyclical variation in testosterone levels associated with nesting. Measured testosterone levels were adjusted for how long the male was in the hand before being bled. Both plumage scores and testosterone levels were arc-sine transformed to normalize the data. Males with redder plumage (lower hue scores) tended to have higher levels of circulating testosterone ($r = -0.20$, $P = 0.069$), but the relationship was not quite significant. There was no significant relationship between plumage saturation and levels of circulating testosterone ($r = 0.08$, $P = 0.55$; Duckworth et al., unpublished data).

pretation is that we rarely recorded the maximum testosterone levels for males, and it may be that such testosterone surges or spikes determine male dominance (Wingfield et al. 1990). Recall that Belthoff et al. (1994) found no relationship between circulating testosterone and dominance status among captive males, but that experiments in which testosterone levels of males were altered clearly showed that testosterone has a strong effect on dominance (Figure 8.4). Such testosterone spikes, which are difficult to measure, could account for the differences in these observational versus experimental dominance studies and could also account for the relationship between plumage coloration, dominance, and measured circulating testosterone among wild males. In the wild, high-condition red males may maintain a higher average level of circulating testosterone, while drab males may have lower average circulating testosterone but may spike the levels of circulating testosterone during aggressive interactions and when displaying to females. This hypothesis does not necessarily invoke any direct influence of testosterone on the color of plumage that is grown; rather, it proposes that males that end up with drab plumage follow a more aggressive behavioral strategy.

Clearly, more studies are needed to reach a better understanding of the inter-relationships between dominance, hormone levels, condition, and disease resistance in the House Finch. Experimental manipulations, in which individuals are assigned high constant levels of testosterone, are the current best approach to getting around the problem of natural fluctuations of testosterone. The development of techniques to measure daily fluctuation of testosterone within an individual would do even more to resolve these important issues. Until we develop a better fundamental understanding of the relationship between testosterone and immunocompetence and testosterone and plumage coloration, we will not be able to explain satisfactorily why drab male House Finches are socially dominant to bright males.

## Are Drab and Bright Males Following Different Reproductive Strategies?

Taken together, these studies suggest an interesting link between plumage coloration, aggressiveness, parental care, and testosterone levels in male House Finches. We can characterize bright red males as non-aggressive individuals that invest in parental care and have relatively low display rates (e.g., low song rates). The association between red plumage and foraging ability (see chapter 5) and investment in offspring make red males preferred as social mates by females. Success within such a reproductive strategy, in which investment is focused primarily toward the social mate and offspring, appears to maximize reproductive output for males. With few extra-pair copulations to be had, there are relatively few opportunities for reproduction outside the interactions with a social mate (see Figure 6.12). In contrast, drably colored males can be characterized as aggressive individuals with high rates of courtship display that are in relatively poor condition and that are poor providers for their offspring and social mates. Consistent with this profile is the observation that drab males are socially dominant to bright males, but they have difficulty attracting social mates, provision their mates and offspring poorly, and have lower reproductive output than do bright males once they are paired.

From what I have presented in these first eight chapters, clearly the best reproductive strategy for a male House Finch is to grow bright red plumage. Red males gain many advantages over drab males, including success at attracting mates, early nest initiation, and the production of more offspring. It would not make sense for any male that had the potential to grow bright red plumage not to do so. Production of drab versus bright plumage almost surely does not represent an equally good reproductive strategy (i.e., it is not an evolutionary stable strategy). Rather, expression of drab plumage coloration and increased aggressiveness are better characterized as conditional reproductive strategies (Dawkins 1980, Johnson et al. 2000). It appears that, for one or more of the reasons outlined in chapter 5—inadequate intake of carotenoid pigments, infection by parasites, or poor nutrition—some males fail to achieve bright plumage coloration. This leaves drab males with a reduced chance of attracting a social mate. Perhaps to compensate for their unattractive plumage ornaments, at least some drab males appear to adopt a secondary, perhaps riskier and costlier, reproductive strategy in which they increase their aggressiveness and display rate. Thus, drab males are dominant to bright males because they have more to gain and are thus willing to take chances and escalate fights. They apparently pay a cost for this increased aggression and social dominance in the form of poor parental care, lower survivorship, reduced immunocompetence, and increased risk of parasitism.

These observations potentially explain why carotenoid-based plumage coloration does not serve as a status signal in House Finches. If only a portion of drab

males in a population followed the strategy of elevating testosterone and aggressiveness, then drab coloration and social dominance would be relatively weakly
linked (as was found in studies of dominance and coloration). Moreover, drab
males are generally in poor condition relative to bright males, making them a threat
to bright males only if they are willing to escalate fights, and displacements at food
sources typically involve no physical contact and probably incur few costs. For all of
these reasons, it would be to the advantage of bright males to assess the aggressiveness, rather than the coloration, of competitors.

A strategy by which drab males elevated testosterone to increase aggressiveness
and display behavior would make sense only if the strategy enhanced the reproductive success of drab males. This is where the trail runs cold, however. The key
prediction of this reproductive strategy hypothesis—that drab males that are aggressive gain higher short-term reproductive success than drab males that are non-
aggressive—has not been tested. We have no data on how testosterone levels and
aggressiveness affect reproductive output. It is not too far fetched, however, to
propose that increased display and social dominance may increase a drab male's
chances of extra-pair matings (Qvarnström and Forsgren 1998), scarce though they
may be (see chapter 6 and Figure 6.12), or may increase its chances of attracting a
social mate if song rate serves as a secondary criterion in mate choice. Also, because
drab males do not have attractive plumage coloration, a social mate is a much more
valuable resource to a drab male than to a bright male. Elevated testosterone and
heightened aggression might enhance the ability of drab males to retain a social
mate if they are able to attract one.

All of these alternative explanations for the observation that drab males are
dominant to bright males remain completely speculative. We have yet to measure
high levels of testosterone in any drab male, so the most basic assumption of this
hypothesis remains to be confirmed. At present, the hypothesis that drab males
adopt an aggressive, risk-taking strategy to compensate for poor plumage ornamentation is the best explanation for the behaviors that we have observed.

## Summary

Explaining the observation that drably colored male House Finches are socially
dominant to brightly colored male House Finches remains one of the most intriguing and challenging topics related to ornamental plumage color in this species.
Carotenoid-based plumage coloration is negatively correlated with dominance rank,
but color-manipulation experiments indicate that plumage coloration does not
function as a status signal in the House Finch. In captive flocks of finches, measured
levels of circulating testosterone of males did not predict the dominance rank of
males, but hormone manipulations had a strong effect on dominance status. Males
that were gonadectomized dropped in social dominance rank, while males
implanted with testosterone rose in social dominance rank. In wild male House
Finches, elevated testosterone increased song rate but decreased parental care.
Males with elevated testosterone also grew drabber plumage than males with low
levels of circulating testosterone during molt. Finally, circulating testosterone was
associated with high body condition, and males that were infected with parasites

had lower levels of circulating testosterone than males that were not parasitized. These observations suggest that drab males adopt an aggressive, risk-taking strategy to compensate for their poor plumage ornamentation. This hypothesis predicts that males with drab plumage coloration would have higher levels of circulating testosterone than males with bright plumage. In the field, however, higher circulating levels of testosterone was associated with brighter plumage. These observations could be reconciled if drab males spiked testosterone during aggressive encounters. The relationship between plumage coloration, dominance, immunocompetence, and testosterone remains to be fully resolved in this or any species.

# The Feeling's Mutual

## Female Plumage Coloration and Male Mate Choice

One of the results of this ability [to live in a wide range of habitats] is a certain tendency to vary, which exhibits itself in off-plumages and freaks, quite independently of association or environment . . . As a further example of freakishness, Swarth cites a case where two young females, caught in the wild, showed distinct traces of red in their plumage.

—W. L. Dawson (1923), part of a description of House Finches in *The Birds of California*

Several very close observers of the habits of animals have assured me that male birds and quadrupeds do often take very strong likes and dislikes to females, and we can hardly believe that the one sex (the female) can have a general taste for colour while the other has no such taste. However this may be, the fact remains, that in the vast number of cases the female acquires as brilliant and as varied colors as the male, and therefore most probably acquires them in the same way as the male does—that is, either because the colour is useful to it, or is correlated with some useful variation, or is pleasing to the other sex.

—A. R. Wallace (1878:129)

*"Daddy, why has mommies got boobs?"* was the innocent question of my daughter Savannah one day when she was about five years old. As seems always to be the case with the questions posed by young minds, this one came out of the blue, unrelated to anything we had been discussing.

*"To nurse their babies," was my answer. I'd learned that in talking to young children (I'd been through this before with my son Trevor, who is four years older) it is best to start with simple direct answers and see where the line of questioning leads.*

*"But Callie doesn't have boobs and she nursed her babies," Savannah answered, referring to our family cat who had had a litter of kittens before we adopted and spayed her. I was slightly taken aback by this reply. It was more insightful than many of the comments I've heard from graduate students over the years.*

*"You're exactly right," I continued, "a mommy doesn't need boobs to produce milk for a baby. The real reason that mommies have boobs is because daddies like boobs. And the reason that daddies have whiskers and loud voices is because mommies like those things."*

*"So God gave daddies boobs and mommies whiskers so everyone is happy," Savannah concluded. I started to correct her: "You mean God gave mommies boobs and daddies whiskers," but then I decided that she was just being insightful again.*

*"Yes," I answered, "those things came to be so that everyone would be happy."*

Since Darwin and Wallace, discussion and investigations of sexual selection in animals have focused almost exclusively on ornamental traits in males. The concept of coy and choosy females and rather indiscriminant, displaying males provided the framework for the development of sexual selection theory (Darwin 1871, Fisher 1915, 1930, 1958). Such a framework was also the basis for most of the work that I conducted with House Finches, as presented in the previous chapters of this book. The problem with this approach is that, although female animals are not generally as highly ornamented as males, in many species the sexes are equally highly ornamented, and in even more species in which males are highly ornamented, females show a subdued version of the same ornamental coloration displayed by males (Amundsen 2000, Andersson 1994, Winterbottom 1929). Such ornamental display in females is as demanding of an explanation as is the ornamental display of males. Once biologists became comfortable with theories of sexual selection that accounted for the evolution and maintenance of male ornamental traits, a few biologists turned their attention to an explanation for ornamental traits in females (Burley 1977, Lande 1980, Parker 1983).

Compared to male plumage coloration, female plumage coloration has received little study. For me, also, the study of female ornamentation has never been a central focus. But, even though I always considered quantifying the plumage coloration of females a low-priority activity compared to the detailed measurements that I recorded for male plumage coloration, from my first day of banding I scored plumage coloration of the females that I captured and banded. Moreover, when I ran various aviary experiments to test the effects of carotenoid access, parasites, or nutrition on expression of male coloration (see chapter 5), I invariably included females in the treatment groups. In the end, I accumulated considerable data on female plumage coloration, and these data provided me with the opportunity to test hypotheses related to the function of ornamental plumage coloration in females.

## Ornamental Coloration in Female House Finches

### Expression of Carotenoid-Based Coloration in Female House Finches

As I have presented throughout this book, House Finches are sexually dichromatic, with males having bright coloration and females being drab. In all field guides that depict House Finches and describe their plumage (e.g., National Geographic Society 1999, Peterson 1980), females are presented as lacking yellow, orange, or red coloration. By the time I had banded my first dozen House Finches at the beginning of my study in 1987, however, it was obvious to me that this characterization of females was wrong. While it was correct that females were never as brightly colored as even the drabbest male, many of the females that I banded and examined closely in the hand had conspicuous carotenoid coloration on their rumps, and a few females even had a wash of carotenoid coloration on their breast and crown feathers. I found that, in my Michigan study population, 56% of females had visible carotenoid pigmentation in their plumage (Hill 1993d) (Figure 9.1). All of these colored females had carotenoid pigmentation on their rump feathers, and a few (about 9%) also had a wash of carotenoid pigmentation on their breast and crown. As in males, females varied in hue of carotenoid-based coloration from yellow through orange to red (Hill 1993d). The proportion of females with carotenoid coloration varied somewhat among populations, but in all populations a substantial proportion of female showed carotenoid-based plumage coloration (Hill 1993d).

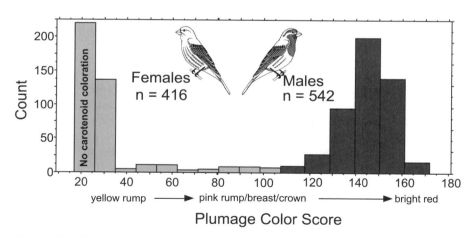

## Plumage Color Score

Figure 9.1. The frequency distribution of plumage scores for female (light gray bars) and male (dark gray bars) House Finches captured in Ann Arbor, Michigan. There was no overlap in plumage scores of males and females. 56% of females showed no detectable carotenoid pigmentation in their plumage. Of those that did have carotenoid-based plumage coloration, most had only a yellow wash on their rump. A few females showed a pink wash on their crown and breast. Modified from Hill (1990) and (1993d).

## Age Effects

One of the first questions that I addressed related to the plumage coloration of females was whether age had an effect. In female Red-winged Blackbirds, one of the few species in which carotenoid-based female plumage coloration had been studied rather extensively, yearling females have less colorful epaulets than older females (Crawford 1977, Miskimen 1980), and older females tend to get brighter as they age (Blank and Nolan 1983, Johnsen et al. 1996). So, using plumage scores from known-age females, I tested to see if there was a similar age effect on expression of carotenoid-based plumage coloration in female House Finches.

To my surprise I found that there was indeed a strong effect of age on expression of plumage coloration in female House Finches, but that it was exactly opposite to that observed in Red-winged Blackbirds and to what I would have predicted. Yearling female House Finches were significantly brighter than females that were two years old or older (Hill 1993d) (Figure 9.2). Almost all the females that showed a wash of red coloration on their breast and crown in addition to their rumps were yearling females.

Why yearling female House Finches tend to grow more ornamented plumage than older females remains unknown. Yearling female House Finches tend to undergo their prebasic molt and grow their ornamental plumage significantly earlier than older females (unpubl. data). It is possible that at the time that yearling females grow feathers, more carotenoid pigments are available in foods than when older females grow their feathers several weeks later (Hill 1993d). Moreover, while yearling females have not bred by the time of their first prebasic molt, older females will have just completed breeding. As a result, older females may be in poorer condition with lower energy reserves at the time of molt than yearling females. It is also possible that yearling and older females are consistent in their expression of plumage coloration, but that brighter yearling females have lower survival than females lacking color and thus are underrepresented in older age cohorts. Finally, any number of proximate control mechanisms (e.g., hormone levels, transport proteins, metabolic enzymes) may differ between yearling and older females, and these proximate factors may account for the difference in plumage coloration. This is an area of research that warrants further study.

It is also possible that yearling and older females pursue different strategies with regard to investment in acquiring caroteᵣ ɔid coloration. Bright ornamental coloration might be more important to young females who have to pair for the first time than to older females who are likely to pair again with their mate from the previous season, or at least to pair with a male with whom they are familiar. The difference in condition between yearling and older females at the time of molt may reinforce such a strategy of reduced pigmentation in older females. All of these hypotheses to account for the age differences in expression of plumage coloration among female House Finches remain to be tested.

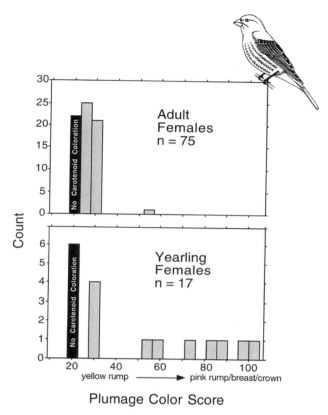

Plumage Color Score

**Figure 9.2.** Age and plumage color of female House Finches captured in Ann Arbor, Michigan. "Adult females" were females known to be in at least their second basic plumage. "Yearling females" were females in first basic plumage. Unlike male House Finches or most passerine birds that have been studied, yearling female House Finches tend to have brighter plumage than older females ($Z = 4.08$, $P = 0.0001$; Mann-Whitney $U$-test). Modified from Hill (1993d).

## Proximate Control of Female Coloration

As was the case with male coloration and female mate choice, to understand the potential signal content of female plumage coloration and the potential benefits to males of assessing female coloration when they choose mates, I had to understand how environmental factors such as access to carotenoids, disease, and nutritional condition during molt affected expression of plumage coloration. It seemed clear from the outset that the proximate control of the plumage coloration of female House Finches was somewhat different than the control of plumage coloration in males. Unlike males, in which all individuals have extensive carotenoid-based color display, nearly half of all females displayed no visible carotenoid coloration. Those females that did have carotenoid-based plumage coloration showed only a shadow of the bright color display of males. Finally, as just presented and unlike what is observed in males, yearling females tended to be brighter than older females.

## Effect of Dietary Access to Carotenoids

In chapter 5, I presented the results of feeding experiments in which I standardized access to carotenoids during molt for captive groups of males and then measured the effect on plumage coloration. In most of those experiments I used hatch-year birds. Because there is no way to determine the sex of House Finches in juvenal plumage by external examination, about half of the birds in these feeding experiments ended up being females. As a result, I recorded the plumage coloration of females as well as males in these feeding experiments. The details of the methods for the feeding experiments that I will present for females are the same as for males and can be found in chapter 5.

Just as in males, dietary access to carotenoid pigments at the time of molt had a significant effect on the expression of carotenoid-based plumage coloration in females. When captive females were fed a seed diet, which provided small quantities of yellow carotenoid pigments, all grew plumage with a wash of yellow coloration on their rumps (Hill 1993d) (Figure 9.3). There was no significant difference between the coloration of adult and yearling females in this treatment (Hill 1993d), but only three yearlings were included and they tended to grow brighter plumage than older females. Despite the fact that only about half the females in wild populations show red/orange/yellow coloration, on a diet supplemented with the red pigment canthaxanthin, all females converged on maximum color expression for females with red rumps and a wash of red over their undersides and crowns. Convincingly this time, there was no significant difference between yearling and older females (Hill 1993d). The mean plumage coloration of females on the plain seed diet was not significantly different than the mean plumage coloration of wild females, but the mean plumage coloration of females fed canthaxanthin during molt was substantially higher than the mean plumage coloration of wild females.

These observations indicate that the variation in expression of plumage coloration in wild females is not due to intrinsic differences among females in their capacity to produce colorful plumage. Rather, all females have the potential for ornamental color display, but in the wild some females achieve brighter coloration than others. Feeding experiments suggest that access to dietary carotenoid pigments at the time of molt is an important factor in determining the color of feathers grown by females. As with males, a role for dietary access in determining plumage expression does not mean that other environmental factors aren't also important in determining female plumage coloration.

It is interesting that supplementing females with abundant red pigments caused them to grow feathers with the maximum amount of carotenoid pigmentation seen in wild female House Finches, but not to converge on male expression of ornamental coloration. Maximum ornamental coloration in female House Finches is still only a shadow of male ornamental coloration, indicating that the proximate mechanisms that control pigmentation—carotenoid absorption, transport, metabolism, and deposition—must be fundamentally different in males and

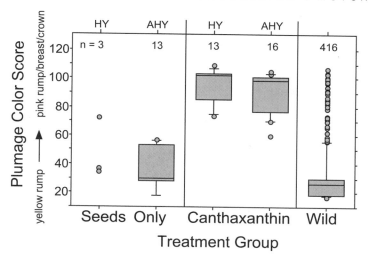

Figure 9.3. The effect of dietary access to carotenoid pigments during molt on the expression of plumage coloration in captive female House Finches. Codes above the box plot indicate whether the group of females was molting from juvenal to first basic plumage (HY) or molting from one basic plumage to another (AHY). Yearling and older females did not differ in plumage coloration when they were maintained on the same diet (seed diet: $t = 1.49$, $P = 0.16$; canthaxanthin: $t = 1.079$, $P = 0.29$). Access to carotenoids during molt had a significant effect on plumage coloration; females maintained on a plain-seed diet grew significantly less colorful plumage than females maintained on a diet supplemented with the red pigment canthaxanthin ($t = 12.03$, $P = 0.0001$). The right box plot shows the distribution of plumage scores of wild female House Finches captured in Ann Arbor, Michigan. Adapted from Hill (1993d).

females. This may seem like a trivial observation, but in Red Crossbills (*Loxia curvirostra*), another species of cardueline finch with sexual dichromatism similar to the House Finch, when females are fed canthaxanthin during molt, they grow red plumage that makes them indistinguishable from males (Hill and Benkman 1995). So, the effect of carotenoid supplementation on expression of carotenoid pigmentation in females is different among different species of cardueline finches. Female House Finches are not as brightly colored as males because they have a sex-specific pigment physiology (although specifically what these sex-specific mechanisms are remains to be determined) that precludes a male-like expression of carotenoid pigmentation. McGraw et al. (2001) observed similar sex-specific patterns of carotenoid display independent of diet in the American Goldfinch. Female Red Crossbills, in contrast, are not as red as males in the wild apparently because they do not ingest sufficient carotenoid pigments at the time of molt. Such differences in female pigment control among closely related species has the potential to provide important insight into the evolution of both male and female carotenoid pigmentation, but it is a topic that has yet to be investigated thoroughly.

*Parasites and Nutrition*

As in males, two additional environmental factors besides carotenoid access have been proposed to affect expression of female coloration—parasites (Potti and Merino 1996) and nutritional condition during molt (Hill 2000). The arguments for how these environmental factors could affect expression of plumage coloration in females are exactly the same as the arguments that I made in chapter 5 for how these variables could affect expression of plumage coloration in males. I will first discuss whether parasites affect expression of female carotenoid-based plumage coloration and then consider nutritional condition.

In the experiment in which captive House Finches were either infected with coccidia or maintained free of coccidia (see chapter 5 and Brawner et al. 2000 for details), we included four females in the inoculated group and six females in the control group. These females were maintained on seed diets supplemented with relatively small quantities of canthaxanthin. When these females had completed their prebasic molt, we used a Colortron to score the coloration of rump feathers. We found a tendency for females that were free of coccidial infection to grow redder plumage than females infected with coccidia, but the difference was not significant (Figure 9.4). When we combined hue, saturation, and brightness into a composite plumage brightness score, we found that, despite the small number of

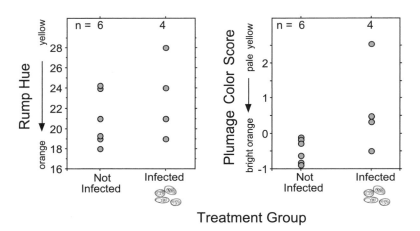

Treatment Group

Figure 9.4. The effect of coccidiosis on expression of carotenoid-based plumage coloration in female House Finches. Yearling females were infected either with 2000 isosporan oocysts prior to molt or maintained on the drug sulfadimethoxine, which reduced coccidiosis to subclinical levels. All females were fed the same seed diet supplemented with canthaxanthin. Females treated with sulfadimethoxine that had no clinical coccidial infection tended to grow redder rump feathers, but the effect was not significant ($Z = -0.99$, $P = 0.17$, one-tailed Mann-Whitney $U$-test). I used Principal Components Analysis to combine hue, saturation, and tone of rump feathers into a single "plumage color score." Plumage color scores of infected and uninfected females were significantly different ($Z = -1.91$, $P = 0.028$, one-tailed Mann-Whitney $U$-test), with infected females growing significantly less brightly colored rump feathers than females that were not infected. See Brawner et al. (2000) for details of infection experiments.

females included in the study, females that were maintained free of coccidia had significantly brighter plumage than females that were infected with coccidia. So, as was true of expression of carotenoid-based plumage coloration in males, coccidial infection inhibited expression of carotenoid-based plumage coloration in females too.

I next tested the effect of nutritional condition during molt on expression of female plumage coloration. Again, females in this study were treated identically to males, so details of this experiment can be found in chapter 5 and in Hill (2000). In the male version of this study, different flocks were fed either canthaxanthin, β-cryptoxanthin, or plain seeds. However, females were included only in the canthaxanthin portion of the experiment. Two groups of females were maintained on the same diet of sunflower seeds and millet with canthaxanthin added to their water, but one flock had its food removed for 38% of daylight hours during molt while the other group had unrestricted access to food throughout the molt period. After they had completed prebasic molt, I scored the plumage coloration of females in each treatment group and compared the hue of feather coloration of females in the food-stressed and ad-lib-food groups. Females in the food-stressed group had significantly less red plumage than females in the group that was not food stressed (Figure 9.5). Moreover, when hue, saturation, and brightness of rump coloration were combined

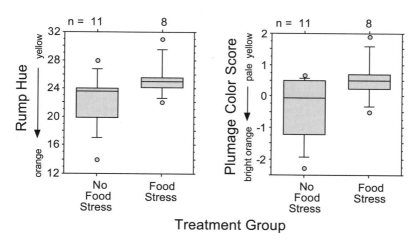

**Figure 9.5.** The effect of food limitation on expression of carotenoid-based ornamental plumage coloration in female House Finches. All females were maintained during molt on the same diet of plain seeds supplemented with canthaxanthin. Females in the food-stressed group had all food removed from their cage for 38% of daylight hours during molt. Females in the group subjected to no food stress had unlimited access to food throughout the molt period. Females that were food-stressed grew rump plumage with significantly higher hue scores (less red plumage) ($Z = -1.90$, $P = 0.029$, one-tailed Mann-Whitney $U$-test). When I used Principal Components Analysis to combine the hue, saturation, and tone of rumps into a single plumage color score, I found that females that were food-stressed grew less colorful plumage than females that were not food-stressed ($Z = -1.81$, $P = 0.035$, one-tailed Mann-Whitney $U$-test). See Hill (2000) for details of experimental design.

into a composite variable, the mean plumage brightness was significantly higher for birds with access to unlimited food than for birds that were food stressed, indicating that these females had not just redder plumage but more saturated and less dark rump plumage as well.

Taken together, these observations indicate that carotenoid-based plumage coloration is a condition-dependent trait in female House Finches just as it is in males. Females apparently must have access to relatively large quantities of appropriate carotenoid pigments during molt to express plumage coloration fully. Moreover, nutritional stress during molt reduces expression of carotenoid pigmentation in females, at least in captivity. Parasites also decrease expression of female plumage coloration. One might conclude from these experiments that female plumage redness serves as an indicator of the females that are of highest quality and would make the best mates for males. The relationship between plumage redness and condition among wild House Finches, however, cannot be as simple as high-condition females always being redder. Recall that yearling females have more colorful plumage on average than older females, and that yearling females nest later, have smaller clutch size, and have lower annual reproductive success than older females (K. J. McGraw et al., unpublished data). It appears that a female must be in good condition to produce maximum carotenoid display, but that age-dependent strategies also influence the degree to which a female will produce colorful plumage. To better understand the factors that control expression of plumage coloration, I looked at condition, age, and plumage coloration among wild females.

### Female Redness and Condition in the Wild

Subjecting females to parasites or nutritional stress during molt in captivity had a clear effect on expression of carotenoid-based plumage coloration. But is expression of female plumage coloration related to relevant aspects of condition in the wild? To address this question, I used field observations from Michigan and Alabama to test for a relationship between plumage coloration and various measures of female condition. Over the years, in Michigan and Alabama, I have assigned studies of female coloration low priority relative to studies of male coloration. Consequently, for some analyses I have data just from Michigan, for others I have data just from Alabama, and for a few I have observations from both study sites.

I first tested the hypothesis that carotenoid-based plumage coloration in female House Finches serves as a signal of individual condition in wild populations by looking at female body condition relative to plumage coloration. "Body condition" is a measure of phenotypic quality that is routinely familiar to humans. When describing each other we associate terms like "scrawny," "emaciated," and even in some contexts "lean" with poor condition, and terms like "muscular," "brawny," and "fleshed-out" with good condition. Such condition is not simply a measure of stored fat; it is a combination of muscle mass and fat. In birds, body condition is typically measured as the mass of a bird adjusted for its skeletal size. This measure ensures that birds are not scored as being in good condition just because they are large, or in poor condition just because they are small. From my banding data in Michigan, I looked for a relationship between plumage coloration and body condition in females. For these analyses, I first looked for patterns among all females banded and then I looked just among females known to be two years old or older. Because there were age-specific patterns of color display I wanted to be certain that potential confounding effects of age did not obscure patterns between female condition and plumage coloration. Contrary to the hypothesis that female coloration is related to female condition, however, I found no relationship between female plumage coloration and body condition, regardless of whether I included yearling females in the analysis (Figure 9.6) (Hill 1993c).

Another indirect measure of condition is over-winter survival. If one assumes that females in better condition survive better than females in poor condition, and if plumage coloration is related to condition, then females with more elaborate plumage coloration should survive better than females with less elaborate plumage coloration. This prediction also was not supported by observations from my Michigan study population. There was no difference in the mean coloration of females that survived versus females that did not survive between years (Figure 9.7) (Hill 1993c). Again, this was only an indirect measure of condition, but the hypothesis that more brightly colored females were in better condition than drabber females was not supported.

So, despite the fact that, in aviary experiments, parasites, nutritional condition, and access to carotenoid pigments during molt all had significant effects on expression of female plumage coloration, in the field female plumage coloration was not related to any of my measures of condition. Certainly not all, or perhaps even the

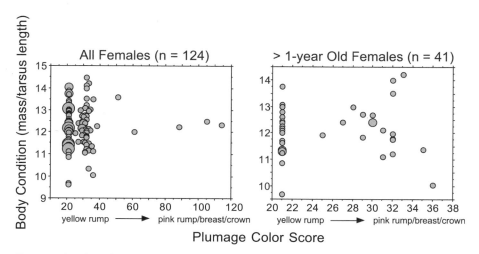

Figure 9.6. The relationship between body condition and the brightness of carotenoid-based plumage coloration for female House Finches captured in Ann Arbor, Michigan. Body condition was calculated by dividing mass by tarsus length. In addition, because the mean body mass of females increased over the sampling period, I adjusted body condition for date (see Hill 1993c for more details on calculations). There was no significant relationship between female body condition and plumage coloration regardless of whether yearling females were included ($r_s = 0.29$, $P = 0.79$) or excluded ($r_s = 0.85$. $P = 0.59$). Adapted from Hill (1993c).

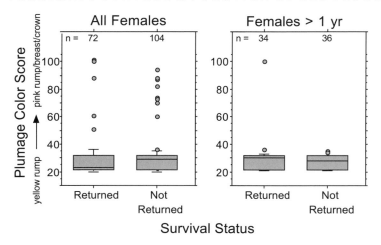

Figure 9.7. Plumage coloration in relation to over-winter survival for female House Finches in Ann Arbor, Michigan. There was no significant difference in the coloration of females who returned versus females who did not between years, regardless of whether yearling females were included ($U = 3369$, $P = 0.23$; two-tailed Mann-Whitney $U$-test) or excluded ($U = 562$, $P = 0.54$; two-tailed Mann-Whitney $U$-test). Adapted from Hill (1993c).

most relevant, measures of condition were included in these field studies. So, although initial tests did not support the idea that female plumage coloration indicates female condition in the wild, it is premature to reject this hypothesis.

## The Function of Female Coloration

Given the prevailing view that sexual selection will act most strongly on males, causing them to evolve elaborate ornamentation to compete for females, ornamental traits in females have always been a bit troubling to evolutionary biologists. If females are not subject to intense sexual selection, why do they need such ornaments? One hypothesis for ornamental traits in females, ignoring species that show sex role reversals such as some shorebirds, is that such ornaments are simply a correlated response to selection for the ornamental trait in males (Lande 1980, Lande and Arnold 1985). By this view, females gain no advantage from ornamentation. Female ornamentation persists because selection for the trait in males is intense, selection against the trait in females is relatively weak, and the genetic linkage between expression of the trait in males and females has not been completely eliminated by selection. I will refer to this as the "correlated-trait hypothesis."

Alternatively, females may be subject to the same sort of sexual selection pressures as males, although not necessarily at the same intensity. The same functional explanations that have been applied to male ornaments—mate choice and intrasexual competition—could then be applied to female ornaments. As was presented in chapter 2, House Finches are socially monogamous, and males contribute greatly to the care of young. Moreover, there is relatively little extra-pair paternity

in House Finch populations. Thus, most of a male's reproductive success is determined by the young that it produces with its social mate. Under such conditions, we
would expect males to be nearly as choosy as females in selecting mates (Burley
1986, Trivers 1972), and females should be subject to sexual selection similar to
that experienced by males. Given that female House Finches use male plumage
coloration as a criterion in mate choice and thereby gain valuable information about
potential mates, a logical question is: do male House Finches assess female coloration in a like manner and thereby gain information about prospective mates? An
answer to this question requires the same sort of data that I collected to address
female mate choice and the function of male plumage coloration. The data available
on male mate choice and female coloration are not as extensive as those available to
address female choice and male coloration, but I have enough observations to test
some basic ideas about the function of female coloration. I will begin by considering
the correlated trait hypothesis. I will then look for evidence that female plumage
coloration is used in male mate choice, beginning with tests of male mate choice
relative to female ornamental display that I conducted with captive birds and then
considering evidence from field data. Finally, I will test the hypothesis that female
plumage coloration serves as a signal in intrasexual aggressive encounters.

## Simply a Correlated Character?

In many ways, the correlated trait hypothesis has become the default explanation
for female ornamentation, and support for the hypothesis has come almost exclusively through the failure of studies to support functional explanations for female
coloration. In a paper on the function and evolution of female plumage coloration
that I published in 1993, I used such a default argument, concluding that the
correlated-trait hypothesis was a likely explanation for bright plumage in female
House Finches because female coloration did not seem to be related to the condition or reproductive success of wild females (Hill 1993c).

   Support for the correlated-trait hypothesis, however, does not need to be
limited to negative evidence from tests of other hypotheses. The correlated-trait
hypothesis can be tested directly. As outlined by Amundsen (2000) the correlated-
trait hypothesis makes specific predictions about the patterns of change in male and
female ornamentation that should be observed among taxa (Figure 9.8). According
to this hypothesis, female ornamentation serves no purpose, and if there is selection
on the trait, it should be only negative selection. Given these conditions, the only
force that could lead to an increase in female ornamentation is selection on expression for the trait in males. The correlated-trait hypothesis is falsified if it can be
shown that, over evolutionary time, female ornamentation has increased without an
associated increase in male ornamentation (assuming that the cost of female coloration remains constant). To perform such a comparative test, one needs a phylogeny
of the taxa most closely related to the species of interest and information on the
character states of the taxa included in the phylogeny.

   Unfortunately, we do not have a robust phylogeny for red finches (genus
*Carpodacus* and their relatives). There is strong support for the hypothesis that
the three North American *Carpodacus* finches are each other's closest relatives,
with Cassin's (*C. cassinii*) and Purple Finches being more closely related to each

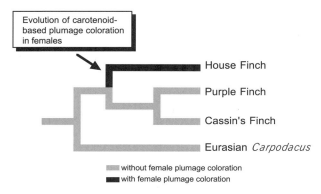

**Figure 9.8.** Expression of female plumage coloration mapped onto a phylogeny of House Finches and other cardueline finches in the genus *Carpodacus*. The relationship between House Finches, Purple Finches, and Cassin's Finches is well supported. The relationship between North American and Eurasian *Carpodacus* finches is less well supported. Female House Finches show carotenoid-based plumage coloration; female Purple and Cassin's Finches entirely lack carotenoid-based plumage coloration. In fourteen of eighteen taxa in the "Eurasian *Carpodacus*" group, females lack carotenoid-based plumage coloration, so this entire group was categorized as having females that lack carotenoid ornamentation. Expression of female plumage coloration in the House Finch appears to be derived from a condition in which females show no plumage coloration. The increase in female ornamental coloration in House Finches was not associated with an increase in male ornamentation. This observation falsifies the hypothesis that female plumage coloration is a correlated response to selection on male coloration.

other than to House Finches (Badyaev 1997, Hill 1994a, 1996b, Marten and Johnson 1986) (Figure 9.8). Beyond this, we can propose that North American *Carpodacus* finches are more closely related to the eighteen species of Eurasian *Carpodacus* finches than they are to other cardueline finches. This gives us a working hypothesis for the historical relationships of the red finches that are likely the closest relatives of the House Finch.

To proceed with a comparative analysis tracing changes in expression of male and female plumage coloration along this phylogeny, we need to know the character states of extant taxa. In the House Finch, as I've documented in this chapter, females have ornamental plumage coloration, but in the Purple Finch and Cassin's Finch, females lack ornamental coloration entirely (Clement et al. 1993, Hahn 1996, Wootton 1996). "Eurasian *Carpodacus* finches" enter into the phylogeny as one lineage, which leaves us with a problem. There is variation in the expression of female plumage coloration across the taxa in this group. Females from fourteen of the eighteen species have no ornamental coloration, but in four species (Three-banded Rosefinch, C. *trifasciatus*; Pallas's Rosefinch, C. *roseus*; Blanford's Rosefinch, C. *rubescens*; White-browed Rosefinch, C. *thura*) females show a shadow of male ornamental coloration (Clement et al. 1993) (personal observation of author based on examination of museum specimens). One way to proceed in this type of analysis is to characterize the lineage by the majority expression (Wiens 1998). Because 78% of Eurasian *Carpodacus* finches show no female ornamentation, I characterized the lineage as having no female ornamentation.

With a phylogeny of *Carpodacus* finches and knowledge of the character expression in extant taxa, it is a simple matter to look at change in male and female plumage ornamentation to test the correlated-trait hypothesis. I found that the likely ancestral state for House Finches is elaborate male pigmentation and no female coloration. In the House Finch lineage, expression of female coloration increased, but there was no corresponding change in male coloration (Figure 9.8). This pattern of change is not consistent with the predictions of the correlated-trait hypothesis, under which the only way that there could be an increase in female ornamentation without a change in male coloration would be if selection against female coloration was relaxed in the House Finch lineage. There is no evidence that selection against female coloration is different in House Finches relative to other *Carpodacus* finches.

This comparative test is built on several critical assumptions regarding the relationships of *Carpodacus* finches, and conclusions could change substantially if the relationships of these finches turn out to be different than I have assumed in my test. Even if it must be viewed as preliminary, however, this approach provides a direct means of assessing the correlated-trait hypothesis.

Assuming that the above analysis is correct and the correlated-trait hypothesis can be eliminated as an explanation for female plumage coloration in the House Finch, that leaves us searching for functional explanations for the trait. Female ornamental traits might function like male ornamental traits, aiding females in attracting higher quality mates or serving as signals in physical contests with other females. I will next consider the evidence that female plumage coloration is used by males as a criterion in choosing mates.

## Laboratory Tests of Male Mate Choice

In the lab, I tested male mate choice relative to female plumage coloration using the same mate-choice apparatus and the same basic procedures that I used to test female mate choice relative to male coloration (see chapter 6). I selected twelve females from my captive flock that had no detectable carotenoid pigmentation and that were known to be at least two years old. Thus, this experiment controlled for age and the natural plumage coloration of females. Each of these twelve females was then assigned randomly to one of four treatment groups: no color; yellow rump; pink rump; or pink rump with a red wash on the crown and breast. I used hair dyes to alter the color of females in much the same way that I used hair dyes to change the plumage color of males (see chapter 6). For treatment of females, however, I used less colorant, and I left the colorant on the feathers for a shorter period before I rinsed them. The result was a faint rather then a bright yellow or red plumage coloration. At the end of the manipulation, these females looked like naturally pigmented female House Finches (Hill 1993c).

In three-hour mate-choice trials I presented thirteen males with sets of four females, with one female from each color type. In these trials, males could move about and associate with females, but females were confined to individual cages and could not see each other. I recorded the time that males associated with females of the various color types, and used these association times to rank the females in each trial according to male preference. I then analyzed the preference ranks for each

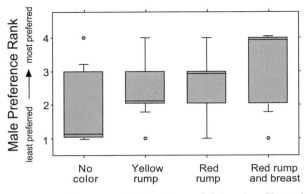

**Plumage Coloration of Stimulus Females**

Figure 9.9. Male mate choice in relation to carotenoid-based plumage coloration of females in captive House Finches. Shown are box plots of the preference ranks of each female plumage type. Three females represented each of the four plumage types, and each male was presented with a set of four females, one from each plumage type. Preference ranks are based on the time males spent in association with the various females. The distribution of preference ranks differed significantly from random ($\chi^2 = 8.04$, $n = 13$, $P = 0.05$), with males showing a significant preference for females with red rumps and breasts ($P < 0.05$). Adapted from Hill (1993c).

female-plumage type across trials. I found that males showed a significant preference for the females with the most elaborate plumage coloration (Figure 9.9) (Hill 1993c). These laboratory mate-choice trials provided clear evidence that, at least when age and other confounding variables were controlled in captivity, male House Finches preferred to associate with more brightly colored females.

*Field Studies of Male Mate Choice*

One way to determine the criteria that males are using in their choice of mates is to compare the traits of paired and unpaired females. In my study populations in Michigan and Alabama, however, virtually all females paired each year, so I could not compare the coloration of paired versus unpaired females as I had done with males. There remained, however, an indirect means of measuring male mate choice in the wild. If we accept the premise that brightly colored males are preferred by females as mates and that brightly colored males, therefore, have their choice of potential mates, then the attributes of those females paired with the most brightly colored males can be concluded to be the attributes preferred by males. Accordingly, if males prefer brightly colored females as mates, then there should be assortative pairing of males and females by plumage coloration. I tested this idea by comparing the plumage coloration of males and females within pairs. In Michigan, I found no assortative mating by color when I considered all females, but I found a significant tendency for males and females to mate assortatively by plumage color when I excluded first-year females from the analysis (Hill 1993c) (Figure 9.10).

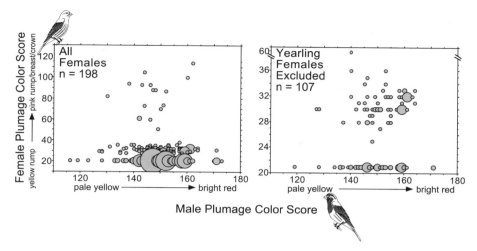

Figure 9.10. Assortative mating by plumage coloration in wild House Finches in Ann Arbor, Michigan. Data are based on observations of color-banded individuals. When all females were used in the analysis, there was no significant relationship between the brightness of a female's plumage coloration and the brightness of her mate's plumage ($r_s = 0.0003$, $P = 0.49$). When yearling females were removed from the analysis, however, there was a significant tendency for brightly colored females to pair with brightly colored males ($r_s = 0.22$, $P = 0.02$). Adapted from Hill (1993c).

The observation that there is assortative mating by plumage coloration only when yearling females are removed from the analysis makes sense if we consider the effect of age on plumage coloration. Yearling females tend to be brighter than older females (see Figure 9.2), but regardless of their plumage color, yearling females are less experienced than older females and should not be chosen over older females as mates. Not surprisingly, there is strong assortative mating by age (Figure 9.11), so yearling females tend to pair with yearling males. Brightly colored males tend to pair with older females (Hill 1993c). Including bright first-year females in the analysis confounds any pattern of association by color; removing first-year females reveals the pattern.

These field data corroborated laboratory mate-choice experiments, supporting the idea that male House Finches prefer to mate with brightly colored females, but with a very important qualification. Age appears to be a primary criterion in male mate choice. Females that are two years old or older are preferred as mates over yearling females, despite the fact that yearling females tend to be more colorful (Hill 1993c). Among females that are two years old or older, however, males do appear to choose more brightly colored females as mates.

### Female–Female Competition

Besides functioning as a criterion in male mate choice, female ornamental coloration has been proposed to function as a signal in female–female contests over resources (Amundsen 2000, Muma and Weatherhead 1991). Female House

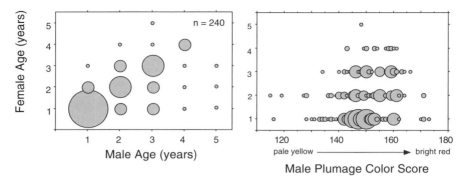

Figure 9.11. The relationship between the age of female House Finches and the age (left figure) and plumage coloration (right figure) of their mates in Ann Arbor, Michigan. House Finches showed a strong tendency to pair assortatively according to age ($r_s = 0.59$, $P = 0.0001$), and older females tended to pair with more brightly colored males ($r_s = 0.20$, $P = 0.001$). Adapted from Hill (1993c).

Finches appear to use aggression in establishing and maintaining nest sites and probably in establishing and maintaining pair bonds with preferred males (McGraw and Hill 2002). At feeding stations and in captive flocks, female House Finches are dominant to males even though males on average have a larger body size (Belthoff and Gauthreaux 1991a, Belthoff and Gowaty 1996, Brown and Brown 1988, Hill 2000, McGraw and Hill 2002). Thus, it seemed plausible that female coloration could function in aggressive encounters with other females.

To test this idea, my graduate student Paul Nolan and I looked at female plumage coloration in relation to dominance in captive flocks of House Finches. We captured females at feeders in early December in Alabama and recorded their masses and the plumage hues of their rumps. At this time of year, we could not age the birds. We housed these females in two large flocks in outdoor aviaries. Each female was given a different combination of colored leg bands, and we determined the dominance rank of each female in each flock by recording winner and losers in squabbles over perch sites and food. The dominance hierarchies for both flocks were linear and stable. We then compared the rump hue of females to their dominance ranks. We found no significant relationship between female plumage coloration and dominance rank in either flock (Figure 9.12). Female mass, in contrast, was significantly positively correlated with dominance rank in both flocks.

Going into our study, there were several reasons to predict that the carotenoid-based plumage coloration of females would not function as a signal of social status in the House Finch. First, carotenoid-based color displays in general do not appear to function as signals of status (Chapter 8; McGraw and Hill 2000a), and carotenoid-based plumage coloration in male House Finches does not serve as a status signal (McGraw and Hill 2000a). In the only other study of status signaling relative to a carotenoid-based female color display, plumage coloration was unrelated to dominance among captive female Red-winged Blackbirds (Muma and Weatherhead 1991). Second, yearling female House Finches tend to be more brightly colored than older females, and it seems unlikely that younger females would tend to be

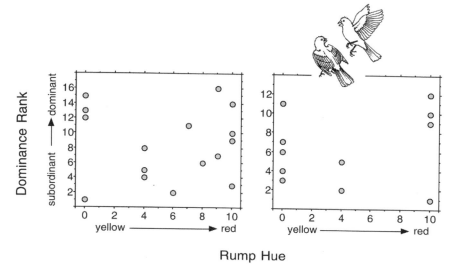

Figure 9.12. The relationship between the natural rump hue and winter dominance status in two flocks of captive female House Finches. In contrast to what was observed for males, there was no significant relationship between carotenoid coloration and dominance rank (left scatterplot: $r_s = 0.15$, $n = 12$, $P = 0.64$; right scatterplot: $r_s = 0.03$, $n = 14$, $P = 0.91$). In both flocks there was a significant positive relationship between female mass and dominance rank (flock 1: $r_s = 0.61$, $n = 12$, $P = 0.04$; flock 2: $r_s = 0.62$, $n = 14$, $P = 0.02$; not illustrated) (P. M. Nolan and G. E. Hill, unpublished data).

socially dominant to older females. Although the hypothesis that female plumage coloration functions in dominance interactions should be tested experimentally before it is discarded, particularly in the breeding season when females compete for nest sites and mates, it seems unlikely that signaling in dominance interactions provides the selective advantage that maintains expression of colorful plumage in female House Finches.

## The Benefits to Males of Choosing Brightly Colored Females

Female House Finches show condition-dependent variation in expression of caro-tenoid-based plumage coloration, and males show a mating preference for the most brightly colored females. These observations lead to the question: What are the benefits to males of choosing brightly colored females as mates? I addressed this question with field data from both the Michigan and Alabama populations.

### Clutch Size and Reproductive Success

Several components of reproductive investment, including nest initiation date, clutch size, and number of clutches per season, are under female control and are likely to affect male reproductive success. If brightly colored females are better than average at initiating early nesting, laying larger clutches, or laying more clutches per season, then males would directly benefit by choosing brightly colored females as mates.

Female House Finches vary substantially on the date at which they begin nesting in the spring, and early nesting leads to greater offspring production (McGraw et al. 2001) (see Figure 6.11). Presumably, females initiate nesting as early as possible in the spring and are constrained by their physical condition in how early they can start. Females in good condition should begin nesting before females in poor condition, and if plumage coloration is related to condition, brightly plumaged females should initiate nesting before drab females. I tested this hypothesis by comparing the nest-initiation dates of females to their plumage coloration in both the Michigan and Auburn study populations. I found no significant relationship between female plumage coloration and nest initiation date whether yearling females were included in the analysis or not (Figure 9.13) (Hill 1993c).

Clutch size may also affect the reproductive output of pairs, and given that eggs are energetically costly for females to produce, if colorful females are in better condition, we would predict that bright females would lay larger clutches than drab females. However, I found no significant relationship between clutch size and female plumage coloration (Figure 9.13). I also found no relationship between female plumage coloration and the time between nest initiations (interclutch inter-

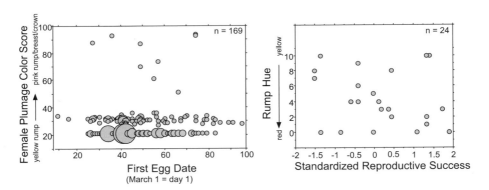

Figure 9.13. Female plumage color in relation to breeding onset (left scatterplot) and reproductive success (right scatterplot). There was no significant relationship between the ornamental plumage coloration of females and either nest initiation date ($r_s = 0.07$, $P = 0.94$; modified from Hill 1993c) or reproductive success measured as chicks produced per season and standardized across years ($r_s = -0.11$, $P = 0.60$; K. J. McGraw et al., unpublished data). Removing yearling females from the analyses did not substantially change the relationships for either comparison ($P > 0.20$). There was also no relationship between female plumage coloration and clutch size ($r_s = -0.05$, $n = 80$, $P = 0.63$; not illustrated; Hill 1993c) or the interval between nesting attempts within a season ($r_s = -0.05$, $n = 12$, $P = 0.41$; not illustrated; K. J. McGraw et al., unpublished data). Note that for the left scatterplot, book scores were used to estimate female plumage coloration, so higher scores correspond to brighter plumage. For the right scatterplot, a Colortron was used to score plumage, so lower scores correspond to brighter plumage.

val). Finally, I compared female plumage coloration to the number of young that were fledged in a breeding season, and found no significant relationship. So, none of these measures of female reproductive success was significantly related to female plumage coloration, and this series of analyses provided no explanation for the benefits that males might receive by pairing with brightly colored females.

## Resource Investment in Relation to Female Coloration

Females benefit from their choice of bright red males as social mates at least in part because redder males tend to provide more resources to them and their offspring during nesting. Thus, a possible benefit to males for choosing to mate with brightly colored females is greater resources provided to offspring. To test this idea, we filmed nests on the Auburn study site and recorded the rates that nestlings were provisioned by females. In the Auburn population, we found that the rate at which females provisioned nestlings was not related to female plumage coloration (Figure 9.14). Interestingly though, male provisioning rate was positively correlated with female plumage coloration. Males paired to redder females fed offspring at a higher rate than males paired to less colorful females. This sort of differential allocation of resources has been observed in several passerine species, but in all previous studies it was females who invested more in offspring sired by attractive males (Burley 1988, Sheldon 2000). The accepted explanation for change in feeding behavior in response to the ornamentation of a mate is a good-genes explanation: young sired by a well-ornamented male are likely to be high-quality offspring with a high

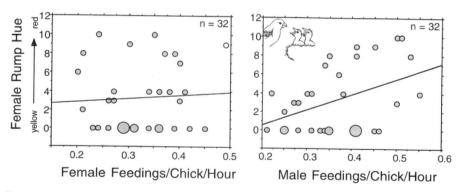

**Figure 9.14.** The relationship between the plumage coloration of female House Finches and their provisioning rate (left scatterplot) and the provisioning rates of their mates (right scatterplot) in Auburn, Alabama. There was no significant relationship between the plumage coloration of females and the rate at which they provisioned young ($r_s = 0.06$, $P = 0.76$), but there was a significant positive relationship between the coloration of females and the rate at which their males provisioned young ($r_s = 0.40$, $P = 0.03$) (K. J. McGraw et al., unpublished data). Provisioning rate was measured by filming nests for approximately 8 h on day 11 after hatching. The total number of provisioning trips made by the male and the female were then divided by the length of the observation period and the number of chicks in the nest.

probability of being successful reproducers, and hence they are worthy of extra resource investment. Our observations of provisioning behavior suggest that, by choosing redder females as mates, male House Finches are not gaining resource investment. The behavior of males, however, suggests that the offspring of brightly colored females are worthy of above-average resource investment.

### Female Plumage Coloration and Maternal Investment

As with other behavioral ecologists, for me female coloration was always an after-thought. My one paper on the topic was compiled primarily from data that I collected in my studies of male coloration. As can be seen by the new studies presented in this chapter, I've been sitting on data related to the proximate control of female coloration (collected again as an afterthought in studies of male colora-tion) for years. The incentive to publish the data on female ornamentation never seemed as strong as the incentive to publish the male data. Moreover, the initial patterns related to female coloration in the House Finch were messy—yearlings brighter than older females, no relationship with body condition, and so on.

Perhaps the problem in my studies, and in most other studies of female colora-tion, has been that the same criteria used in studies of male coloration have been applied to studies of female coloration. By not focusing on those aspects unique to female reproductive biology, we failed to grasp the relevance of ornamental traits in females. As maternal effects have become a greater focus of interest among evolu-tionary biologists, it is perhaps time to approach the study of female ornamentation from a truly female perspective.

I think that the recent focus on maternal investment in the egg introduces a new and potentially powerful approach to the study of ornamental traits in females. Recent studies have shown that females control investment of testosterone as well as other hormones in the eggs (Birkhead et al. 2000, Gil et al. 1999, Lipar et al. 1999, Schwabl 1993). Testosterone deposited in eggs has been shown to have several beneficial effects on developing birds (Schwabl 1996a,b), but the basis for variation among females in testosterone investment in eggs has scarcely been stu-died. Female birds also impart immunity to the embryo that must carry it through its first weeks of development (Hassan and Curtiss 1996, Smith et al. 1994). This component of maternal investment is almost completely unstudied in birds outside of poultry. Finally, and most interesting relative to carotenoid-based ornamenta-tion, females pass substantial amounts of carotenoid pigments to offspring—this is what makes the yolk of eggs of virtually all species of birds orange or yellow (Blount et al. 2000). As outlined above and in chapter 5, carotenoids are important as free radical scavengers. The metabolic rate of a developing chick is extremely high and free radical buildup and disposal may be an important factor limiting the rate at which chicks can grow and develop (von Schantz et al. 1999). It has been proposed that carotenoids deposited by females during egg formation may play a key role in helping chicks deal with this free radical buildup (Blount et al. 2000).

One interesting possibility is that, through their carotenoid displays, females might signal their ability to provide testosterone, antibodies, or especially carote-noid pigments to offspring. Males could then assess female quality relative to maternal investment by assessing plumage coloration. By this idea, the eggs laid by more colorful females would tend to be "better" and worthy of more resource investment than the eggs laid by less colorful females. If female plumage coloration is associated with more resource investment by females, then the greater color display of yearling versus older female House Finch becomes difficult to explain. It may be that signaling condition is more important for yearling females, who are pairing for the first time and are unknown to any males, than to older females, who pair within a community of males to whom they are known. Older females may use more of their carotenoid resources for maternal investment and hence display less carotenoid pigment in their plumage.

If female plumage coloration predicted carotenoid or testosterone investment in eggs, this could explain why males choose more brightly colored females as mates and invest more in the offspring of brightly colored females. These ideas remain completely untested.

## Summary

Ornamental traits expressed by females in species in which males are more ornamented have been a focus of curiosity since the studies of Darwin and Wallace. Yet, the study of female coloration has remained a backwater, having been viewed as worthy of a sentence or two of speculation or perhaps a side project when the real work was completed, but not as a topic worthy of being the focus of a research program. The patterns that have emerged to date provide an intriguing but confusing framework for understanding female ornamentation. Preliminary comparative analyses suggest that female plumage coloration in the House Finch cannot be dismissed simply as a correlated response to selection for colorful plumage in males. In further support of the hypothesis that female plumage coloration is maintained by current selection is the observation from lab and field studies that male House Finches prefer to mate with brightly colored females. As with expression of male plumage coloration, expression of female plumage coloration appears to be condition dependent; in the lab, access to carotenoid pigments, exposure to parasites, and access to nutrition during molt all had a significant effect on expression of female plumage coloration. In contrast to male plumage coloration, however, in field studies female plumage coloration was not related to female condition or reproductive investment. Although, female plumage coloration was not related to female provisioning rate, it was related to the provisioning rates of their mates, suggesting that the offspring of brightly colored females are worthy of above-average investment by males. Many questions related to ornamental coloration in female House Finches and in female birds in general remain to be answered.

# PART 3

*Biogeography and the Evolution
of Colorful Plumage*

# 10 From the Halls of Montezuma to the Shores of Tripoli (New York)

## Populations, Subspecies, and Geographic Variation in Ornamental Coloration

If in some way the American House Finch could be induced to come east, and the English Sparrow could be given papers of extradition, the exchange would be a relief and a benefit to the whole country.

—L. S. Kenyon (1902)

The House Finches are definitely "on the march."
—R. T. Moore (1939), one year before House Finches were introduced to Long Island

It is often assumed that these mainland birds were descended from the colonists on Long Island. There is not one shred of evidence for this.
—G. Cant and H. P. Geis (1961), commenting on the origin of House Finches in New York, New Jersey, and Connecticut that gave rise to the eastern population of House Finches

*In the spring of 1983 I was a senior at Indiana University in Bloomington working part time in the ornithology lab of Val Nolan, Jr. and Ellen Ketterson. In April, as I entered Morrison Hall to go to work, I heard a strange bird singing in the courtyard. It sounded to me something like a vireo, but not specifically like any of the eastern vireos whose songs I knew. The bird was hidden in the foliage of a tree so I couldn't see what it was. I went up to the ornithology lab and mentioned the odd bird song to the graduate students who were present—Jim Hengeveld, Liecia Wolfe, and Chris Rogers. We could clearly hear the bird through the lab window. The graduate students knew the sound from various times spent in California. It was a House Finch, a species of bird that I had heard was spreading west from a population that had been introduced around New York City. Over the course of that spring, the song of the House Finch became routinely familiar to me on the Indiana University campus as they went from never-before-seen on campus to seemingly everywhere. Although I had only vague appreciation of it at the time, we had witnessed the passing of the front edge of the wave of finches that was washing over eastern North America.*

Those who assert that humans are destroying all life on Earth are taking a narrow view of life. There can be no doubt that some species suffer utterly in the wake of human activities and that loss of these selected species leads to loss of diversity. Loss of species diversity is a tragedy of our time, and future generations will not forgive us for what we are doing. However, there is a non-trivial set of organisms that give a collective cheer as forests and prairie give way to strip malls and suburbs. For these species the opening of the dense forests and the creation of structure on the barren plains provide new opportunities—a potential for life where none existed before. And no species of North American bird has benefited more from such change than the House Finch.

## The Colonization of a Continent

Before the colonization of North America by Europeans, the House Finch was found exclusively in Western North America (Figure 10.1). As Europeans filtered into the great western wilderness, they encountered the warbling song of the House Finch from the western edge of the Great Plains to the Pacific Coast. Spanish settlers found House Finches virtually throughout Mexico, from the Sonoran and Chihuahuan Deserts south to the edge of tropical rain forest in Oaxaca. It is interesting that the House Finch was not described in the scientific literature until 1776, based on a bird collected in "The Valley of Mexico" (Muller 1776), even though it must have been a familiar bird to the Spanish in Mexico and the southwestern United States for 200 years before this date of official "discovery." At the time of European colonization of North America, however, no House Finches sang north of what would become the Canadian border, or even in what would become states that border Canada. There were no House Finches in what would be Washington, Idaho, Montana, or Wyoming, all part of the present range of the House Finch.

Many people who are familiar with the House Finch and its incredible spread in eastern North America assume that the western population, like the populations of most other western bird species, has remained more or less stable through recent decades or centuries. The notion of a stable western range for the House Finch is wrong, however. The distribution of House Finches began to change rapidly in the early twentieth century, as human activities reshaped the landscape across broad regions of western North America. Beginning in the early decades of the twentieth century, they spread up the west coast, colonizing, over the course of four or five

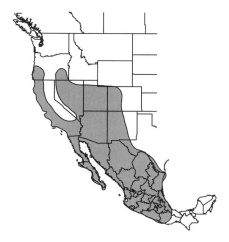

Figure 10.1. The approximate range of the House Finch before 1900. Distribution in the United States for this time period is relatively well documented. The distribution in Mexico is based on the current distribution of the species.

decades, the northern half of the Oregon coast, the entire length of the Washington coast, and finally southern coastal British Columbia (Figure 10.2). At the same time they expanded north in the Rocky Mountain and Great Basin regions reaching the Okanagan Valley of British Columbia within a year (1937) of reaching coastal British Columbia (1938) (Cannings et al. 1987, Cowen 1937). Thus, Canada went from having no breeding populations of House Finches to almost simultaneously becoming home to two populations of finches that were the result of two independent range expansions. A third breeding population for Canada would come later. In the early decades of the twentieth century, yet another finger of expansion spread up through Wyoming into central Montana and then west to the Idaho border (Hill 1993b). In the southern portion of their United States range House Finches also spread east, reaching the Oklahoma panhandle by the 1920s (Tate 1925) and the Edward's Plateau in Texas by the 1930s (Oberholser 1974). Expansion in Mexico is poorly documented, but it seems likely that House Finches expanded their range throughout Mexico as human activities created House Finch habitat in areas that had previously been unsuitable.

   Looking back, this extensive expansion of the western range seems like a modest preamble to the astonishing range expansion that began in about 1940 with the liberation of House Finches in the east. The establishment of the House Finch in the eastern United States was a direct consequence of the fact that they are excellent caged birds. Throughout Mexico, it is still very common to go into a shop or house and see a male House Finch vigorously singing from its small cage. To my knowledge, House Finches are not bred in captivity in Mexico, but rather the caged birds that one encounters are captured from the wild. Bird catching is a traditional profession in Mexico, and was once also a common profession in the United States. Because House Finches are such excellent caged birds, they were primary

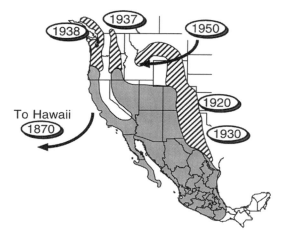

Figure 10.2. The expansion of the range of the House Finch in western North America in the nineteenth and twentieth centuries. The gray region is the approximate range of the House Finch before European colonization; the hatched regions show areas into which House Finch populations spread. In contrast to the general assumption of a constant western range, House Finches expanded their western range substantially north and east. Dates give the approximate years at which House Finches reached various locations. (See text for details.)

targets of United States bird catchers. By the thousands, House Finches were trapped in the late nineteenth and early twentieth centuries, primarily in coastal California, and shipped to eastern urban centers to be sold as pets. The passage of the Migratory Bird Treaty Act in 1918 may have slowed the trade in House Finches, but it did not stop it. In the early decades of the twentieth century, tens of thousands of House Finches were shipped from the west and held in cages in the eastern United States (Cant and Geis 1961).

Elliot and Arbib (1953) give the most thorough account of the sequence of events surrounding the establishment of a wild population of House Finches in the east. In early 1940, Dr. Edward Fleisher discovered House Finches being sold in a pet shop in Brooklyn, New York. Dr. Fleisher knew that such sales were illegal under the Migratory Bird Treaty Act, and he alerted the United States Fish and Wildlife Service. In response to a letter Dr. Fleisher wrote concerning the illegal trade in House Finches, wildlife law enforcement authorities contacted him saying, "We have been able to stop the trapping and transportation from California, and have stopped sales throughout the United States" (cited in Elliot and Arbib 1953).

The assumption by most ornithologists has been that in response to the threat of prosecution by wildlife enforcement officers, pet dealers released their stocks of House Finches into the wild in the New York City area. These liberated caged birds established populations that eventually spread across the entire eastern portion of the continent. The timing of the discovery of wild populations in the New York City area supports the assertion that this specific effort by law-enforcement officers led to the establishment of wild populations of House Finches. Soon after the crackdown on illegal sales, House Finches began to be seen around New York City. In April 1941, a male House Finch was observed at Jones Beach, Long Island, the first eastern record of a wild male House Finch. A small flock was discovered in 1942 in Babylon, Long Island, 12 miles (20 km) from Jones Beach. In 1944, a nest was found in this colony, establishing the first breeding record for the eastern United States. Seventy birds were counted in the vicinity of this breeding area in the winter of 1949–50. The first record of a wild House Finch in Eastern North America away from Long Island was in 1948, in Tarrytown, New York, followed soon after by observations in Ridgewood, New Jersey (1949) and Greenwich Township, Connecticut (1951) (summarized from Elliot and Arbib 1953). By the early 1960s House Finches were breeding from Boston to Washington, D.C. They reached Ontario in the late 1970s, and by the early 1980s they had reached what would be my study sites in Michigan and Alabama (Figure 10.3). And, at the beginning of the third millennium, they have spread west through the Great Plains so that it is now impossible to tell if finch populations in towns in the

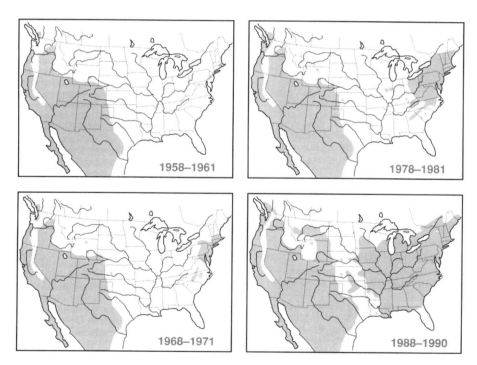

Figure 10.3. Expansion of the range of the House Finch in eastern North America from birds released around Long Island, New York, in approximately 1940. Distributions are based on data from Christmas Bird Counts. As of 2000, birds expanding east at the edge of the western range and birds expanding west at the edge of the eastern range had met in the Great Plains so that it is no longer possible to assign an origin to populations breeding in Oklahoma, Kansas, Nebraska, and the Dakotas. Maps adapted from Hamilton (1992) and figure reproduced from Hill (1993b) with permission.

Great Plains are derived from eastern or western range expansions (e.g., Tyler 1992). The colonization of the continent is essentially complete.

The colonization of eastern North America was not the first time that House Finches had colonized a completely new region. In the mid-1800s, House Finches were released on the Hawaiian Islands by American settlers. The details of this introduction are much sketchier than the events of the introduction into the eastern United States. At the beginning of the twentieth century, Grinnell (1911) reconstructed the introduction of House Finches to the Hawaiian islands as best he could, but by then House Finches were long established in Hawaii, and it was already too late to find first-hand accounts of the introduction. Based on his correspondence with biologists who had lived on Hawaii and who had inquired about the origin of House Finches among older inhabitants of the islands, Grinnell deduced that they had been on the major Hawaiian islands since at least the 1870s. It is presumed that House Finches were introduced purposefully to Honolulu on the island of Oahu and that they subsequently colonized the other major islands in the chain. Because there are no accounts of the introductions, however, where and how many birds were released is entirely speculative. The origin of the birds released on Hawaii is also unknown. Most ships sailing to Hawaii in the middle of the nineteenth century left from San Francisco, and so Grinnell (1911) proposed that the House Finches that gave rise to the Hawaiian population were captured in the Bay area. Since biologists began recording the distribution and abundance of birds on Hawaii, House Finches have been an abundant breeder on all of the major islands.

*How Many Founding Fathers?*

How many birds founded the eastern North American population of House Finches? The presumption has always been that the eastern population was established by a single introduction of a small number of birds on Long Island in 1940. This account is probably correct in linking the establishment of some Long Island populations to a specific crackdown by wildlife officers. This may not have been the only introduction of House Finches to the east, however. An enormous number of House Finches, estimated at over 100,000 birds, were shipped from California to the eastern United States in the early part of the twentieth century (Cant and Geis 1961). As Cant and Geis (1961) argue, with so many House Finches in pet shops and homes, it seems likely that they were being released into the east periodically during the early decades of the twentieth century, and that some House Finches were free-flying in the east before 1940.

Small populations could have been present in various locations around eastern population centers for some time, yet remained unreported—either because no one was paying attention, or because the birds were mistaken for Purple Finches. It wasn't until House Finches were discovered breeding on Long Island, and the establishment of eastern House Finches became widely known, that bird watchers started looking for and finding populations in New Jersey, Connecticut, and mainland New York. Everyone then assumed that these "new" populations had spread from the populations on Long Island.

It is not unprecedented for the introduction and spread of a conspicuous urban bird to go unnoticed in the United States. It happened in the 1970s and 1980s with another great avian colonist and disperser, the Eurasian Collared-Dove, *Streptopelia decaocto* (Smith 1987). As its name suggests, the Eurasian Collared-Dove is native to Europe and Asia. The species was not recorded in the wild in North America until the mid-1980s, but by the time it was "discovered," its population in south Florida was already estimated in the tens of thousands (Smith 1987). The problem was that Eurasian Collared-Doves are almost identical to Ringed Turtle-Doves, which are a domestic race of the African Collared Dove (*Streptopelia roseogrisea*). Ringed Turtle-Doves are common caged birds throughout North America. For decades, they have been released into the wild in North America during wedding ceremonies, so there were records of these birds throughout the United States. Bird watchers had learned to pay them little notice. Domestic doves do poorly in the wild, and breeding populations of Ringed Turtle-Doves through the twentieth century were limited to small populations in Los Angeles and a couple of locations in Florida (Smith 1987).

The Eurasian Collared-Doves that founded the Florida and eastern United States populations apparently escaped from an aviary on the Bahamas in 1974 (Smith 1987). The escaped Eurasian Collared-Doves rapidly established a large (>10,000 birds) breeding population in the Bahamas and then spread north to Florida in the late 1970s. Remarkably, despite the fact that they are large, conspicuous birds that are most abundant around cities and suburbs, the establishment and subsequent spread of Eurasian Collared-Doves in south Florida went unnoticed and unrecorded. Observers simply assumed that the "ringed doves" they were seeing were escaped Ringed Turtle-Doves. By 1986, when a few keen observers realized that the doves that were becoming increasingly abundant in South Florida and spreading north were not Ringed Turtle-Doves but a new species for Florida and North America, Eurasian Collared-Doves were already breeding three-fourths of the way up the Florida peninsula (Smith 1987).

It seems likely that a similar sequence of events occurred with House Finches in the eastern United States. With so many finches being imported from the west, some House Finches were surely being liberated periodically in the east in the early part of the twentieth century (Cant and Geis 1961). To all but the most experienced observers, when these birds were observed in the east, they were dismissed as the similar Purple Finch. Once people were alerted to begin looking for House Finches in the east, the overlooked birds started to be noticed.

## Founder Effects

Why does the number of birds that established the eastern population of House Finches matter? If a new population of an organism is founded by a small number of individuals, this founder population will have less than the total genetic diversity of the parent population. This is known as the "founder effect," and the smaller the founding population, the greater will be the loss of genetic diversity. Loss of genetic diversity can make populations more vulnerable to parasites and pathogens. Moreover, by definition, there is less genetic variation in such a founder population, so there would also be less chance for ornamental traits to reflect male genetic quality. It is therefore potentially very important whether the eastern House

Finch population was established from a founding population of five, ten, or fifty or more birds.

Two graduate research projects have been devoted to determining the degree of genetic diversity in the introduced eastern and Hawaiian populations of House Finches compared to the parent population in the western United States. For his master's thesis, Manuel Vazquez-Phillips (1992) looked at allozymic variation in proteins from 568 House Finches collected in California, Colorado, New Mexico, and Arizona in the west, several locations throughout the east, and several of the Hawaiian islands. Allozymes are different forms of a protein that are coded for by the same genetic locus. An individual bird can have one allozyme of a given protein (homozygous) or two allozymes (heterozygous). Within a population all individuals might have only one allozyme for a given protein, or there might be many forms of the protein. Usually, in a population of birds there are one or a few common allozymes for each locus and other allozymes, if they occur, are found in only a few individuals. Vazquez-Phillips found that the mean heterozygosity (whether loci coded for one or two allozymes averaged across the thirty-one loci examined for each individual bird averaged across the population) was about the same for eastern, western, and Hawaiian populations (Vazquez-Phillips 1992). He did find, however, that the Hawaiian population had only 29.2% and the eastern population only 51.8% of the rare alleles found in the western population.

In another master's thesis, William Benner (1991) conducted a similar comparison between native western and introduced eastern and Hawaiian House Finches, but instead of comparing protein allozymes, he compared gene sequences of mitochondrial DNA. He found that eastern House Finches have maintained most of the genetic diversity found in the native western population, while Hawaiian House Finches had lost a substantial amount of mitochondrial gene diversity (Benner 1991). Both the study by Benner and the study by Vazquez-Phillips therefore support the idea that the eastern population of House Finches was founded by enough birds that most of the genetic diversity of the western population was maintained. The Hawaiian population, on the other hand, appears to have been founded by a smaller number of birds and to have gone through a significant genetic bottleneck at the time of the establishment of the population.

Interestingly, both Vasquez-Phillips and Benner found little genetic or morphological differentiation among birds collected throughout the western United States (Colorado, New Mexico, Arizona, California), supporting the idea that all House Finch populations north of the Rio Grande comprise one population and should be grouped into one subspecies, although House Finches from the Great Basin were not included in this comparison. The distinctiveness of different populations of House Finches is our next topic.

## House Finch Taxonomy and Systematics

### Deeper Divisions

Thus far in this chapter, I have focused on populations of House Finches north of the Rio Grande. Most of the really interesting population variation, however, occurs south of the United States border and on islands off the coasts of southern California and Baja, where House Finches exist in a number of geographically isolated populations with diagnostic morphological features. These distinctive populations are currently afforded subspecies status, although the number of subspecies into which House Finches should be divided is somewhat unclear. In 1874, Baird, Brewer and Ridgway (Baird et al. 1874) recognized three subspecies of House Finches—an interior United States subspecies, a coastal California subspecies, and a southern Mexican subspecies (Table 10.1). By 1901, with a much more extensive series of skins to examine, Ridgway recognized nine taxa of House Finches, including three species—the House Finch (C. *mexicanus*), which included all mainland and most island forms, the Guadalupe Island House Finch (C. *amplus*), and the San Benito House Finch (C. *mcgregori*). The latter two species were found only on the Pacific islands of Guadalupe and San Benito, respectively. Within the mainland populations, Ridgway (1901) recognized seven subspecies, and he placed all finches north of the Rio Grande into one subspecies.

In what remains the only comprehensive treatment of House Finch taxonomy, Moore (1939) recognized eighteen distinctive populations, including four species and fifteen subspecies. Moore recognized the same three species as Ridgway (1901), but following Grinnell (1911, 1912), he also recognized the Hawaiian population as a distinct species, C. *mutans*. The fifteen subspecies of *Carpodacus mexicanus* in Moore's taxonomy included four mainland United States subspecies north of the Rio Grande Valley, three subspecies in southern Mexico referred to as the "Sur Group," a subspecies on the islands off the coast of southern California, and seven subspecies in Mexico between the Sur group and United States subspecies (Moore 1939) (Table 10.1; Figure 10.4). Moore's taxonomy has been modified somewhat in subsequent editions of The A.O.U. Check-list of North American Birds. The fifth edition (A.O.U. 1957), which is the last edition to include subspecies, followed Moore's taxonomy except that (1) all taxa are given subspecies status, so C. *amplus* becomes C. *m. amplus*, and C. *mcgregori* becomes C. *m. mcgregori*; (2) C. *mutans*, the Hawaiian population, is no longer recognized as a species or subspecies, but simply as an introduced population of coastal California birds; (3) all mainland United States House Finches north of the Rio Grande are lumped into a single subspecies C. *m. frontalis*; and (4) some of the questionable Mexican subspecies that were based on one or a few specimens are no longer recognized (Table 10.1; see also Howell et al. 1968). Thus, by current convention, all House Finches now belong to one species that contains eleven subspecies, one of which—the San Benito House Finch—is sadly now extinct (Jehl 1971).

The purpose of this chapter is not to exhaustively review the merits of the subspecies status of different populations. Whether there are eleven or eighteen subspecies is of no particular importance to understanding why House Finches have red plumage. An appreciation that there are several populations of House Finches that have had a unique evolutionary history for many thousands of years, however, is critically important to my investigation of the evolution of plumage ornamentation. For the remainder of this and the next chapter, I will use the conservative taxonomy of the A.O.U., except that I will follow Moore in recognizing a twelfth subspecies—the Oaxaca House Finch (C. *m. roseipectus*)—based on a recent review by Binford (2000) and on my own examination of museum skins. Also, I will treat

Table 10.1. The taxonomy of the House Finch group (subgenus Burrica) over the last 125 years. No modern molecular techniques have yet been used to better resolve the distinctiveness of the various populations.

| Latin name | Common name | Range |
|---|---|---|
| **Baird et al. (1874)** | | |
| C. frontalis frontalis | House Linnet | United States western interior |
| C. frontalis rhodocolpus | Crimson-fronted Finch | Pacific Coast United States and Mexico |
| C. frontalis haemorhous | | Southern Mexico |
| **Ridgway (1901)** | | |
| C. mexicanus mexicanus | Mexican House Finch | Southern Mexico |
| C. mexicanus roseipectus | Oaxaca House Finch | Oaxaca |
| C. mexicanus rhodocolpus | Cuernevaca House Finch | Southwest Mexican Plateau |
| C. mexicanus sonoriensis | Sonoran House Finch | Sonora and Chihuahua |
| C. mexicanus ruberrimus | Saint Lucas House Finch | Southern Baja Penninsula |
| C. mexicanus frontalis | House Finch | Western United States |
| C. mexicanus clemintis | San Clemente House Finch | Santa Barbara Islands |
| C. mcgregori | San Benito House Finch | San Benito Island |
| C. amplus | Guadalupe House Finch | Guadalupe Island |
| **Moore (1939)** | | |
| C. mexicanus frontalis | Pueblo House Finch | New Mexico and S. Colorado |
| C. mexicanus smithi | Dusky House Finch | N. Colorado and Wyoming |
| C. mexicanus solitudinis | Desert House Finch | Great Basin |
| C. mexicanus grinnelli | Grinnell House Finch | Coastal California |
| C. mexicanus clemintis | San Clemente House Finch | Santa Barbara Islands |
| C. mexicanus nigrescens | Tamaulipas House Finch | Tamaulipas, Miquihauna |
| C. mexianus potosinus | San Luis House Finch | S. Texas to San Luis Potosi |
| C. mexicanus centralis | Guanajuato House Finch | Guanajuato and N. Michoacan |
| C. mexicanus coccineus | Scarlet-breasted House Finch | S.W. Pacific Slope of Mexico |
| C. mexicanus altitudinis | Sierra Madre House Finch | Border of Sinaloa, Durango, Chihuahua |
| C. mexicanus rhodopnus | Sinaloa House Finch | Pacific coast Sinaloa |
| C. mexicanus ruberrimus | Saint Lucas House Finch | S. Baja and adj. Sonora and Sinaloa |
| C. mexicanus mexicanus | Mexican House Finch | Southern Mexico |
| C. mexicanus roseipectus | Oaxaca House Finch | Oaxaca |
| C. mexicanus griscomi | Guerrero House Finch | Guerrero |
| C. mcgregori | San Benito House Finch | San Benito Island |
| C. amplus | Guadalupe House Finch | Guadalupe Island |
| C. mutans | Hawaiian House Finch | Hawaiian Islands |
| **A.O.U. (1957)** | | |
| C. mexicanus frontalis | | United States and S. Canada, Hawaiian I |
| C. mexicanus clemintis | | Santa Barbara Islands |
| C. mexianus potosinus | | S. Texas to San Luis Potosi |
| C. mexicanus centralis | | Guanajuato and N. Michoacan |
| C. mexicanus coccineus | | S.W. Pacific Slope of Mexico |
| C. mexicanus rhodopnus | | Pacific coast Sinaloa |
| C. mexicanus ruberrimus | | S. Baja and adj. Sonora and Sinaloa |
| C. mexicanus mexicanus | | Southern Mexico |
| C. mexicanus griscomi | | Guerrero |
| C. mexicanus mcgregori | | San Benito Island |
| C. mexicanus amplus | | Guadalupe Island |

the Great Basin House Finch population (part of C. *m. frontalis*) as a separate taxon, as explained below (Figure 10.4).

## A House Finch Phylogeny

What are the phylogenetic relationships of the House Finch subspecies and populations? We know the relationships among populations of *frontalis* finches in the lower forty-eight states, southern Canada, and Hawaii because we have a written history of the establishment of these populations (Figure 10.5). The populations that have given rise to the subspecies of House Finches diverged many thousands of years ago, however, so the relationships among these taxa have to be deduced. To derive a phylogenetic hypothesis for the twelve subspecies, I first conducted a simple cladistic analysis based on plumage characteristics that I could observe on study skins of the birds (Hill 1994a). A cladistic analysis simply groups taxa based on derived traits that they share and then minimizes the number of independent evolutionary events needed to account for all the traits of all the taxa (see Kitching et al. 1988). Cladistic analysis is based on the principle of parsimony—the simplest hypothesis is the hypothesis most likely to be true. Because most of the subspecies are similar in plumage characteristics, this analysis provided minimal resolution of the relationships of the taxa, but it did separate the Sur group—the three southernmost House Finch subspecies—from other House Finches (Hill 1994a) (Figure 10.6), which supports the suggestion of several authorities that the Sur group has been isolated from other House Finch populations for a long time (Moore 1939) and may be distinct at the species level (Sharp 1888).

To this phylogeny derived through cladistic analysis, I added three additional groupings among House Finch lineages. First, based on a study of morphology and biogeography, Power (1979) proposed that birds in the subspecies C. *m. clementis* from the northern California Islands are derived from the mainland *frontalis* population, and that the subspecies C. *m. mcgregori* and C. *m. amplus* from the southern California Islands are derived from the mainland *ruberrimus* population. I include Power's hypotheses in my phylogeny. Second, I assume that birds from the Great Basin *frontalis* population are a sister lineage to the *frontalis* populations that surround it on the west, south, and east. Because finches in these *frontalis* populations are so similar that they do not warrant taxonomic distinction, treating them as sister taxa to other *frontalis* populations seemed reasonable. Finally, based on evidence from molecular studies (Marten and Johnson 1986), I placed the two North American outgroups, Cassin's Finch (*Carpodacus cassinii*) and Purple Finch (*Carpodacus purpureus*), as being more closely related to each other than either is to any House Finch population. As a second outgroup, I use Eurasian *Carpodacus* finches combined into one group. The result is a fairly well resolved phylogeny for populations of House Finches (Hill 1996b) (Figure 10.6).

This composite phylogeny based on a simple cladistic analysis of a few morphological traits and the hypotheses of other authors should certainly be viewed as preliminary. A better approach to resolving the relationships of these populations of House Finches would be to conduct a molecular systematic study, but this has not been attempted. So, although this hypothesis must be viewed cautiously, it is

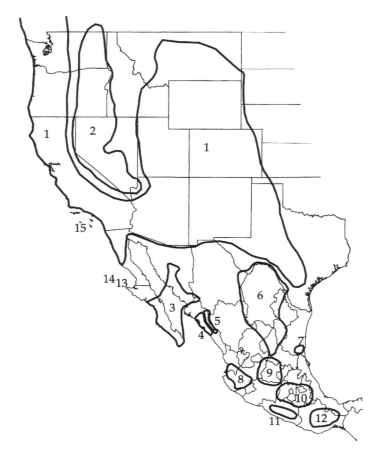

Figure 10.4. Approximate ranges of subspecies of House Finches in western North America. Ranges are redrawn from Moore (1939) except for that of *C. m. frontalis*, which has undergone extensive range expansion in the United States since 1939. Also, the range of the Great Basin population of *C. m. frontalis* is here proposed to extend to British Columbia. Eastern and Hawaiian ranges of *C. m. frontalis* are not shown. Many of the gaps between the depicted ranges have populations of House Finches of unknown subspecies affiliation. (see Moore 1939). 1: *C. m. frontalis*; 2: Great Basin *C. m. frontalis*; 3: *C. m. ruberrimus*; 4: *C. m. rhodopnus*; 5: *C. m. altitudinus*; 6: *C. m. potosinus*; 7: *C. m. nigrescens*; 8: *C. m. coccineus*; 9: *C. m. centralis*; 10: *C. m. mexicanus*; 11: *C. m. griscomi*; 12: *C. m. roseipectus*; 13: *C. m. mcgregori*; 14: *C. m. amplus*; 15: *C. m. clementis*. *C. m. altitudinus* and *C. m. nigrescens* are no longer recognized as valid subspecies. Reproduced from Hill (1996b) with permission.

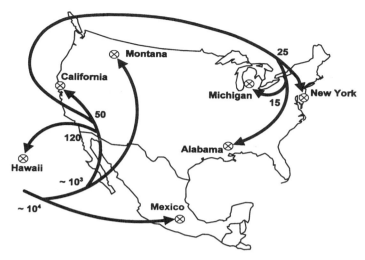

Figure 10.5. The known phylogenetic relationships of populations of House Finches in the subspecies *C. m. frontalis*. Numbers at the nodes give the years since divergence of the lineages. Birds in the population in Mexico belong to the subspecies *C. m. griscomi*, and it is assumed that *frontalis* finches split from *griscomi* finches more than 10,000 years ago. Reproduced from Badyaev and Hill (2000b) with permission.

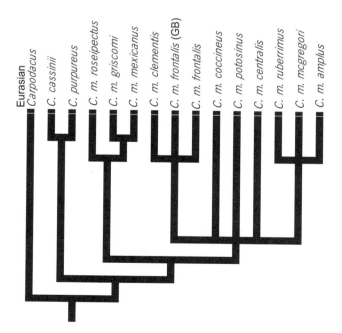

Figure 10.6. A composite phylogeny of House Finch lineages based on a cladistic analysis and biogeographical and morphological evidence. This phylogeny shows only the hypothesized ancestry of lineages—branch length is arbitrary. From Hill (1996b). *C. m. frontalis* (GB) is the Great Basin population of *C. m. frontalis*.

consistent with the biogeography of the taxa, and I'm confident that the main features of the topology will persist as more detailed phylogenies are derived.

## Time Scale

How deep are the divisions among these House Finch lineages? Following a molecular clock assumption, Marten and Johnson (1986) estimated that House Finches split from a Purple Finch/Cassin's Finch ancestor about eight million years ago. Moore (1939) speculated that the three southern Mexican populations (the Sur Group) were isolated from other House Finch populations during the Pleistocene, about 1.5 million years ago. Presumably, the divergence of populations within the Sur and northern finch populations occurred since that time. Power (1979) speculated that, following the Pleistocene and until about 6000 years ago, C. m. rubberrimus had a continuous distribution throughout the Baja Peninsula, across the top of the Gulf of California, and south along the mainland shore of the Gulf of California. Climatic changes that led to the desertification of the northern Gulf of California caused the ruberrimus population to contract south along both sides of the Gulf, leading to the present disjunct distribution of this population (Figure 10.4). If this scenario is correct, it would mean that the subspecies' characteristics of ruberrimus have been maintained by two populations in isolation for 6000 years. A reanalysis of House Finch populations using improved molecular techniques is needed before the age of any of these taxa can be stated with confidence, but there is very good evidence that at least some of the populations of House Finches identified as subspecies have had a unique evolutionary history for at least tens of thousands of years, and some may have been isolated for hundreds of thousands of years.

## Geographic Variation in Ornamental Display

House Finches in the unique populations described above live in very different environments with very different plant communities, parasites, predators, and competitors. So, the strength of natural and sexual selection acting on the phenotypes of birds in these isolated populations should have varied considerably, and as a result of such selection, birds in these populations would be expected to have diverged in characteristics, including sexually selected traits. Indeed, both Moore (1939) and Power (1979), and more recently Badyaev and Hill (2000b), documented substantial geographic variation in the size and shape of House Finches across their range. Of greater interest was the substantial variation in expression of plumage coloration recorded among males in the different subspecies, among *frontalis* males in California, and in the recently introduced populations in the eastern United States and Hawaii.

## Variation in Plumage Redness Among Populations of
C. m. frontalis

That males in different populations of House Finches vary substantially in expression of carotenoid-based plumage coloration was recognized long ago. In 1911,

Grinnell documented that populations of House Finches on the Hawaiian islands had a much higher frequency of males with yellow and orange coloration than he observed in California. Grinnell (1911) attributed this difference in coloration to a genetic change in the Hawaiian birds, and he proposed that the Hawaiian population be given species status because of its different coloration (Grinnell 1912; see taxonomy section above).

As part of my doctoral thesis, I captured and examined House Finches on Oahu and Hawaii islands, in the San Francisco Bay area of California, and in Ann Arbor, Michigan, and Long Island, New York, in the eastern United States. At each location, I carefully quantified the ornamental plumage coloration of the males that I captured. I found that the mean plumage color scores (which combined plumage hue, saturation, and tone) were significantly lower (less red and less saturated) for males from both Oahu and Hawaii compared to males from Michigan or New York (Hill 1993a) (Figure 10.7). Surprisingly, in California I found substantial variation in the plumage coloration of males at two nearly adjacent banding stations. One of these sampling sites was the backyard of Dr. Dick Mewaldt in the suburbs of San Jose (hereafter San Jose), where males tended to be bright red and had a mean plumage color score much like eastern United States populations. The other sampling site was Coyote Creek Riparian Station near Alviso (hereafter Alviso), where males tended to be yellow and orange and had a mean plumage score much like Hawaii. Despite the very different expression of plumage coloration by males at these California banding stations, the sites were only 12 km apart and connected by continuous finch habitat.

Interestingly, the mean size of the ventral patch of carotenoid pigmentation displayed by males did not differ significantly among the populations (Figure 10.7). Whether they were from bright or drab populations, expression of patch size was similar among the six populations. So one component of color varied substantially among populations, but another component did not.

The variation in expression of carotenoid-based ornamentation among these populations is interesting given that the Hawaiian and eastern United States populations are recently derived from birds taken from the coastal California population (see Figure 10.5), and that all birds in coastal California belong to one gene pool (Benner 1991, Power 1979, Vazquez-Phillips 1992). The large difference between the two populations in the San Francisco Bay area is particularly striking considering that males in these populations live within a few minutes flight of each other, with no dispersal barriers between them. So, this great variation in expression of plumage coloration would seem to have arisen over a very short time period and without the populations undergoing genetic differentiation. These factors clearly implicated an environmental basis for the differences in plumage coloration among these populations.

## Temporal Stability of Population Coloration

During the four years in which I monitored the House Finch population in Ann Arbor, Michigan, I found that the mean and standard deviations of ornamental male plumage coloration was remarkably stable, with no significant changes between years (Figure 10.8). In three other populations, I was able to compare the plumage

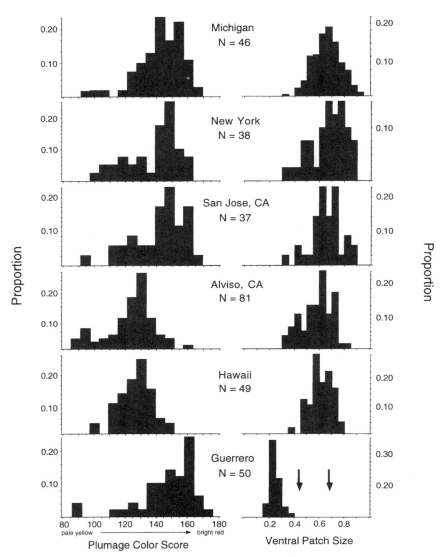

Figure 10.7. Frequency distributions of plumage color scores and ventral patch size (proportion of underside with carotenoid pigmentation) for several populations of males in the subspecies *C. m. frontalis* (upper ten histograms) and one population in the subspecies *C. m. griscomi* (lower two histograms). There were significant differences in the mean plumage coloration of various populations ($F = 52.35$, d.f. $= 5$, 802, $P = 0.0001$, one-way ANOVA) with males from New York, Michigan, San Jose, and Guerrero being significantly brighter on average than males from Alviso or Hawaii ($P < 0.05$, Scheffe's test). There was a significant difference in patch size only between males from *griscomi* and *frontalis* populations ($F = 125.10$, d.f. $= 5$, 798, $P = 0.0001$ ANOVA with Scheffe's test, $P < 0.05$). There were no significant differences in patch size among *frontalis* populations ($P > 0.05$, Scheffe's test). The arrows in the lower right histogram indicate the ventral patch size of the single *frontalis* × *griscomi* hybrid male (left arrow) and the three *frontalis* × *frontalis* males (right arrow) raised in captivity. Adapted from Hill (1993a).

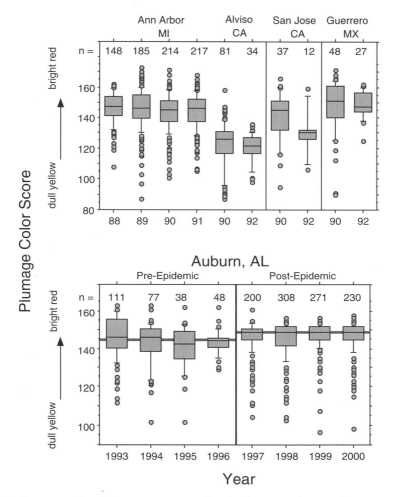

Figure 10.8. Temporal stability in expression of male plumage coloration for various populations of House Finches. Male plumage coloration did not change significantly across years in Ann Arbor, Michigan ($H = 7.22$, d.f. = 3, 760, $P = 0.07$), Alviso, California ($z = -1.52$, $P = 0.13$), or Guerrero, Mexico ($z = -0.75$, $P = 0.46$). It did change significantly between 1990 and 1992 in San Jose, California ($z = -2.01$, $P = 0.04$). In Auburn, Alabama, male plumage coloration was stable for four years until the onset of a mycoplasmal epidemic. Plumage coloration increased significantly between 1996 and 1997 ($z = -4.29$, $P = 0.0001$) and has not changed significantly since. The horizontal gray bars running across the lower figure indicate the mean plumage coloration of males in the pre- and post-epidemic periods, respectively. Mann-Whitney $U$-tests were used for paired comparisons, and a Kruskal-Wallis test for the multi-year comparisons in Alabama and Michigan.

coloration of males during bouts of sampling two years apart. In two of these populations, Alviso and Guerrero, the mean and distribution of the plumage coloration of males were similar between sampling times, with no significant differences. In the San Jose population, however, the plumage coloration of males dropped significantly between the first time I captured finches there in 1990 and second time I sampled finches in 1992. During these two sampling periods, I trapped birds in the same yard in San Jose, using the same techniques. This dramatic change in the mean plumage coloration of males over two years at one sampling site, as well as the large difference in plumage coloration between the San Jose and the nearby Alviso population suggests that some environmental variable in this area of California is causing the mean coloration of males to fluctuate substantially over brief intervals of time and space. I've implicated access to carotenoid pigments in microgeographic variation in expression of plumage coloration (see below), but local outbreaks of a parasite such as avian pox (Zahn and Rothstein 1999) or some combination of carotenoid access and parasite infection could also be responsible.

In the Auburn population, my students and I recorded the coloration of males from 1993 to 2000. Between 1995 and 1996, we switched from scoring plumage by visual comparison to *The Methuen Handbook of Colour* to scoring coloration using a Colortron (see chapter 3). Colortron plumage scores and book scores are related in a linear fashion (Hill 1998a), so I was able to convert Colortron plumage scores to book scores for comparisons across years. I found that the plumage coloration of males in the Auburn population was stable until the onset of the mycoplasmal epidemic in late 1995 and early 1996, which, as I discussed in chapter 7, disproportionately killed drab males and caused the mean plumage coloration of males in the population to increase (Nolan et al. 1998) (see Figure 7.6). Mean plumage redness rose between 1996 and 1997 and has remained stable and high since (Figure 10.8).

The overall conclusion from these observations of change in plumage score over time is that, although the mean plumage coloration of males within a population can remain very similar across years, expression of male plumage coloration can also change dramatically between years if environmental conditions change. As I documented in chapter 5, male plumage coloration is affected by access to carotenoid pigments, exposure to parasites, and nutritional condition, and changes in any of these environmental factors over time can lead to substantial changes in plumage coloration within a population, as we were able to document with the mycoplasmal epidemic in the Auburn population.

## Patch Size Variation Among Subspecies

The patterns of variation in expression of plumage coloration described above concern coastal California House Finches (subspecies C. *m. frontalis*) and their recent descendents. Among subspecies, there is also some variation in expression of plumage hue and saturation as will be described below, but the most striking variation among subspecies is in the expression of patch size (Figure 10.9). The patch size of most male House Finches in most subspecies is the patch size seen in the familiar House Finches of coastal California origin, with carotenoid pigmentation covering an average of about 60 to 65% of the underside of the bird (Hill 1993a, 1996b, Moore 1939). In birds with this medium expression of patch size, the lower boundary of the carotenoid patch is fuzzy, fading slowly from intense pigmentation to no carotenoid pigmentation in the lower breast region. In addition, there is often a wash of red outside the patches of concentrated coloration on the

Figure 10.9. The extent of ventral carotenoid pigmentation (patch size) of males in three subspecies of the House Finch. Illustrated are typical males from: (top) the small-patched subspecies C. *m. griscomi*; (middle) the medium-patched subspecies C. *m. frontalis*; and (bottom) the large-patched subspecies C.*m. ruberrimus*. Of the thirteen subspecies and populations of House Finches that I recognize in this book, four have males with small patches (C. *m. frontalis* (Great Basin), C. *m. mexicanus*, C. *m. griscomi*, and C. *m. amplus*), two have males with large patches (C. *m. ruberrimus* and C. *m. rhodopnus*), and seven have males with medium patches (C. *m. frontalis* (typical), C. *m. potosinus*, C. *m. coccineus*, C. *m. centralis*, C. *m. roseipectus*, C. *m. mcgregori*, and C. *m. clementis*). Drawing by Phillip C. Chu.

upper crown as well as on the back and on some of the primary coverts (Hill 1993a, Moore 1939). This medium-patched carotenoid display is found in seven subspecies and populations of the House Finch (Hill 1993a, 1996b).

In contrast to the medium expression of patch size displayed by males in most subspecies of House Finches, in southern Mexico there are two subspecies, *mexicanus* and *griscomi*, with much reduced patch size (Hill 1993a, 1996b, Moore 1939) (Figure 10.9). Males from these two subspecies have carotenoid pigmentation covering an average of only 24% of the ventral plumage area (Hill 1993a, 1996b). In males from these subspecies, there is no "bleeding" of carotenoid pigmentation outside the clearly demarcated patches; rather, there is an abrupt transition between feathers with intense carotenoid pigmentation and feathers lacking carotenoid pigmentation, so the boundaries of the ventral patches of these males are sharp (Hill 1993a, Moore 1939).

House Finches of the subspecies *griscomi* and *mexicanus* are the most distinctly marked of the small-patched populations of House Finches—they are the only populations that have the clearly demarcated small patches—but two other populations of House Finches also have small patches of ventral coloration. First, males from Guadalupe Island (*C. m. amplus*) have an average of 32% of their ventral plumage with carotenoid pigmentation, which is a significantly smaller patch size than *frontalis* males, but not significantly different than *griscomi* males (Hill 1996b). Second, males in the Great Basin of the interior, western United States also have consistently smaller patches of ventral carotenoid pigmentation than males from coastal California, Arizona/New Mexico, or the Rocky Mountain Region (Moore 1939, Van Rossem 1936) (pers. obs.). In the Smithsonian Museum of Natural History, I measured the patch size of thirty-five males collected in the Great Basin, and observed that they had on average 33% of their ventral surface with carotenoid coloration. This patch size is much like that of Gaudalupe Island House Finches, and significantly smaller than typical *frontalis* males. Except for their small patches of coloration, *amplus* and Great Basin *frontalis* male are like coastal California *frontalis* males in having fuzzy boundaries to their ventral carotenoid coloration and bleeding of carotenoid coloration onto their upper crown, back, and wings.

Perhaps the most interesting subspecies with regard to patch size is the Oaxaca House Finch, *C. m. roseipectus*, which is the third subspecies, along with *griscomi* and *mexicanus*, in the "Sur" group and the most southerly population of the House Finch (Figure 10.4). *C. m. roseipectus* males are similar to *mexicanus* males in having a small, clearly demarcated patch of bright carotenoid pigmentation on the throat and upper breast. Below this small patch of bright carotenoid pigmentation, however, is a wash of carotenoid coloration that extends the patch so that it covers an area of the underside like that of the medium-patched populations described above. *C. m. roseipectus* males are like *griscomi* and *mexicanus* males in that, except for the diffuse extension of carotenoid pigmentation on the breast, there is no "bleeding" of pigmentation on the crown, back, or wings. Thus, even though this population of House Finches is very much like the two other southern subspecies of House Finches in plumage coloration, I classify it with the medium-patched populations.

Finally, in two subspecies of House Finches, *ruberrimus* and *rhodopnus*, virtually the entire underside of males has carotenoid pigmentation. I measured the patch size of thirty-nine *ruberrimus* males in the Field and Smithsonian Musuems of Natural History. These *ruberrimus* males had an average of 71% of the ventral plumage with carotenoid coloration, which is a significantly larger patch size than *frontalis* males from coastal California. Not only are males from these populations extensively bright red on their undersides, but there is also a prominent wash of carotenoid coloration on the upper crown, back, and wing coverts of males of these subspecies. Both of these large-patched populations of House Finches are found on the Pacific coast of Mexico (Figure 10.4).

## Variation in Hue and Saturation Among Subspecies

Most males in most subspecies of House Finches have reddish plumage hue, although in all subspecies, yellow and orange males occur at least occasionally (Moore 1939; pers. obs.). Overall, there is more variation in plumage hue within *frontalis* males, as described above, than between any subspecies. Next to the drab populations of *frontalis* males in Hawaii and California, House Finches on Guadalupe Island (*amplus*) seem to have the lowest mean plumage coloration of any subspecies of House Finch, with many yellow and orange males (Moore 1939; pers. obs).

Males of some subspecies seem to have characteristically more saturated carotenoid pigmentation than males of other subspecies. For instance, I found that, independent of plumage hue, the plumage saturation of *griscomi* males was significantly higher than *frontalis* males (Hill 1993a) (Figure 10.10). And, in a study of the carotenoid pigments in the plumages of *griscomi* and *frontalis* males (see chapter 4 for details), we found that the quantity of carotenoids was twice as high in the plumage in *griscomi* males compared to the plumage of *frontalis* males (Inouye et al. 2000). This suggests that the mechanisms for pigment processing and deposition are fundamentally different for some populations of House Finches. I will return to this topic in the next chapter.

## The Proximate Bases of Population Variation

Why is there so much variation among populations and subspecies of House Finches in the expression of male plumage coloration? Do they differ in the genes coding for plumage coloration as proposed by Grinnell (1912)? Do they face different environmental challenges in different regions of the continent that cause them to express plumage in different ways? Or, is a more complex, gene–environment interaction responsible for the differences?

Male House Finches differ in ornamental plumage coloration in two primary ways: they differ in plumage hue within and among populations, and they differ in patch size among subspecies. I tested for the proximate factors that might be responsible for this geographic variation in plumage coloration using the same type of feeding experiments that I used to investigate within-population variation in plumage coloration among males (see chapter 5). In controlled feeding experi-

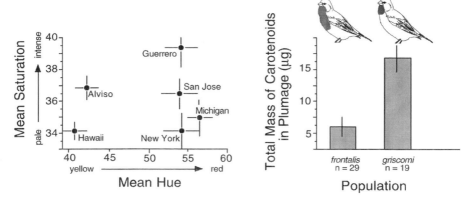

Figure 10.10 Variation among populations and subspecies of House Finches in mean saturation of plumage coloration. On the left is shown a plot of the mean ± standard errors plumage hue scores versus plumage saturation scores for males sampled in five populations of the *frontalis* subspecies and one population (Guerrero) of the *griscomi* subspecies. *griscomi* males had significantly more saturated plumage than males from any of the *frontalis* populations ($F = 7.48$, $P = 0.0001$ ANOVA, $P < 0.05$ post hoc Scheffe's test comparing populations). Adapted from Hill (1993a). On the right is shown the means ± standard deviations of concentration of total carotenoid pigments in the plumage of *griscomi* males sampled in Guerrero and *frontalis* males sampled in coastal California. *griscomi* males have significantly higher concentration of carotenoid pigments ($t = -14.3$, $P < 0.0001$, two-tailed *t*-test), which is the biochemical basis for the greater color saturation of males in these populations. Adapted from Inouye et al. (2001).

ments in which access to carotenoid pigments was standardized among males, I compared the plumage coloration and patch size of (1) males from different *frontalis* populations, and (2) males from the subspecies *frontalis* and *griscomi*. Let me first present my tests of the effect of access to carotenoid pigments on expression of plumage hue among populations of *frontalis* House Finches.

## Proximate Control of Color Differences Among frontalis Populations

Given that standardizing access to carotenoid pigments at the time of molt elimi- nated most of the variation in plumage coloration among males from the Ann Arbor population, I predicted that standardizing carotenoid access would also eliminate variation in expression of plumage coloration among populations of *frontalis* males. To test this idea, I first had to bring birds from various populations around North America and the Hawaiian islands into my aviaries at the University of Michigan. Looking back, I was extremely fortunate to have been able to capture sufficient numbers of male House Finches in California, in the Hawaiian islands, and in Guerrero, Mexico, and to transport these birds alive back to my aviaries in Michigan. I was often working with small groups of birds from the various locations so that one or a few deaths would have eliminated a population from my study. The birds did amazingly well in transport and in my aviaries, and I ended up with enough finches to test the effect of diet on male plumage coloration for a drab California population (Alviso), a bright California population (San Jose), the Hawaiian islands population, the *griscomi* population in Guerrero, Mexico, and of course the population in Michigan.

I found that, just as it had essentially eliminated variation in expression of plumage coloration within a population, standardizing access to carotenoid pigments at the time of molt essentially eliminated variation in expression of plumage coloration among males from the different *frontalis* populations. When they were supplemented with the red pigment canthaxanthin added to their water, all males from all *frontalis* populations converged on a similar bright red plumage coloration, with no significant differences in expression of plumage coloration among males from the different populations (Hill 1993a) (Figure 10.11). The birds from Hawaii, where red males are rarely encountered, were just as red as males from Michigan, where most males in the wild are red. Likewise, males from the drab Alviso population grew feathers that were just as red as those grown by males from the bright San Jose population. I observed a similar convergence of color display when males were maintained on a seed diet with no carotenoid supplementation; on such a low-carotenoid diet, males from all populations grew drab yellow plumage (Hill 1993a). Males from the Michigan population

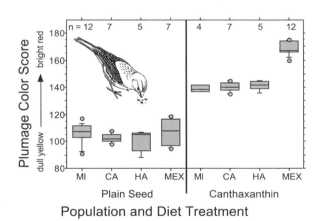

Figure 10.11. The effect of controlled access to carotenoid pigments during molt on expression of plumage coloration for male House Finches from different locations in North America. Shown are the post-molt, plumage color scores of males from Ann Arbor, Michigan (MI), San Jose, California (CA), and Pohakuloa, Hawaii (HA), which are in the subspecies *C. m. frontalis*, as well as males from Guerrero, Mexico (MEX), which are in the subspecies *C. m. griscomi*. There were no differences in mean plumage coloration among any of the groups fed plain seed ($F = 1.00$, $P = 0.41$, ANOVA). In the canthaxanthin treatment there were no differences among the *frontalis* populations, but *griscomi* males had significantly brighter plumage coloration than males from any of the *frontalis* populations ($F = 98.36$, $P < 0.0001$, ANOVA; post hoc Scheffe's test showed sigificant differences between all *frontalis* population and the *griscomi* population $P < 0.05$, but not among *frontalis* populations). Adapted from Hill (1993a).

were just as drab after molt as were males from the Hawaiian population, and San Jose males were as drab as Alviso males.

So, observations from this feeding experiment supported the idea that differences in expression of plumage coloration among *frontalis* populations are the result of differences among these populations in mean male condition, including access to carotenoid pigments at the time of molt. This should hardly be a revelation. The males that I captured in Michigan, Hawaii, San Jose, and Alviso are all derived from the same coastal California gene pool. The San Jose and Alviso males were still part of that gene pool at the time of their capture, and the Hawaiian and Michigan males were only a few dozen generations removed. To have found genetic-based differences in expression of plumage coloration among these populations would have been very surprising. It makes much more sense that changes in environmental conditions led to rapid and local changes in male coloration among populations or even neighborhoods of *frontalis* males. Even though it was what I predicted going into the feeding experiments, it was important to demonstrate experimentally the plastic nature of plumage expression for males from all populations.

Differences in the expression of patch size by males in the various *frontalis* populations were also eliminated by the standardized diets. The populations started out more similar in mean patch sizes (differences were not significant) from mean plumage coloration, but the mean patch sizes of males from the various populations became even more similar after the controlled feeding experiments (Hill 1993a) (Figure 10.12).

### Proximate Basis for Color Differences Among Subspecies

Male House Finches from the *griscomi* population have much smaller, more discrete patches of coloration with more intense carotenoid pigmentation than do *frontalis* males. To see how these subspecies' differences are affected by standardized diets, I fed *frontalis* and *griscomi* males during molt either (1) plain seed diets with few carotenoid pigments, or (2) the same diets supplemented with the red pigment canthaxanthin. On a plain seed diet, males from both the *griscomi* and *frontalis* populations grew very similar patches of pale yellow plumage, and there were no significant differences between the subspecies in plumage color scores (Hill 1993a) (Figure 10.11). On a diet supplemented with canthaxanthin, males from both subspecies grew bright red plumage, but the red plumage coloration grown by *griscomi* males was significantly more saturated than the red plumage grown by *frontalis* males (Hill 1993a). So, for *griscomi* and *frontalis* House Finches, dietary access to carotenoid pigments can greatly affect expression of plumage coloration, but on the same supplementation of canthaxanthin, *griscomi* males are able to grow significantly more saturated plumage than *frontalis* males. These observations suggest that *griscomi* males are able to concentrate available carotenoids in their colored patches more effectively than *frontalis* males.

In contrast to the large effect of dietary carotenoids on expression of plumage coloration, the carotenoid content of diet had no significant effect on expression of patch size by *griscomi* males (Hill 1993a) (Figure 10.12). Whether they were fed red pigments and grew red plumage or were fed only seeds and grew yellow plumage, *griscomi* males invariably grew the same small patches of ventral carotenoid pig-

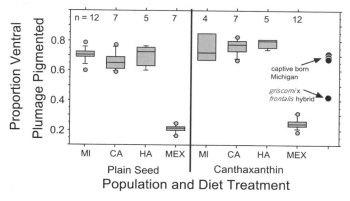

Figure 10.12. The effect of controlled access to carotenoid pigments during molt on expression of ventral patch size for male House Finches from different locations in North America. Shown are the post-molt proportions of ventral pigmentation for males from Ann Arbor, Michigan (MI), San Jose, California (CA), and Pohakuloa, Hawaii (HA), which are part of the subspecies *C. m. frontalis*, as well as males from Guerrero, Mexico (MEX), which are in the subspecies *C. m. griscomi*. The points on the right edge of the figures are the patch sizes for the three captive-bred males from Michigan parents and one captive-bred male from a *frontalis* × *griscomi* (Michigan × Guerrero) cross. There were no differences in mean patch size among *frontalis* males in either of the diet treatments ($P > 0.05$, Scheffe's test). In both diet treatments, *griscomi* males had significantly smaller patches than any of the *frontalis* populations (seed diets: $F = 133.16$, $P < 0.0001$; canthaxanthin-supplemented: $F = 199.14$, $P < 0.0001$). The hybrid male was intermediate in patch size to the two parental subspecies. Redrawn from Hill (1993a).

mentation. In both diet treatments the mean patch size of *griscomi* males was significantly smaller than the mean patch size of *frontalis* males. The persistence of differences in patch size between *frontalis* and *griscomi* males across dietary treatments suggests that these differences are due to fixed genetic differences between the subspecies.

Not only did *frontalis* and *griscomi* males held on a canthaxanthin-supplemented diet differ in patch size and pigment saturation, they also differed in the extent to which carotenoid pigmentation colored feathers outside their discrete patches of color. In all *frontalis* males maintained on canthaxanthin, there was a heavy bleeding of coloration across the upper crown and down the back of the bird. In *griscomi* males, in contrast, the red pigmentation was restricted entirely to the well-defined patches of color. There was no bleeding of carotenoid pigmentation outside the patches.

## Characteristics of a Hybrid

I further tested the idea that patch-size differences between *griscomi* and *frontalis* males represented genetic differences between the populations by cross-breeding a *frontalis* female with a *griscomi* male in my aviary. At the same time, I also paired Michigan males with Michigan females and allowed them to breed in captivity.

The *griscomi* × *frontalis* pair produced one male fledgling and the *frontalis* × *frontalis* pairs produced three male fledglings. All four of these males that fledged survived through fall molt, during which they were maintained on a canthax-anthin-supplemented diet. After molt, I scored the patch size of these four captive-hatched males. The males from the *frontalis* × *frontalis* cross had patch sizes and plumage color scores like wild-caught *frontalis* males maintained on canthax-anthin during molt (Hill 1993a) (Figures 10.7 and 10.12). The hybrid male, however, was intermediate in patch size between *frontalis* males and *griscomi* males (Hill 1993a). Also, unlike *griscomi* males, the hybrid male had some crown and back feathers with carotenoid pigment outside the discrete carotenoid patches, but this was restricted to only a few feathers and was much less extensive than that which is displayed by all *frontalis* males. Clearly, one must be careful in drawing too many conclusions from a single hybrid male, but the intermediate patch expression of this hybrid male further supports the idea that the patch size differences between *frontalis* and *griscomi* males represent fixed genetic differences between these subspecies.

## Delayed Plumage Maturation

Not only does the definitive plumage of males differ among subspecies of House Finches, but males in these subspecies also vary in the developmental sequence (ontogeny) leading to brightly colored plumage. In the two finch taxa most closely related to the House Finch, the Cassin's and Purple Finches, it is well known that males do not acquire bright carotenoid pigmentation until their second prebasic molt (Samson 1977, Wootton 1996). As a result, males of these species spend their first potential breeding season in a female-like plumage. Despite their drab plumage, these first-year males are sexually mature; they produce sperm and they can and do breed (Samson 1977, Wootton 1996). This pattern of plumage development is known as delayed plumage maturation (Rohwer et al. 1980).

As I have already discussed in chapter 2, *frontalis* males tend to be drabber in their first year than in subsequent years, but virtually all males in familiar United States *frontalis* populations grow definitive basic plumage through their first fall molt (see Figure 2.1). Based on this pattern of plumage development of the most familiar populations, House Finches in general have been classified as not having delayed plumage maturation (Rohwer and Butcher 1988, Rohwer et al. 1980, Studd and Robertson 1985a; but see Lyon and Montgomerie 1986). However, the pattern of plumage development observed in most *frontalis* populations is not necessarily the pattern in all House Finch populations.

Van Rossem (1936) noted that the population of House Finches in the Great Basin of southern Nevada had a plumage ontogeny like Cassin's and Purple Finches; in their first potential breeding season, males from this population had female-like plumage coloration with little or no carotenoid pigmentation. This was the first suggestion that not all populations of House Finches were necessarily like the familiar Coastal California House Finches in the ontogeny of plumage expression.

In January 1990, I traveled to Guerrero, Mexico, to capture and sample males in the small-patched subspecies endemic to that state, *C. m. griscomi*. I was surprised to find that about 20% of the males that I captured had female-like plumage coloration, with only very small patches of carotenoid pigmentation on their throat and crown plumage (Hill 1996b). Two years later, when I returned with colleagues to Guerrero in the fall to collect birds for a study of plumage pigments (Inouye et al. 2000), I was able to examine sixty-three molting hatch-year males that I sexed by gonadal examination. I noted that female-like plumage of these males resulted from the growth of feathers lacking carotenoid pigmentation, and not from the retention of juvenal feathers (Hill 1996b). Finally, based on examination of museum specimens, I also concluded that the Guadalupe Island House Finch (*C. m. amplus*) and the Great Basin population of the *frontalis* House Finch have delayed plumage maturation (Hill 1996b). The typical pattern of male plumage development in other subspecies and populations is for males to acquire definitive basic plumage by their first winter. Based on written descriptions and examinations of museum specimens, I concluded that subspecies other than *griscomi*, *amplus*, and the Great Basin *frontalis* populations do not typically show delayed plumage maturation. The evidence is most convincing for males in the subspecies *ruberrimus* and *frontalis* for which many museum specimens exist (Hill 1996b).

## Geographic Variation in Female Coloration

### Population Variation

As described in the last chapter, some female House Finches show a wash of carotenoid-based plumage coloration. Female coloration is generally restricted to the rump, but occasionally there is also a wash of coloration on the crown and upper breast. The degree of female ornamentation varies somewhat across populations and subspecies, but at least some females in all populations show ornamental coloration (Hill 1993d). Most of my observations of geographic variation in female coloration have been of various *frontalis* populations and of the *griscomi* population near Chilpancingo, Guerrero, Mexico. In museum collections I have also looked at expression of plumage coloration in females in several subspecies, including the large-patched subspecies *ruberrimus*, the small-patched subspecies *amplus*, and the small-patched Great Basin *frontalis* population.

Females in *frontalis* populations in which male coloration averages bright red tend to have redder and more extensive carotenoid pigmentation than females from *frontalis* populations in which males average drabber in coloration (Figure 10.13; see also Figure 11.11). This pattern of association between female and male plumage coloration clearly does not hold when females from the *griscomi*, *ruberrimus*, or Great Basin *frontalis* populations are considered. Although *griscomi* and Great Basin *frontalis* males average bright red in coloration, females in those population show little ornamental coloration; conversely, although *ruberrimus* males have a mean color score like Michigan *frontalis* males, females in their population tend to have extensive carotenoid ornamentation. The significance of this difference between the populations will be discussed in the next chapter.

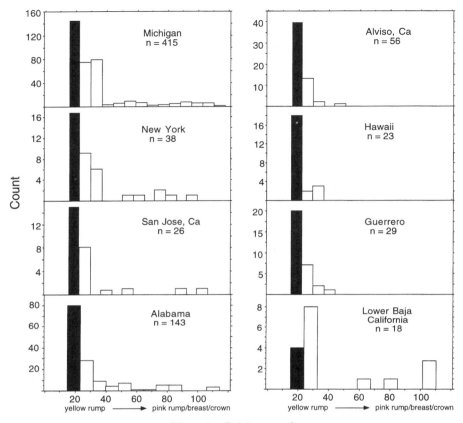

**Plumage Brightness Score**

Figure 10.13. The distribution of female plumage color scores in various populations of House Finches. Black bars in each of the histograms indicate females with no detectable carotenoid coloration. Males in Auburn, Alabama; Lower Baja, California; Ann Arbor, Michigan; New York, New York; and San Jose, California, averaged bright in coloration while males in Pohakuloa, Hawaii; Chilpancingo, Guerrero; and Alviso, California, averaged drab in coloration. The mean plumage coloration of females was significantly different between but not within these two color groups ($H = 31.05$, $P < 0.0001$ with post hoc tests). Adapted from Hill (1993d) with Auburn, Alabama, and Lower Baja, California, data added.

## Proximate Basis for Population Variation in Female Coloration

I tested the effect of access to dietary carotenoids on expression of plumage coloration for females from different populations using the same type of feeding experiments that I described previously (see chapters 5 and 9). In this case, I maintained females from Michigan, Hawaii, Alviso (California), San Jose (California), and Guerrero during fall molt on either plain seed diets or seed diets supplemented

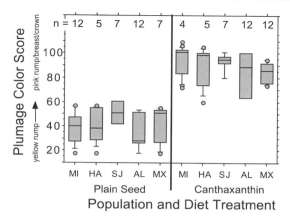

Figure 10.14. The effect of access to carotenoid pigments during molt on expression of plumage coloration for captive female House Finches from different areas of North America. Females were either fed the red carotenoid pigment canthaxanthin, which was added to their drinking water, or they were maintained on a plain-seed diet, which provided few carotenoid pigments. For all populations, the plumage of females maintained on plain-seed diet was significantly less colorful than females maintained on canthaxanthin ($P < 0.01$). MI = Ann Arbor, Michigan; HA = Pohakuloa, Hawaii; SJ = San Jose, California; AL = Auburn, Alabama; MX = Chilpancingo, Guerrero. Adapted from Hill (1993d) with Auburn data added.

with canthaxanthin. As I described in chapter 6, I found that supplementing the diets of females with canthaxanthin caused them all to grow a red rump with a pink wash on their breasts and crowns. Maintaining females on plain seed diets caused them to grow plumage with a yellow rump. Most importantly for the topic of this chapter, there were no significant differences among the populations or subspecies within either diet group in expression of carotenoid pigmentation (Figure 10.14) (Hill 1993d).

So, despite the fact that females from some populations average much drabber in expression of plumage coloration than females from other populations, all females from all populations appear to possess the potential for maximum female ornament display. This observation suggests that some environmental constraint such as access to sufficient dietary carotenoids during molt is responsible for the differences in expression of female plumage coloration among various populations of House Finches.

## Summary

Across North America, House Finches exist not as one panmictic population but as many populations that are isolated and differentiated to varying degrees. Since the middle of the nineteenth century, House Finches of the familiar northern subspecies, C. *m. frontalis*, have undergone a remarkable expansion of their original range in western North America. From a handful of birds introduced in the vicinity of

New York, House Finches spread through most of the eastern North America. They also spread throughout the Hawaiian islands after being introduced there. The *frontalis* subspecies now occupies an area from the Canadian Maritimes to Florida to the Pacific Northwest to the northern Baja Peninsula.

Populations of *frontalis* finches vary substantially in the mean plumage coloration of males. Males in introduced eastern populations in New York and Michigan averaged bright in coloration; males in an introduced population on the island of Hawaii averaged drab in coloration. Males from two sampling sites in central coastal California separated by only 12 km averaged drab and bright, respectively. South of the United States border and through the islands off the southern Pacific coast of North America, House Finches exist in about thirteen well-marked subspecies and populations. These populations have had a unique evolutionary history for many thousands of years, and they show fascinating variation in size, shape, and intensity of carotenoid-based plumage coloration. Perhaps most interesting of all, there is substantial variation among the subspecies in extent of male carotenoid pigmentation, and subspecies can be divided into those in which males have small patches, medium patches, and large patches of ventral pigmentation. Controlled feeding experiments with males from a small-patched subspecies from Guerrero, Mexico (C. *m. griscomi*) and with males from various populations of the medium-patched *frontalis* subspecies indicated that variation in expression of plumage coloration among *frontalis* populations is a result of different environmental conditions, not genetic differences for plumage expression among the populations. In contrast, differences in patch size among the subspecies appear to reflect fixed genetic differences for expression of patch size. Finally, in some subspecies of House Finches males have a female-like plumage during their first potential breeding season—a condition called delayed plumage maturation. This tremendous variation in the color, patch size, and ontogeny of ornamental plumage pigmentation provides a unique opportunity to conduct comparative tests of hypotheses related to the evolution of carotenoid-based plumage coloration. Such comparative tests will be the focus of the next chapter.

# 11  Why Red?

## The Evolution of Color Display

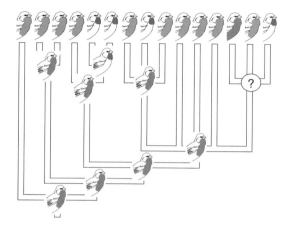

The most significant interrelation of colors, however, and the one which I believe to be of wide application in the explanation of bird colors, is that between red and yellow. . . there is a high degree of probability that red is simply an intensification of yellow. There is much evidence to show that yellow is a more primitive stage, and that the latter has always, or nearly always, been developed from the former.

> —C. A. Keeler (1893), in the first public speculation
> of which I am aware of the evolutionary sequence
> leading to red coloration in birds

"Whence," it may be asked, "has this extremely uniform and definite taste for a particular detailed design of form and colour arisen?" Granted that while this taste and preference prevails among the females of the species, the males will grow more and more elaborate and beautiful tail feathers, the question must be answered "Why have the females this taste? Of what use is it to the species that they should select this seemingly useless ornament?"

> —R. A. Fisher (1915), summarizing the great
> challenge to the idea of sexual selection
> driven by female choice

Perhaps even more fanciful are some of the theories designed to explain the strikingly marked or brilliantly colored plumages, which obviously have little or no concealing value. Darwin's theory of sexual selection, which presupposes that females select the most

strikingly colored males, has not stood the test of time (mating is
not that simple). . . . Many special markings on birds, however, are
probably primarily for species recognition rather than being strictly
epigamic.

—G. J. Wallace (1955), presenting the consensus
opinion on the evolution of plumage coloration in
the middle of the twentieth century

*I met Jorge Serano in the central market of Chilpancingo, Guerrero, in January 1990.
My friend Michael Nachman and I were camping in the mountains above the city
trying, with little success, to catch House Finches to ship back to Michigan for my studies.
We had come to town to get supplies and were drawn to the man who was sitting next to
a tall stack of cages, each of which contained a House Finch. I couldn't help but notice
that he was knitting a mist nest as he sat in the busy market.*

*"How much for the finches?" I asked Michael to inquire. Embarrassingly, I speak
no language but English, and Michael (who was a doctoral student in the mammal
division at Michigan) had come along in exchange for travel and living expenses as a
companion and translator. I don't remember the amount in pesos but it was about
$10 US per bird. Michael explained that we were scientists from the United States
and that we'd be catching our own House Finches for our research. Jorge smiled and
said, "Come back if you need help." Three days and a pathetic few netted House
Finches later, we drove back to town to see what sort of help the grizzled old bird
catcher could provide.*

*As it turned out, Jorge Serrano was extremely knowledgeable about "los gorriones"
as he called them, and much better at getting them into a net than I was! At home on my
study sites, I establish feeding stations and easily catch finches in traps. In the field,
away from established feeding stations, my approach had been to put up nets in House
Finch habitat and hope that a bird would hit the net. Jorge took a much more active
approach to catching finches. The first day out with two of his sons, he guided us out of
town into an area with large agricultural fields, some of which had gone fallow. We
drove until, with his one good eye, Jorge spotted a flock of House Finches feeding out in
fields. Jorge quickly assessed the situation—how many birds and which way they were
moving. He then directed us like the captain of a SWAT team. Under Jorge's command,
we quickly and quietly moved a hundred yards ahead of the advancing flock of finches
and began to erect a line of mist nets.*

*My standard mist net set requires rebar, a mallet, and metal poles. At the very
least, in the field I cut saplings or bamboo for net poles. For Jorge this was frivolous
extravagance. He used the weed stems themselves as poles for the mist nets. Within
fifteen minutes he and his sons (despite our best efforts, Michael and I mostly got in the
way) had a line of five nets (a hundred and fifty feet of net) across the path of the
advancing finches. Then, directing us mostly with hand signals, Jorge indicated that we
should move in two groups around either flank of the finch flock. Once in position behind
the flock, Jorge directed us to move forward at a modest pace—fast enough to drive the
birds toward the nets, but not so fast as to flush them from the field. As I walked to the
net lane, I was astonished to see it full of birds. We had twenty-two House Finches in 90*

*minutes of work, after I had captured only ten finches in five full days of hard work in the mountains.*

*Jorge and his sons helped me catch birds for five days, keeping our captive birds alive and healthy at his house. He asked for no pay—I had none to give.*

*Speaking through Michael, he told me: "I've loved birds my entire life and I know that I am a bastard for what I do to them."*

*"Tell him," I said, "that I feel the same way."*

Why did various lineages of birds evolve brilliant patches of plumage coloration? This was a focus of discussion and investigation for Darwin and Wallace, Poulton and Huxley, O'Donald and Maynard-Smith and a thousand other biologists over the last 150 years. It was the question that initiated my studies of plumage coloration fifteen years ago, and it is the question that has fueled the fires of my curiosity ever since. Early on, I realized that even complete understanding of the current function of ornamental plumage coloration does not necessarily reveal the process by which these traits evolved. To investigate the evolution of a trait, one must adopt a historical perspective, assessing changes in the expression of traits across lineages and comparing the patterns of change in the character of interest to changes in other traits or to changes in the environment. This approach requires a phylogeny, which is a hypothesis of the history of relationships of the populations of birds under study, and it requires knowledge of the character states of the taxa being investigated. Enumerating the character states of taxa is relatively simple if the character of interest is plumage coloration—one need only to look at color illustrations in a book or to examine study skins in a museum. Deducing the character states across taxa for a trait such as female mate preference or for an environmental variable such as access to carotenoid pigments poses a bit more of a challenge. In this chapter, I will present the efforts to date to use a historical (comparative) approach to test hypotheses for the evolution of carotenoid-based plumage coloration.

Before I can present comparative studies that have tested hypotheses for the evolution of carotenoid-based ornamental plumage coloration, I first have to review the current hypotheses for how plumage traits have evolved. In chapter 1, I side-stepped the issue of explicit models for the evolution of ornamental traits through female mate choice because the focus of the first two sections of the book was on the present function of plumage coloration. This final chapter, however, deals explicitly with the evolution of plumage displays, and it is time to review the models of sexual selection that have been proposed since Darwin. Here, I present a brief verbal description of the principal models for the evolution of ornamental traits. For more detailed reviews of these models, including reviews of attempts to formalize these models mathematically, see Andersson (1994) or Møller (1994).

## Models of Trait Evolution via Female Choice

### Fisherian Model of Sexual Selection

As I reviewed in chapter 1, after Darwin (1859, 1871) proposed his idea that female mate choice drove the evolution of ornamental traits in animals, there was a century of almost universal denial that female choice was a selective force in the evolution of such traits. Much of the blame for this initial failure of the idea of sexual selection driven by female choice must lie with Darwin himself, for although he focused on an aesthetic sense in females that selected for ever more gaudy, beautiful, and extravagant traits in males, Darwin never explained the origin or maintenance of such a female preference (Blaisdell 1992, Cronin 1991). Without a logical basis for a female mate preference for ornamental traits, it didn't seem reasonable to propose that female choice was the selective force behind the evolution of ornamental traits (e.g., see Wallace 1889). Much as the lack of a mechanism hampered the acceptance of evolution as an explanation for species diversity in the first half of the nineteenth century, so too did the lack of a mechanism by which female choice could drive the elaboration of ornamental traits cause scientists to look elsewhere for an explanation for display traits.

In 1915, R. A. Fisher took the challenge of working through the logic of the evolution of ornaments from an initial rudimentary appearance to a peacock's tail or a bellbird's display. Fisher's initial effort was sketchy, but for the first time he proposed that genes for female preference for a specific male trait could increase and become fixed in a population if that trait was related to increased survivorship of males. He further outlined that, once female mate preference for a male trait became established in a population, the male trait could be elaborated simply because of the mating advantages gained by ornamented males. Furthermore, once a mating advantage became the primary benefit for having the trait, the trait could be elaborated far beyond that which would be optimal for simple survival and fecundity.

Fisher presented his model in more detail in his book, *The Genetical Theory of Natural Selection* (Fisher 1930, revised in 1958). The traditional trait to use as an example in explaining Fisher's process is tail length, and so I will use tail length in this discussion. As in Fisher's original model, the process begins with a trait that is associated with increased male survivorship; for instance, males with slightly longer tails fly slightly better than males with shorter tails, and therefore survive slightly better. Females in the population have a range of genetically based preferences for males, and a very small proportion of females happen to prefer the males with slightly longer tails. (This is the part of sexual selection models that, from Fisher onward, has been generally glossed over, but a predisposition to respond to a trait that does not yet exist in a population is often termed "sensory bias"; we'll return to this idea of a sensory bias driving the evolution of traits in a couple of paragraphs.) The females that prefer males with long tails make a better-than-average choice among males because long-tailed males survive better than average and pass their long-tailed genes to offspring who also survive better than average. As a result, the genes for longer tails and the genes for choosing longer-tailed males would increase in the population due to natural selection. Thus, Fisher proposed that the establishment of female preference for at least the antecedent to the ornamental trait, the sticking point for acceptance of female choice as a selective force, was a result of natural selection.

To this point, all Fisher had outlined was a simple process of natural selection, but Fisher understood that this simple scenario set the stage for the sort of trait evolution that Darwin had proposed but never justified. Once female preference for the trait is established in the population, males begin to benefit not just from the enhanced flying ability associated with longer tails but also by having a trait that causes them to be chosen as mates. Moreover, females who chose long-tailed males would benefit not just by having offspring with this survival advantage, but also through the mating advantage gained by long-tailed offspring. Thus, sexual selection—selection for a mating advantage—would begin to drive the evolution of the trait. The more the genes for preferring long tails spread through the population, the greater would become the mating advantage for longer tailed males (i.e., the stronger would become sexual selection on the trait). As Fisher (1958) stated, ". . . it is easy to see that the speed of development will be proportional to the development already attained, which will therefore increase with time exponentially, or in a geometric progression." Once this runaway stage of sexual selection was reached, the trait under selection would be elaborated rapidly. As the trait was driven by sexual selection further and further from a natural selection optimum, it would eventually reach a point at which the survival cost imposed by the elaborated trait outweighed the mating advantage created by the trait. At that point, further trait elaboration would cease. This model for trait evolution has been called the Fisherian or Runaway Model of Sexual Selection.

In his Runaway Model, Fisher provided for the first time a logical framework for how female mate choice could evolve and drive the evolution of ornamental traits in males. But despite the fact that Fisher was among the greatest population geneticists of the twentieth century, he never formalized his model of runaway sexual selection. That is, he never constructed the mathematical models needed to test whether and under what conditions his hypothesized runaway process would work. Shortly after the publication of Fisher's revised *The Genetical Theory of Natural Selection* (Fisher 1958), however, others took up the challenge of formally modeling the process. Peter O'Donald (1962) constructed the first population genetic model in which he looked at change in gene frequencies over generations under a runaway process, allowing not just the frequency of genes for male display to evolve but also the frequency of genes for female mate choice. O'Donald found that if he stuck strictly to the process outlined by Fisher, that "female preference . . . will have only a slight effect on the rate of sexual selection."

What really pushed the process forward, O'Donald discovered, was genetic linkage between the preference genes and the ornament genes. When a female chose to mate with long-tailed males, not only did her offspring inherit the long-tailed genes of their father but they also inherited the preference genes for long tails from her. Consequently, female choice for longer tails would lead to both an increase in tail length and an increase in preference for tail length. In turn, selection for increase in preference for tail length would lead to both an increase in the preference genes and an increase in the tail length itself. O'Donald pointed out that such assortative mating caused by female mate preference would lead to linkage disequilibrium—the genes for long tails and the genes for preference for long tails would begin to assort together. Even more than the mating advantages proposed by Fisher, the genetic association between preference genes and ornament genes would lead to an exponential rate of elaboration of ornamental traits.

Since Fisher and O'Donald, this process has been modeled in much more sophisticated ways. In general, the basic result has remained the same—on paper, with a number of simplifying assumptions, the runaway process works to drive the evolution of ornamental traits. One problem that has been identified in models of the runaway process is that if there is a cost to females of choosing males, then the runaway process will not proceed (Lande 1980). Because there should generally be some cost in the form of search time or increased risk of predation associated with virtually any female choice, some have viewed the cost of choice as a serious problem for the Runaway Model of Sexual Selection (Pomiankowski 1987).

## Honest Advertisement Models

The model laid out by Fisher and modified by O'Donald was a beautiful explanation for how female mate choice could lead to the evolution of extravagant ornamental traits. For the first time it provided a mechanism for how female mate choice could drive the evolution of ornamental traits, and it resurrected an interest in sexual selection after a century of neglect. Once Fisher's model was appreciated fully by the biological community and formalized by population biologists, it quickly became THE explanation for the evolution and maintenance of ornamental traits. To some theoreticians, the problem was solved. As Andersson wrote in 1986: "One might therefore argue that there is no need to ascribe importance to another genetic mechanism: The Fisherian one can explain what needs to be explained. Although Maynard-Smith does not express this view, there seem to be signs of it in other authors . . ." (Andersson 1986b).

Beginning with Wallace (1889), however, biologists had discussed the tendency for extreme expression of ornamental traits to be associated with individual health and vigor. In a more modern context, the idea that expression of ornamental traits might be associated with male condition had been briefly discussed by Trivers (1972) and Williams (1966). Fisher himself proposed that natural selection often led to female choice for male traits that indicated condition: "Consider then what happens when a clearly-marked pattern of bright feathers affords, in a certain species of birds, a fairly good index of natural superiority. A tendency to select those suitors in which the feature is best developed is then a profitable instinct for the female bird and the taste for this 'point' becomes firmly established among the female instincts" (Fisher 1915:187). Fisher proposed that females would use many such indicators of male quality in their assessment of mates, but he did not propose an explicit model by which a trait could be elaborated into an ornament and remain a reliable indicator of male quality. Clearly, however, the idea that expression of ornamental traits is associated with male health and condition had been batted around by evolutionary biologists since they first started to discuss the evolution of ornamental traits.

Indeed, to the many biologists observing animals with ornamental traits in the wild, by the 1980s the Runaway Model of sexual selection was being viewed as an incomplete explanation of ornamental traits. What field biologists were repeatedly observing was that the males with the most elaborate ornamentation were also the biggest, strongest, least parasitized, and most fit males in a population. There was no reason to expect such an association under the Runaway Model proposed by Fisher. The notion that the ornamental traits were simply arbitrary markers of attractiveness did not seem to fit a majority of observations of ornamental traits in nature.

The first real champion of the idea that ornamental traits arise, are elaborated, and are maintained because they serve as honest signals of male condition was Amotz Zahavi (1975, 1977). He proposed that ornamental traits are handicaps to male survival. According to Zahavi, only males in the best condition with the best genotype will be able to bear the burden of a well-developed ornament and still function in the population. By choosing males with well-developed ornaments, then, females choose as mates the fittest males in the population. This idea was widely appealing to empirical biologists who saw evidence for such honest advertisement in their observations of animals, but Zahavi's "Handicap Principle" met with initial harsh criticism from theoretical biologists who found in their genetic models that the process did not work (Bell 1978, Davis and O'Donald 1976, Maynard-Smith 1976, 1978b, 1985). The problem with Zahavi's idea that was identified in these initial mathematical models was that any benefit gained by the female in mating with a high-quality male would be outweighed by the cost of the handicap that would be imposed on offspring. As Zahavi (1977) pointed out, however, the handicap did not inflict a uniform cost on all males; it would, he proposed, be less of a burden to high-quality males than it would be to low-quality males. This is what Maynard-Smith (1991) and others have termed a "conditional handicap." With this adjustment in the assumptions of the model, the handicap model could be shown to work in mathematical models (Andersson 1986a).

Finally Grafen (1990a,b) developed a much more complex game-theory model in which males have variable quality, ornament display is variable and related to quality, males pay a cost for advertising that is related to their quality, females benefit by mating with a high-quality male, and female choice is variable with respect to ornament expression (they cannot detect quality directly). In this model, Grafen clearly showed that the concept developed by Zahavi, of ornamental traits evolving and being maintained because they provided information to females about the quality of males, could work. Although empiricists interpreting observations of wild animals had maintained that ornamental traits were signals of individual quality, it was not until the process was shown to work in formal models on paper that the idea became widely embraced among evolutionary biologists.

In retrospect, one of the primary problems with Zahavi's original model was that he considered only the cost of maintaining ornamental traits; an ornamental trait increased the risk of death for males that carried the trait. When discussing plumage coloration, for instance, Zavahi suggested that bright plumage increased predation risk so that by choosing bright males, females were choosing males who had survived despite this handicap of conspicuousness (Zahavi 1975:210). Zahavi's original presentation of the handicap model, which for a decade was the definition of honest signaling, did not explicitly consider the costs associated with the production of ornamental traits. This may seem like a small point, but a focus on maintenance costs rather than production costs of ornamental traits framed the model in a way that made the idea unworkable in early theoretical models.

What one typically observes in nature, however, is that the costs of ornamental traits are incurred primarily during production—carotenoids must be ingested for feathers to be pigmented (see chapter 5); large dipteran prey must be captured for long tails to be grown (Møller 1991); and good nutrition must be available for the development of a well-structured brain that can produce elaborate song (Nowicki et al. 2000). The condition of an individual constrains the production of such ornaments, making the signal honest. Moreover, most ornamental traits appear to pose, at worst, a minimal burden once they are produced, so the term "handicap" becomes somewhat misleading. Although "conditional handicaps" are often proposed to have a viability cost associated with their maintenance, they can also work by having minimal maintenance costs but high production costs. This view of honest advertisement, with a focus on production rather than maintenance costs, was first articulated in a verbal model by Kodric-Brown and Brown (1984).

A final, often overlooked problem with the handicap model developed by Zahavi as well as the modified Honest Advertisement or Indicator Models that have followed is that they provide no mechanism for the elaboration of ornamental traits. When a trait arises that has variable expression and for which expression is associated with male quality, it would be expected to spread in the population, as described by Zahavi (1975) and Kodric-Brown and Brown (1984). This would be the antecedent to an elaborate ornamental trait. However, continued selection on such a trait with non-heritable variation would affect the good genes correlated with the expression of the character (if good genes and not simply resources are being signaled by the trait), but there could be no evolutionary response to selection by the display character itself (Dominey 1983). The question, then, is how can such an antecedent ever evolve into a peacock's tail or a House Finch's color display? In the models of Andersson (1986a), Grafen (1990b), and others, a single allele for trait expression spreads through the population. There is no mechanism for further trait elaboration once this "ornamentation allele" is fixed. Because most models of honest advertisement provide no mechanism for trait elaboration, they are not legitimate alternatives to the Runaway Model of Sexual Selection.

I attempted to provide a mechanism for trait elaboration under the Honest Advertisement Model (Hill 1994c). I proposed that ornamental traits would not be elaborated from their initial rudimentary forms if signaling remained honest. (This is very much an echo of Fisher's (1915) idea that most signals of condition used by females in mate choice, such as his examples of human complexion or breath odor, remain simple and reliable signals of quality that are not elaborated.) However, I proposed that the evolution of dishonest signaling strategies by males would cause female choice to drive the evolution of the display trait into more complex and costly forms: ". . . once female mate preference for such a viability-indicating trait and male potential to display such a trait are fixed in a population, there will also be strong selection on males to find shortcuts to trait expression (i.e., to express the trait regardless of their quality). Whenever an allele arises that allows for expression of a viability-indicator trait at reduced cost to males, it will spread rapidly through the population. Thus, just as with predator–prey, host–parasite (Van Valen 1973), and even other male–female interactions (Trivers

1972, Eberhard 1993), one could imagine an evolutionary race in which females are selected to choose mates by assessing traits that correlate with male fitness, and males are selected to find ways to express these characters at reduced cost" (Hill 1994c). By this model, traits are elaborated to retain links between trait expression and male condition, and traits remain reliable indicators of quality throughout the process of elaboration.

Runaway and Honest Advertisement Models have often been discussed as diametrically opposed views of how ornamental traits evolve, but in my view as these models become better developed they begin to describe the same process of ornament evolution. Fisher's original model begins with a trait that is related to male fitness. Honest Advertisement Models, if they are to include a mechanism for trait elaboration, proceed through a series of evolutionary steps, each of which resembles a runaway process (Figure 11.1). Males evolve new or more elaborate display traits in response to the mating advantage created when females alter their mate preferences in response to a loss of information content in the current male trait (Hill 1994c). Runaway models can be formulated so that expression of the ornamental trait is linked to direct benefits for females (Kirkpatrick 1985) making the runaway process very much like the process envisioned by Zahavi. The more Runaway and Honest Advertisement Models of sexual selection are modified and made to better account for what is observed in nature, the more they seem to converge in their proposed mechanisms of action.

### Species Recognition and Sensory Bias Models

Runaway and Honest Advertisement Models of sexual selection remain the primary models to explain the evolution of ornamental traits. There are, however, a few alternative models that bear mentioning. Wallace (1889) first proposed that ornamental traits and female preference for the traits evolve because they enhance the ability of females to recognize males within their species and hence avoid non-adaptive matings outside the species (see also chapter 1). This idea became the standard explanation for ornamental traits through the first two-thirds of the twentieth century and was advocated by Dobzhansky (1937), Huxley (1942), and Mayr (1942). With the acceptance of female mate choice as a viable selective agent in wild populations, however, species recognition has all but been forgotten in recent discussion of sexual selection. There is good reason to reject species recognition as a complete explanation for extremely elaborate structures or behaviors. As Maynard-Smith (1991) stated: "it would be absurd to suppose that a male nightingale must sing like that [very elaborate song] in order for a female to tell that he is not a willow warbler. Indeed, species recognition would be better served by a single song type, and not by a varied repertoire." But there are other cases where species recognition may play a role in plumage evolution, and this is certainly a hypothesis that warrants testing when the data to do so are available.

More recently, an entirely different mechanism for the evolution of female choice has been proposed. This new model proposes that some ornamental traits may arise through biases in the sensory systems of females. For instance, there might be strong natural selection for individuals of a species to search for red foods, resulting in an acute ability to detect red and a predisposition to respond

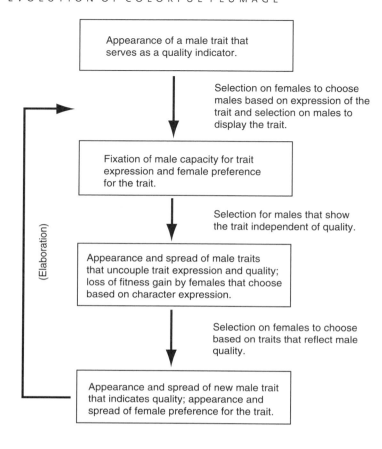

Figure 11.1. A mechanism by which selection for honest signals of male quality might lead to the evolution of elaborate display traits. By this model, trait elaboration is driven by a basic conflict of interest between males, who gain by having well-developed ornaments regardless of quality, and females, who gain only if ornamental traits reliably signal quality. Evolutionary events are indicated in the boxed portion of the figure, and the arrows connecting boxes show forces of change. From Hill (1994c).

in a positive manner when red is detected. Once these detection and response systems are in place, they predispose individuals seeking mates to be attracted to red displays (Endler and Basolo 1998, Ryan and Keddy-Hector 1992). This Sensory-Bias Hypothesis proposes truly arbitrary mate choice with respect to fitness attributes of males because the mate preferences of females evolve prior to the trait in males. The Sensory-Bias Hypothesis has been proposed as a model for trait evolution that competes with Runaway and Honest Advertisement Models. In my opinion, however, the Sensory-Bias Hypothesis is more productively viewed as a proximate mechanism that establishes female mating preferences in populations (Hill 1995a). In all models of sexual selection, at the initial stages of the evolution of a new display trait, some females in a population must have a predisposition to respond to the male trait. This is where sensory bias plays a key role in the evolution of ornamental traits by establishing an initial preference on which sexual selection can act. By itself, however, the Sensory Bias Model is not a complete explanation for the evolution of elaborate ornamental traits.

## The Color Equivalent of a Long Tail

Before I proceed with comparative tests of models of sexual selection, it is important that I establish what makes one carotenoid-based color display more or less elaborate than another. In studies that deal with morphological traits that differ in size, like the tail length of Long-tailed Widowbirds, it is obvious what is a greater and what is a lesser expression of the ornamental trait. The issue becomes much less clear, however, when one is dealing with a color display. Across a suite of potential color hues, which is the equivalent to a long tail?

A basic premise throughout this book has been that more saturated plumage coloration is a more elaborate form of ornamentation than less saturated plumage and that red is more elaborate than yellow. The assumption that more saturated color is a more elaborate form of pigment display is relatively easy to justify. Saturation is a function of the concentration of pigment put into feathers (see chapter 4). The logical premise of most hierarchies of ornament expression is that more is better; longer tails are more elaborate than shorter tails; larger repertoires are more elaborate than smaller repertoires; and, more pheromones are more elaborate than fewer pheromones. It follows then, that more carotenoid molecules concentrated in plumage comprise a more elaborate ornamental display than fewer carotenoid molecules.

But what about hue? How can I justify claiming that, as a general principle, red carotenoid coloration is a more elaborate condition than yellow carotenoid coloration? First, as presented in detail in chapter 3, red carotenoid pigments are often the metabolic derivatives of yellow carotenoid pigments. Red pigments are rarer in most environments than yellow carotenoid pigments (Goodwin 1980, 1984), so to display red carotenoid pigmentation an organism must either seek out scarce red pigments or convert yellow pigments to red pigments at some metabolic cost (Hill 1996a). Strictly from a biochemical and physiological standpoint therefore, there is reason to assume that red is a more costly display than yellow (Hill 1996a).

The basic patterns of color display that I've observed in House Finches also support the assertion that red is a more elaborate form of color display than yellow. Hue and saturation are correlated such that redder males tend to be males with more saturated plumage (see Figure 3.6). The plumage of males tends to become redder between their first and subsequent breeding seasons (Figures 2.3 and 2.4) (Hill 1992), just as the tails of Barn Swallows get longer and the antlers of red deer get larger as these animals age (Clutton-Brock et al. 1982, Møller 1988). Redder male plumage is also associated with better condition (see chapter 5). And, as was presented in detail in chapter 6, females prefer to mate with red rather than yellow males (Hill 1990, 1991). Observations of House Finches clearly support the idea that red is a more exaggerated form of color display than orange or yellow (Hill 1996a).

A comparative analysis across cardueline finches also provided support for the assertion that red is a more elaborate form of color display than orange or yellow. For morphological traits in which the direction of elaboration is unambiguous, juvenal males almost invariably show reduced expression of ornamental traits relative to adult males and, with rare exceptions that are generally associated with polyandrous mating systems, females show equal or reduced ornamentation relative to adult males (Ligon 1999). For example, relative to adult males, juvenal males and females typically show reduced tail streamers, horns, antlers, wattles, and song production (Andersson 1994). If red carotenoid pigmentation is an inherently more costly display than orange or yellow carotenoid pigmentation, then I predicted that in cases in which hue varies among age/sex classes, juvenals and females would show plumage that was less red than adult males. I tested this prediction by systematically tallying the plumage hue of adult male, juvenal male, and female cardueline finches. I found that, without exception, when adult males and adult females or adult males and juvenal males differed in plumage redness, females were less red than males and juvenal males were less red than adult males (Hill 1996a).

As a final test of the hypothesis that red is a more elaborate expression of ornamental plumage coloration than is yellow, Alex Badyaev and I compared patterns of redness and sexual dichromatism across species of cardueline finches. Cardueline finches vary substantially in the degree of sexual dichromatism, and this variation in sexual dichromatism is largely a function of the amount of ornamentation displayed by females (rather than a loss of coloration by males) (Hill 1996a). Assuming that expression of female ornamental coloration is a function of a balance between the mating advantages of displaying the trait and the costs of producing the trait, we predicted that, across species, the degree of female ornamentation, as measured by degree of sexual dichromatism, should correlate with the cost of producing the ornament. In other words, if ornamental plumage becomes more costly to produce as it becomes redder, then sexual dichromatism should increase as male plumage gets redder. Thus, we predicted a positive correlation between the degree of sexual dichromatism and redness of plumage. Across cardueline finches we found a significant relationship—as plumage gets redder, sexual dichromatism increases (Figure 11.2) (Badyaev and Hill 2000a, Hill 1996a). This pattern persisted when we corrected for phylogeny, but, interestingly, there was no equivalent pattern when we compared extent of melanin-based ornamental coloration to sexual dimorphism. Thus, patterns both within House Finches and among species of cardueline finches support the idea that redness is a measure of the production costs of carotenoid-based ornamental traits.

## Female Mate Preferences Among Populations and Subspecies

Tests of models for the evolution of ornamental plumage coloration focus on how male traits co-evolve with female mate preferences. To conduct effective comparative tests of the models of trait evolution, therefore, I had to know the preferences of females from various populations and subspecies of House Finches relative to

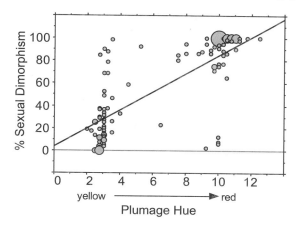

Figure 11.2. The relationship between the hue of male carotenoid pigmentation and the degree of sexual dichromatism of carotenoid ornamentation for 107 cardueline finch species with carotenoid-based plumage coloration ($r_s = 0.70$, $P = 0.02$). This relationship suggests that plumage redness is a measure of the cost of trait production in cardueline finches. Adapted from Hill (1996a).

male ornament display. Unfortunately, you cannot simply open a field guide or peruse a tray of study skins to determine the mating preferences of females. You have to ask the living females what they prefer by presenting them with a range of male phenotypes in a mate-choice experiment. So, I spent several years capturing females from different populations, transporting them back to my aviaries at the University of Michigan, and testing their mate preferences relative to variation in male plumage coloration and patch size.

For mate-choice trials, I captured females from five populations: (1) Ann Arbor, Michigan; (2) San Jose, California; (3) Alviso, California; (4) Hawaii Island, Hawaii; and (5) Chilpancingo, Guerrero, Mexico. Males in the Michigan, San Jose, and Guerrero populations averaged bright red in plumage coloration, while males in the Hawaiian and Alviso populations averaged drab yellow/orange (see Figure 10.7). House Finches from Michigan, San Jose, Alviso, and Hawaii belong to the *frontalis* subspecies, in which males have medium patches of ventral carotenoid pigmentation. House Finches from Guerrero belong to the *griscomi* subspecies, in which males have small patches of ventral pigmentation (Figure 10.9). I will begin by describing my tests of the preferences of females from the various bright and drab *frontalis* populations relative to the plumage redness of prospective mates. I will then describe my tests of the preferences of females from *frontalis* and *griscomi* populations relative to both plumage coloration and patch size of prospective mates.

## Mate Preferences Related to Plumage Hue Among frontalis *Populations*

As I documented in the last chapter, various populations of *frontalis* House Finches vary substantially in the mean expression of male plumage coloration. The obvious but not-so-easily-answered question related to this geographic variation in male plumage coloration is: do the mate preferences of females co-vary with the expression of male plumage among these populations? Answering this question was a critical first step in understanding how carotenoid-based plumage coloration evolved in the House Finch.

In previous research, I had established that laboratory tests of female mate choice based on female association with males was a reasonable way to assess female mate preferences (see chapter 6). So, I used this approach in my tests of the preferences of female House Finches from different populations. I presented females from Michigan, Hawaii, San Jose, and Alviso with four stimulus males that had medium ventral patches ranging in coloration from drab yellow to bright red (Hill 1994a). This experiment followed the protocol outlined in detail in chapter 6. Briefly, twelve different Michigan males were used as stimulus males. These twelve males were assigned randomly to color types such that three males represented each color type. In this way, I was able to uncouple plumage coloration from age, size, or other male traits that might typically be associated with color expression.

I found that, regardless of the appearance of males in their population, females from all populations showed a significant preference for the reddest male presented. Specifically, females from Michigan and San Jose, where males average bright red in coloration, showed a preference for bright red males; likewise, females from Hawaii and Alviso, where males average much drabber in coloration, showed a significant association preference for the reddest male presented to them (Hill 1994a) (Figure 11.3). So, despite the fact that expression of male plumage coloration has changed from red to yellow/orange in some *frontalis* lineages, females in these populations have retained a preference for red males. These results are perhaps not surprising given that finches in the two drab populations are recently derived from a bright coastal California population. But the observation that females were consistent in their preference for red males, even when males in their populations rarely display red plumage, will be important when we begin to unravel how patterns of colorful plumage evolve.

## Mate Preferences Related to Patch Size Within and Between Subspecies

So, females from all populations of *frontalis* finches, regardless of whether males averaged bright or drab in coloration, showed a preference for males with bright red plumage. The hue and saturation of plumage coloration is a conspicuous component of the color display of males, but males also vary substantially in patch size—the extent of red coloration on the ventral plumage. Patch size varies somewhat within populations, but, as described in the previous chapter, the most striking variation is between subspecies of House Finches. In some subspecies, males show much reduced expression of ventral pigmentation. Would females from these small-patched populations prefer males with small or large patches?

To answer this question, I determined the mate preferences of females from Hawaii, where males average drab with large patches, from Michigan, where males average bright with large patches, and from Guerrero, where males average bright

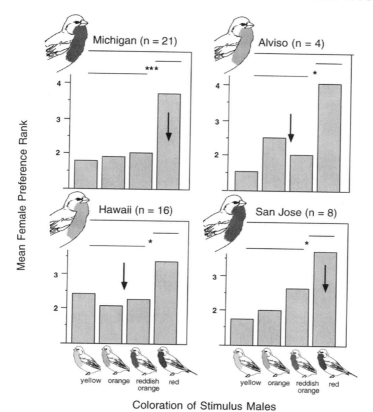

Coloration of Stimulus Males

Figure 11.3. The mate preferences of captive female House Finches from different populations in the subspecies *frontalis* relative to the ornamental coloration of prospective mates. Samples sizes are the number of females whose preference was tested in each experiment. Males in the Ann Arbor, Michigan, and San Jose, California, populations averaged bright red in coloration; males in the Pohakuloa, Hawaii, and Alviso, California, populations averaged drab orange/yellow (as indicated by the arrows in the figures). Regardless of the appearance of males in their population, females from all four populations showed a significant preference for males with bright red coloration. In each experiment, three males were used to represent each of the four plumage types presented to females. Lines above the bars indicate which males share statistically similar ranks, where successive levels are statistically different from one another. *P < 0.05, ***P < 0.001. Adapted from Hill (1994a).

with small patches, relative to both patch size and plumage coloration. To keep the main points of these experiments clear, I'm going to refer to the males from the Hawaiian and Michigan populations as large-patched, and the males from the Guerrero population as small-patched. Technically, the Hawaiian and Michigan populations have medium patches (see Figure 10.9), but using the term "medium-patched" in describing males in these experiments can leave the reader with the impression that females chose an intermediate ornament expression, among the ornamented types presented. Within the context of these experiments, the patch size of males from the *frontalis* populations in Michigan and Hawaii were the largest.

I gave these females a choice of males with ventral patches that were either large bright, small bright, large drab, or small drab (Hill 1994a). One problem with manipulating plumage coloration of male House Finches is that I know of no way to make the patch size of a male smaller. For patch-size manipulations, the only way to proceed is to start with small-patched males and then extend patch size. By necessity, then, the stimulus males in all experiments addressing patch size were *griscomi* males, and in experiments testing the preferences of females from Hawaii and Michigan, *frontalis* females were presented with *griscomi* males. As I mentioned in chapter 10, however, I was able to get *frontalis* females to pair and mate with *griscomi* males in captivity. The greatest risk of presenting females with males of a foreign subspecies is that the females would not respond to the males, and hence make no mate choice. Fortunately, *frontalis* females presented with *griscomi* males showed as strong of an association preference as did *frontalis* females presented with *frontalis* males.

When I presented females from these three populations with males that varied in patch size and coloration, I found that females from all populations showed a preference for the most brightly plumaged, largest-patched males (Hill 1994a) (Figure 11.4). Most significantly, even though *griscomi* males displayed small

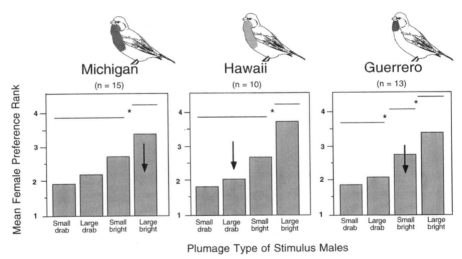

Figure 11.4. Mate preferences of captive female House Finches from different populations relative to the plumage redness and patch size of males presented in choice trials. Sample sizes are the number of females whose preference was tested in each experiment. Hawaiian and Michigan females belong to the subspecies C. m. *frontalis* and males in their populations have large drab and large bright patches of color, respectively (as indicated by the arrows in the figures). Guerrero females belong to the subspecies C. m. *griscomi*, and males in their populations have small bright patches of color. Females from all three populations showed a significant preference for males with large bright patches of color. Lines above the bars indicate which males share statistically similar ranks, where successive levels are statistically different from one another. *P < 0.05. Adapted from Hill (1994a).

patches of coloration, *griscomi* females showed consistent preferences for large, bright patches of coloration. From these experiments I concluded that the preferences of females did not necessarily match the display of males—*griscomi* females showed a preference for a pattern of color display typical of *frontalis* males rather than the pattern typical of their own *griscomi* males.

To further test female preference in relation to patch size and coloration, I wanted to simplify the number of variables being considered, so I conducted two-choice trials. In these experiments I tested females from the same populations—drab with large patch size (Hawaii), bright with large patch size (Michigan), and bright with a small patch size (Guerrero). First, I held plumage coloration constant by giving females a choice of males with either small bright or large bright patches. Females from all three populations showed a preference for large bright over small bright patches (Hill 1994a) (Figure 11.5). Again, this was particularly interesting

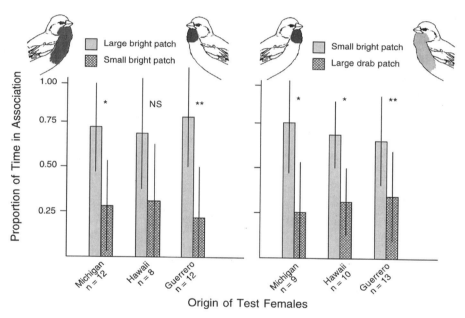

**Figure 11.5.** Mate preferences of female House Finches relative to the plumage redness and patch size of the two males presented to them in choice trials. Shown are the mean (± SD) proportion of times spent by females in association with stimulus males. Hawaiian and Michigan females belong to the subspecies *C. m. frontalis*, and males in their populations have large drab and large bright patches of color, respectively. Guerrero females belong to the subspecies *C. m. griscomi*, and males in their populations have small bright patches of color. Females from all three populations showed a preference for large over small patches of carotenoid pigmentation when hue and brightness were held constant (left figure). When females were forced to choose between large drab or small bright patches of coloration, females from all three populations chose small bright patches (right figure) (*P < 0.05, **P < 0.01; Wilcoxon matched-pairs, signed-ranks test). These results suggest that plumage coloration is the primary criterion and patch size a secondary criterion in female mate choice. Adapted from Hill (1994a).

with regard to females from the Guerrero population because males in their population displayed small patches of coloration, but Guerrero females showed a clear preference for large patches of color. Finally, I made females choose between patch size and coloration by giving them a choice between males with small bright versus large drab patches. Females from all populations showed a significant preference for small brightly colored patches over large drab patches (Hill 1994a). These experiments suggest that plumage coloration is the primary criterion used by female House Finches in choosing mates, and that patch size is a secondary criterion.

## Assessing Models for Trait Evolution

How can these observations of female mate preference and male ornament display in various populations of House Finches be used to test hypotheses for the evolution of ornamental coloration? The Honest Advertisement Model proposes that ornamental traits evolved specifically as honest signals of quality, where signal honesty is maintained by the costs of producing or maintaining the trait. I've already presented evidence that, within populations of House Finches, expression of carotenoid pigmentation is tied to the costs of producing the color display (see chapter 5). The Honest Advertisement Model proposes that, once an ornamental trait like bright coloration exists, subsequent changes in the color display will evolve in response to changes in the cost of trait production (Hill 1994c). If the costs of trait production decrease (e.g., if access to carotenoid resources increase or exposure to parasites decrease), then the ornament will have to become more elaborate to remain honest. Conversely, if the costs of trait production increase, the trait will have to become less elaborate for males to meet the costs of production (Hill 1994c). It follows that if changes in patch size or hue of color display in lineages evolved through an honest advertisement process, then such changes should be linked to changes in the cost of trait production.

While the predictions of the Honest Advertisement Model are clear, testing this hypothesis is difficult because it is difficult to assess the relative cost of trait production for different populations of wild House Finches. In my comparative studies, I focused on carotenoid access as the key element in the cost of carotenoid display. I tested whether the elaborateness of carotenoid display among House Finch populations was related to the availability of carotenoids in different environments.

### Color Expression Among frontalis Populations

Let's begin with what I observed among populations of *frontalis* House Finches. For these populations, we have a written record of the introductions of the various populations, and therefore we know the phylogenetic relationships of these House Finch populations (Figure 11.6). We even have a known ancestral condition for mean redness because Grinnell (1911) looked at mean redness of male House Finches in coastal California at the end of the nineteenth century and the beginning of the twentieth century, before or at about the time that various populations split

from the parent population. Grinnell found that in this ancestral California population males were on average bright red in coloration. So we have a situation in which
independently in two populations, mean male coloration shifted from red to yel-
low/orange. In both of the drab populations, however, females showed a strong
mate preference for bright red males. Thus, a reduction in male ornamentation in
the Hawaiian and Alviso populations has occurred independent of any change in
female preference.

This comparison among *frontalis* populations demonstrates that there can be
substantial change in male ornament expression independent of any change in
female preference. In the case of these *frontalis* populations, the change in mean
male appearance is entirely phenotypic; there appears to have been no genetic
change in male capacity to express bright coloration among populations (see chapter 10). Moreover, because males in all of these populations have the capacity to
express bright red plumage coloration, I interpret the failure of males in some
populations to be red as evidence that the costs of producing bright red plumage
has increased in those populations. Such a change in the cost of trait production
would exert a selective pressure on males, and this is where the plumage coloration/
patch size hierarchy becomes important. If it is more important for a male to
display a brightly pigmented plumage than to display a large patch of pigmentation,
then in environments such as Hawaii and Alviso, where the costs of carotenoid
production have increased, males would benefit if they had smaller patches that
allowed them to more efficiently concentrate available pigments. Males in Hawaii
and Alviso apparently have patch sizes that are larger than would be optimal in
their environments, but House Finches have been in these environments for too
short of a time for a smaller patch size to have evolved.

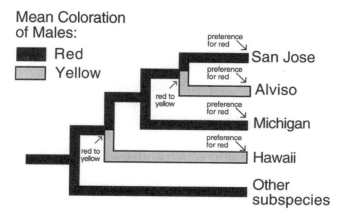

Figure 11.6. The known historical relationships of populations of House Finches in the
subspecies *C. m. frontalis*. In two lineages, male coloration has changed from the known
ancestral condition in which males average red to a condition in which males average drab
yellow/orange. Females in four populations, including the two populations in which males
are drab, prefer as mates males with bright red plumage coloration. The reconstruction
indicates that expression of male ornamental coloration can change independent of any
change in female mate preference.

What about female mate preference and male ornament display in populations where there has been ample time for an evolutionary response to the local environment?

## Subspecies Comparisons

Phylogenetic analysis indicates that the small patch size displayed by males from the *griscomi* population is derived from a larger-patched ancestral state (Hill 1994a, 1996b) (Figure 11.7). Moreover, unlike the phenotypic changes in expression of mean male plumage coloration among populations of *frontalis* males, the patch-size differences between *frontalis* and *griscomi* males are due to fixed genetic differences between the subspecies (see chapter 10). The reduction in mean male patch size in the *griscomi* lineage, therefore, represents an evolutionary change in ornamental display by males. Laboratory mate-choice experiments, however, show that females

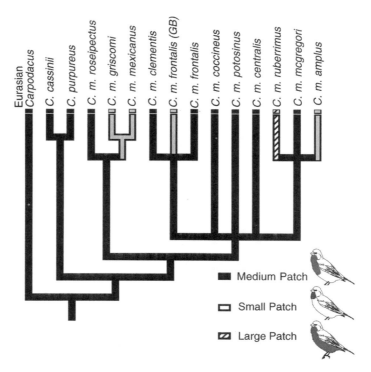

Figure 11.7. A composite phylogeny of House Finch lineages based on a cladistic analysis and biogeographical and morphological evidence (Hill 1996b). *C. m. frontalis* (GB) is the Great Basin population of *C. m. frontalis*. Evolution of the character patch size (extent of ventral carotenoid pigmentation) was traced on the phylogeny and the states of interior nodes were optimized according to Farris (1970). Small patch size has evolved independently from a larger-patched ancestral state three times. Females in the *griscomi* lineage retain the preference for large patches of red (see Figures 11.4 and 11.5) despite the reduction in patch size in males in their population.

from the *griscomi* population retain a preference for larger patch size, the ancestral condition. The reduction in ornamentation in males in the *griscomi* population occurred independent of any change in female mate preference in this population.

The lack of congruence between what females prefer and what males display is not consistent with basic predictions of either Sensory Bias or Species Recognition Models of sexual selection. Sensory bias provides no mechanism for anything other than simple unidirectional elaboration of a male trait: females have a predisposition to respond to a specific sensory stimulus (e.g., red) and this causes the trait to spread in the population if and when the trait arises. By this model, one should either see a female preference for a male trait that does not yet exist or see a close link between what is preferred by the female and what is displayed by the male (Basolo 1990, Ryan et al. 1990, Ryan and Keddy-Hector 1992). The model cannot account for a reversal in ornament elaboration, such as I observed in three populations of House Finches (Hill 1995a). The patterns of mate preference relative to male ornamentation that I observed among populations also falsify the Species Recognition Model. This model proposes that ornamental traits evolve to facilitate recognition by females of members of the same species or population (Wallace 1889), and I observed *griscomi* females preferring the phenotype of *frontalis* males.

Identifying clear predictions of the Runaway Model of Sexual Selection has always presented a challenge (Andersson 1994, Bradbury and Andersson 1987). However, a basic component of this model is that female preference and male ornamentation co-evolve as a result of a genetic correlation between the traits. Although lags in the response to changes in female preference may occur, in general, one would expect to see congruence between the mean preference of females and the mean appearance of males (Hill 1994c, Houde 1993, Houde and Endler 1990). The evolution of display traits need not always be toward more elaborate expression; the runaway process can lead to a reduction as well as an increase in the size or elaboration of traits (Kirkpatrick 1982, Lande 1981), but male appearance should track female preference (Houde 1993). Most importantly for this study, under the Runaway Model, one would not expect to see a decrease in the elaborateness of an ornamental trait without a corresponding change in female ornament preference (Hill 1994a, c, Houde 1993). Thus, the reduction in the patch size of males in the *griscomi* lineage without a corresponding change in female mate preference is not consistent with the process of trait evolution proposed by the Runaway Model (Figure 11.8).

So, three models—Sensory Bias, Species Recognition, and Runaway—do not seem to explain the observed patterns of change in male ornamentation independent of a change in female preference. What about the Honest Advertisement Model? Under the Honest Advertisement Model, female mate preference drives the evolution of display characters in males, and, generally, the mean appearance of males is expected to track the mean preferences of females (Andersson 1982b, 1986a, Kodric-Brown and Brown 1984, Nur and Hasson 1984). However, fundamental to the Honest Advertisement Model is that signal honesty is maintained by the costs of producing or maintaining ornamental traits. Trait expression in males can be reduced or enhanced in response to changes in the costs of acquiring the display trait (e.g., a change in carotenoid abundance) independent of any change in female preference (Hill 1994a, c, Kodric-Brown and Brown 1984). Thus, at least

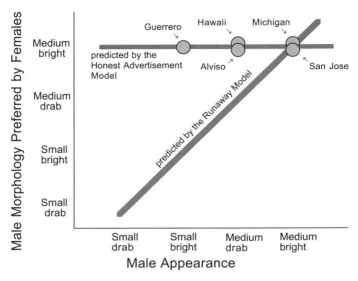

Figure 11.8. A graphical representation of predictions of the Honest Advertisement and Runaway Models of Sexual Selection. The Runaway Model predicts congruence between the ornamental traits displayed by males and the traits preferred by females. The Honest Advertisement Model predicts that females should always choose the most elaborate expression of the ornamental trait. The circles in the figure show the observed preferences of females from populations of House Finches in which males vary in carotenoid ornamentation. The observed pattern fits the predictions of the Honest Advertisement but not the Runaway Model.

under some circumstances, the Honest Advertisement Model predicts a reduction in the magnitude of male ornamentation without a change in female mate preference. This prediction is unique to the Honest Advertisement Model; it provides an opportunity to distinguish between Runaway and Honest Advertisement Models (Hill 1994a, c); and, it is consistent with the reduction in plumage ornamentation independent of any change in female mate preference that I observed in House Finches (Figures 11.4 and 11.7).

## Evidence for High Production Costs of Red Plumage in the Small-Patched Populations

The pattern of expression of male plumage coloration among *frontalis* populations suggested that the costs of producing ornamental coloration varied across environments such that the high cost of trait production in some environments resulted in most males in those environments being poorly ornamented. With such condition-dependent ornaments, females always benefit by choosing males with the most elaborate expression of the trait (Hill 1994c). There is no need for female mate choice to change in response to the change in male plumage expression. However, under the same conditions, there may be strong selection on males to change their expression of plumage ornamentation in response to changes in the cost of trait production. If a small area of intense carotenoid pigmentation is more stimulating to females than a large area of diffuse pigmentation, and if males trade off using carotenoids to make a patch of color more intense versus using carotenoids to make a patch larger, then males should respond to increased production costs of carotenoid pigmentation by reducing patch size and concentrating their carotenoid resources into a smaller patch of coloration. As described above, it appears that males in the Alviso and Hawaiian populations are displaying a patch size that is suboptimal for the environments they currently inhabit—males in these populations would do better if they concentrated pigment into small patches of color. Presumably, in the past, males in southern Mexico, Guadalupe Island, and the United States Great Basin were subjected to similar selective pressure to reduce patch size and, in response to such selection, the mean patch size of males decreased in these populations.

The hypothesis that a change in the production costs of carotenoid display has driven the evolution of patch size in subspecies is testable. In the next section, I present four independent lines of evidence—the carotenoid content of diet samples, patterns of carotenoid deposition, expression of carotenoid coloration in females, and patterns of delayed plumage maturation across subspecies—that support the assertion that the cost of carotenoid ornamentation is higher for populations with reduced patch size.

## Carotenoid Content of Diets

In chapter 5, I presented data on the plumage pigments of male House Finches with a focus on individual variation. At that time, I did not explain why we went all the way to Guerrero, Mexico, to collect birds for this study instead of focusing on much more accessible populations in the United States. Given the previous discussion on the hypothesized effects of the cost of trait production on the expression of colorful plumage, the reasons for sampling the diets of small-patched *griscomi* males as well as much more accessible larger-patched *frontalis* populations should now be more obvious. Part of our goal was to test the idea that males from the small-patched *griscomi* population had access to fewer carotenoid pigments than males from the larger-patched *frontalis* population. As described in chapter 5, we collected the gut contents of males in both Guerrero and central California and Caron Inouye determined the concentration of total carotenoids in the gut samples (see Inouye 1999 for methods).

We found that, as predicted, *frontalis* males had a significantly higher concentration of carotenoid pigments in their food than did *griscomi* males (Hill et al. 2002) (Figure 11.9). This observation supported a basic prediction of the Honest Advertisement Model that the cost of carotenoid pigmentation would be higher in the population with reduced ornament expression. Moreover, recall that on a standardized diet *griscomi* males grew more intensely colored feathers than did *frontalis* males (see Figure 10.11). Recall also that despite the low carotenoid content of their diets, wild *griscomi* males had more intensely pigmented feathers than wild *frontalis* males (Figure 10.10). These observations support the idea that there is a trade-off between having larger patches of coloration and having more intensely

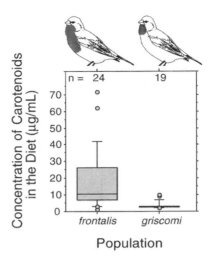

Figure 11.9. The mean concentration of carotenoid pigments in diets of males from two subspecies of House Finches. Males from the medium-patched *frontalis* population had a significantly higher concentration of carotenoids in their gut contents than did males from the small-patched *griscomi* population ($Z = -4.43$, $P = 0.0001$, Mann-Whitney $U$-test). This pattern supports the hypothesis that change in patch size is driven by a change in the cost of producing the trait. Adapted from Hill et al. (2002).

pigmented patches of color. Moreover, these observations suggest that by having smaller patches of coloration, males in the *griscomi* population are able to produce brightly colored feathers despite having access to low quantities of carotenoids in their diet. These data provide key support for the central prediction of the Honest Advertisement Hypothesis. Unfortunately, this diet analysis is a relatively weak test of the prediction that change in resource abundance drove the change in expression of ornamental plumage coloration in male House Finches because we were able to compare only two populations.

*Depositional Patterns*

If carotenoid pigments were more limiting for males in some populations than males in other populations, we would expect males in the carotenoid-limited populations to be economical in their use of carotenoids. It is interesting, therefore, that one of the most striking features of carotenoid deposition among *griscomi* males, which distinguished them from males in all of the medium- and large-patched populations, is the discreteness of their plumage patches (Hill 1993a, Moore 1939). In *frontalis* males and males from all of the medium- and large-patched populations, the patches of colored feathers are not very sharply demarcated. They have fuzzy borders with carotenoid pigment "bleeding" onto feathers outside the bright patches of coloration (Figure 11.10). It is as if excess carotenoids are allowed to wash across the plumage, extending the surface area with red when abundant red carotenoids are available.

Figure 11.10. Differences in the discreteness of colored feather patches in male House Finches from the small-patched *griscomi* population (left) and the medium-patched *frontalis* population (right). In males from the smalll-patched populations *griscomi* and *mexicanus*, patches of carotenoid pigmentation are discrete with no carotenoid bleeding. In *frontalis* males and males from all medium-patched populations, carotenoid pigmentation bleeds out of colored patches. The discrete patches of *griscomi* males are consistent with the idea that they have access to limited carotenoid resources and use their carotenoid resources efficiently.

The plumage pattern of the "Sur" House Finches, including the *griscomi* population, is quite different. The patches of ornamental coloration on the crown, breast, and rump of males in these populations are sharply demarcated. All carotenoid pigmentation of feathers occurs within the bounds of these patches—there is no "bleeding" of carotenoid pigments (Hill 1993a, Moore 1939) (Figure 11.10). Moreover, the discreteness of the patches of males in these populations appears to reflect a fixed genetic difference between these populations and large- and medium-patched populations to the north. There is no variation among males within these populations in expression of patch discreteness (all *griscomi* males display discrete patches regardless of diet (see chapter 10) and all *frontalis* males with red plumage show carotenoid bleeding). The hybrid *frontalis/griscomi* male that I raised in captivity showed an intermediate amount of pigment bleeding (Hill 1993).

Mechanisms to prevent such bleeding of carotenoid pigments would lead necessarily to more efficient use of carotenoids for pigmenting colored patches of feathers and hence to more intense pigmentation. It is interesting that this adaptation is found only in small-patched populations. The presence of discrete patches of carotenoid pigmentation in small-patched populations was not an *a priori* prediction that I made from the Honest Advertisement Model, but it is an observation that supports the idea that the production costs of carotenoid display are higher for males in southern Mexico.

## Ornamental Coloration in Females

As discussed in chapters 9 and 10, some female House Finches have a wash of carotenoid pigmentation on their rumps, crowns, and undersides, and the proportion of females with plumage coloration varies among subspecies (Figure 10.11). Across populations of *frontalis* finches, there is a clear pattern of association between the mean plumage brightness of males in a population and the mean plumage brightness of females (Figure 11.11); females average brighter in plumage coloration when males average brighter. Presumably, this pattern is a result of the same environmental constraints affecting the plumage coloration of both sexes (Hill 1994a).

This pattern of association between male plumage brightness and female color display does not hold when House Finches from the small-patched *griscomi* and the large-patched *ruberrimus* subspecies are included in the analysis (Figure 11.11). Males in the *griscomi* population are brightly colored on average, but females in this subspecies show very little plumage coloration. Conversely, males in the *ruberrimus* population are only moderately bright on average, but most females in this subspecies (75%) show carotenoid coloration and the mean female plumage score is by far the highest of any population. This apparent inconsistency makes sense if the lack of plumage coloration among *griscomi* females reflected a low availability of carotenoid pigments (as presumably do the low plumage scores of *frontalis* females from Hawaii and Alviso), but the small patches of *griscomi* males allowed them to concentrate smaller quantities of carotenoid pigments in achieving bright plumage. Similarly, the high plumage score of *ruberrimus* females could be explained if it

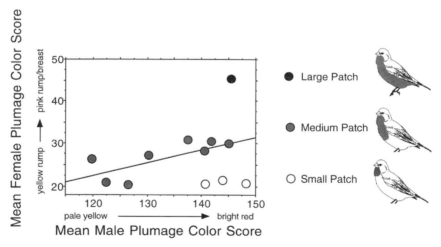

Figure 11.11. The relationship between the ornamental plumage coloration of male and female House Finches in different populations. Among populations of medium-patched House Finches in the *frontalis* subspecies, there is a significant positive relationship between the plumage coloration of males and females ($r_s = 0.74$, $n = 8$, $P = 0.05$). Females in the small-patched *griscomi*, *amplus*, and Great Basin *frontalis* populations are less colorful than expected given the coloration of males in these populations. Conversely, females from the large-patched *ruberrimus* subspecies are more colorful than expected given the coloration of males in their population. This pattern supports the idea that carotenoid resources are more abundant in the environment of large-patched populations compared to small-patched populations. Adapted from Hill (1994a).

reflected a high availability of carotenoid pigments. In *ruberrimus* males, however, carotenoid pigments are dispersed across a large patch of pigmented feathers so *ruberrimus* males are no brighter than *frontalis* males. The important point is that reduced expression of coloration in *griscomi* females is consistent with the idea that the costs of trait production have increased for males in the *griscomi* lineage. Thus, these patterns of variation of female plumage coloration within and among subspecies support the idea that small patches of plumage coloration evolved where the cost of carotenoid display are the highest, and large patches of coloration evolved where the costs are the lowest.

### Patterns of Delayed Plumage Maturation

A fourth line of evidence that supports the idea that reduced patch size evolved in some populations of House Finches in response to an increased cost of trait production is the pattern of delayed plumage maturation among populations. In the last chapter I presented evidence that in three populations of House Finches—C. m. *griscomi*, C. m. *amplus*, and the Great Basin population of C. m. *frontalis*—males have delayed plumage maturation. It is striking that these are three of the four populations that also have reduced patch size (Figure 11.12) (Hill 1996b). The only population that has one trait without the other is C. m. *mexicanus*, which has reduced patch size but no delayed plumage maturation. The chances of reduced patch size and delayed plumage maturation occurring together in almost exactly the

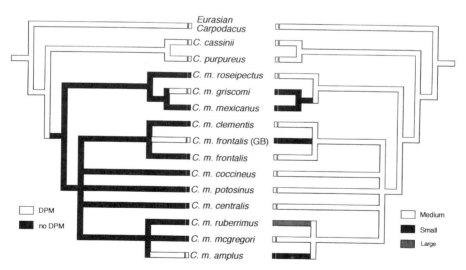

Figure 11.12. The evolution of delayed plumage maturation (left) and patch-size (right) traced on the phylogeny of House Finch lineages. Three of the four taxa that have delayed plumage maturation also have small ventral patches of carotenoid pigmentation. The probability of this degree of concordance of traits occurring by chance is $P < 0.01$ (concentrated changes test; Maddison and Maddison 1993). Reproduced from Hill (1996b) with permission.

same taxa by chance alone is less than one in a hundred (Hill 1996b). What does this co-occurrence of delayed plumage maturation and reduced patch size mean?

Phylogenetic analysis indicates that delayed plumage maturation is the ancestral state for the *Carpodacus* finches, that it was lost in the ancestral House Finch population, and that it was regained independently by the three taxa in which it is now observed (Figure 11.12) (Hill 1996b). Thus, both delayed plumage maturation and lack of delayed plumage maturation are derived traits at different levels in the House Finch phylogeny. This suggests that delayed plumage maturation is evolutionarily labile in House Finch lineages, and that we might reasonably assume that the pattern of acquisition of ornamental plumage in males in different populations has been shaped by local selection (Hill 1996b).

The best explanation for why male House Finches in some populations evolved delayed plumage maturation while males in other populations have not is that it is beneficial for males in some, but not all, populations to delay the costs of ornament production (Hill 1996b). At the end of their first summer, males can either attempt to molt directly into an ornamented breeding plumage, in which case their success at attaining bright plumage will be related to their condition, or they can defer such an attempt at growing ornamented plumage and instead grow a plain brown plumage that resembles female or juvenal plumage (Figure 11.13). By delaying attainment of ornamented plumage, young males would reduce their chances of first-year reproduction, but presumably they would avoid the costs associated with ornamental plumage coloration, and increase their chances of surviving to their second year when they would be more competitive. Such life-history strategies will be shaped by natural selection based on the costs and benefits of producing ornamental plu-

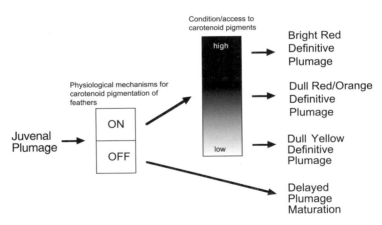

Figure 11.13. Flow chart illustrating the potential patterns of plumage development of yearling male House Finches. Males in most subspecies including *frontalis* and *ruberrimus* molt directly into definitive plumage and the brightness of their first-year plumage is determined by their condition. Males in populations with delayed plumage maturation, including males in the *griscomi* subspecies, shut down the potential for pigmenting growing feathers. A strategy of delayed plumage maturation is presumably adopted when yearling males have little chance of growing bright plumage because of the high costs of ornament production. Reproduced from Hill (1996b) with permission.

mage in the first year (Lack 1968, Selander 1972). It seems reasonable to suggest that the costs of the production of ornamental plumage coloration are higher in populations with delayed plumage maturation than in populations lacking delayed plumage maturation. Thus, the coincidence of delayed plumage maturation and small patch size supports the idea that reduced patch size evolved in response to increased cost of ornament production.

### Summary of Assessment of Sexual Selection Models

The observed patterns of female mate preference relative to male ornament expression across populations are not consistent with Sensory Bias, Species Recognition, or Runaway Models of Sexual Selection. The patterns are consistent with the Honest Advertisement Model, with changes in the cost of trait production in local environments driving the evolution of ornamental traits. As in virtually any test aimed at reconstructing the history of taxa, the evidence in support of an honest advertisement process is indirect. However, the observations that, in small-patched populations of House Finches, there are reduced quantities of dietary carotenoid pigments, that males have discrete patches of carotenoid pigmentation, that females show reduced coloration, and that males have delayed plumage maturation support the idea that increased cost of trait production drove the reduction in ornament size in these populations. With these observations, I have been able to support a specific hypothesis for the evolution of the particular expression of an ornamental display.

### Future Comparative Studies of Carotenoid Pigmentation

The time is ripe for comparative studies that do more than categorize bird species as "bright" or "drab," lumping together plumage that results from carotenoid pigmentation, melanin pigmentation, and structural coloration. Even comparative studies that have focused on one type of ornamental plumage coloration, such as carotenoid or melanin pigmentation, have focused on plumage hue or overall plumage brightness. Thanks to the work of Riccardo Stradi (Stradi et al. 1995, 1996, 1997), we now have detailed data on the pigment composition of plumage coloration for many taxa of cardueline finches. In addition, detailed phylogenies for two major subgroups of cardueline finches (genera *Carduelis* and *Serinus*) have recently been published (Arnaiz-Vellena et al. 1998, 1999), and there are less detailed phylogenies for all cardueline finches (Badyaev 1997, Groth 1998, Marten and Johnson 1986).

Using these resources, for the first time we can begin to look at the transitions between plumage displays, such as from yellow to red, across taxa from the standpoint of changes in the carotenoid composition of plumage. We can look at red plumage not necessarily as one trait with one origin, but rather as a trait that may have evolved independently more than once through different means. For instance, in different lineages, red coloration may have been achieved using different enzymes for carotenoid conversions and with different red carotenoid pigments deposited in feathers. Linking changes in the pigment composition of plumage displays to changes in the environment, including carotenoid content of diet and the social environment of taxa, would provide unprecedented insight into the selective pressures that drive the evolution of ornamental coloration. This sort of comparative study might not only reveal the selective pressures shaping plumage display, but it could also indicate how labile are the mechanisms for pigment uptake and conversion and hence suggest areas for future study regarding the physiology of carotenoid pigmentation.

## Summary

Since Darwin and Wallace, the primary question that has driven scientific investigation of ornamental plumage coloration is why and how such apparently maladaptive traits evolved. Decades after Darwin proposed the idea that female mate choice drives the evolution of ornamental traits, population biologists identified two primary evolutionary mechanisms that could lead to the evolution of female choice for ornamental traits. The Runaway Model of Sexual Selection proposes that once female preference for a trait is established in a population, such as through natural selection, females benefit by choosing the trait because they produce sons that have the trait and are attractive to other females. This leads to a runaway process in which an ornamental trait can be elaborated far beyond a natural selection optimum. The main alternative model, the Honest Advertisement Model, proposes that ornamental traits evolved as reliable signals of male health and condition. By this model, female choice for ornamental traits evolves because choice for such males enables females to choose better-than-average mates. Distinguishing between these hypotheses in studies of ornamental traits in wild animals has proven challenging for behavioral ecologists.

To investigate the evolution of colorful plumage, I adopted a comparative approach, tracing changes in plumage coloration on a phylogeny of House Finch populations and subspecies. Specifically, I looked for the patterns of change in expression of plumage coloration and female mate preference for plumage display. In two populations of the familiar northern subspecies of the House Finch, C. m. frontalis, male plumage coloration is known to have changed from bright red to drab orange/yellow. Despite this change in male phenotype, females from the populations in which males were drab showed a strong mate preference for males with bright red coloration. Considering all subspecies, phylogenetic reconstruction indicated that a medium patch size (red extending about halfway down the underside) was the primitive condition in the House Finch lineage and that, independently in lineages, patch size was reduced to a small area of red. Mate-choice trials with females from small-patched populations showed that they preferred males with large patches.

These observations are not consistent with the idea that colorful plumage evolved as a mechanism for species recognition or in response to a sensory bias. A reduction in the elaborateness of male coloration independent of a change in female preference is also not consistent with the Runaway Model of Sexual Selection. Four sets of observations are consistent with the idea that carotenoid-based plumage coloration evolved as an honest signal of male quality—compared to populations with medium or large patches of color, populations with small patches have few carotenoids in their diets, have discrete patches of carotenoid pigmentation, have drab females, and have males with delayed plumage maturation. Similar comparative studies on a broader range of taxa are needed to test whether the patterns observed in House Finches may describe a general pattern for the evolution of ornamental traits.

# 12 Epilogue

In this work an attempt is made to give a detailed explanation of how and why the land birds of North America have acquired their tints and markings. The subject chosen is thus one that might well be selected for the crowning work of a long life of special research instead of the maiden effort of one who has still his spurs to win in the field of zoological investigation.

—J. A. Allen (1893), in a brutal (and largely unfounded) critique of Keeler's *Colours of North American Land Birds*. Keeler never recovered from the thrashing that his first and only book received.

Whenever I have found out that I have blundered, or that my work has been imperfect, and when I have been contemptuously criticized, and even when I have been overpraised, so that I have felt mortified, it has been my greatest comfort to say hundreds of times to myself that "I have worked as hard and as well as I could, and no man can do more than this."

—C. Darwin (1882), *Autobiography*, p. 126

*When I was a teenager, one of my favorite hobbies was spelunking—exploring caves. Southern Kentucky has hundreds of extensive caves that are known to only a handful of amateur explorers. One of the favorite caves of my little spelunking group was Sloan's Valley, a huge labyrinth with over twenty miles of passages. These twenty miles of passageways were not, primarily, wide corridors with smooth even flooring—maybe 1% of the cave was like that. Most of the cave was a stoop walk, a hands-and-knees crawl, or a belly crawl. We entered the Sloan's Valley cave system through one of several obscure, unmarked entrances with names like the Post Office, Scowling Tom's, the Garbage Pit, or Screaming Willie's. Each of these entrances posed its own challenges and rewards, but usually we entered at the Post Office, a vertical tube, five feet in diameter, that dropped fifty feet from where it opened under a rock overhang behind the Sloan's Valley Post Office. The Post Office was our preferred entrance because it put us near the center of the cave, only an hour or so of crawling from the Grand Central Spaghetti, the nexus of the five major regions of the cave.*

*The majority of this cave had been mapped by volunteer cave enthusiasts. For anyone unfamiliar with a wild cave, however, it is difficult to comprehend the incredible complexity of three-dimensional air space within rock or the difficulty of depicting such complexity as a two-dimensional map. As one stood in the Grand Central Station at the center of the Grand Central Spaghetti, a room perhaps eighty feet by eighty feet with floor and walls that were a jumble of giant limestone slabs, the map said that there were eleven passages that led from the room. But as we would sweep our dim carbide lamps around the room, we would see at least a hundred dark recesses that appeared to be passages. When we began to check out the recesses, we would find that some were simply spaces between rocks that extended only a few feet. Others were passages that could be followed but that ended abruptly and went nowhere. Our expeditions were spent checking leads, getting lost sometimes for hours at a time, finding a landmark, moving on, getting lost, finding a landmark, and eventually finding the destination.*

*I think I learned a lot about science in the hours that I spent underground. When I first began caving, the experience was overwhelming—at once horrifying and exhilarating. The entire time I was underground I would pray for nothing but the surface, swearing I would never return, but within minutes of leaving the cave I would be planning the next trip with my friends, giddy at the thought of an even more adventuresome trip. When we first started caving without the adult who introduced us to the activity, we frequently found ourselves in "hopeless" situations. We would get lost, the first few leads would fail to check out, and I would be in despair—how will we ever find our way out of the cave? Invariably, one of the next few leads would check out, and we would again be progressing forward. So it has been with my studies of bird coloration. During a project, when trouble arises, as it always does, it is easy to begin to despair—this is impossible, it cannot work. But invariably, a path around the difficulty is found, a new approach is invented, and a goal is achieved. As with caving, it leaves me at once terrified and exhilarated, and at the end of every experiment I am giddy at the prospect of the next study or the next experiment that will bring me—er, I mean Science—closer to understanding why birds have colorful plumage.*

Male House Finches have colorful plumage that impresses potential mates and thereby increases their mating success. Few males achieve maximum expression of red plumage because growing brilliant red feathers is challenging. To display red plumage coloration, males must ingest substantial quantities of appropriate carotenoid pigments, but carotenoids are scarce in food. Even if a male forages efficiently and ingests large amounts of appropriate carotenoids, poor nutrition or the effects of parasites may inhibit it from using ingested carotenoids as feather pigments. Thus, encoded in the ornamental coloration of a male House Finch is reliable information about its foraging success, health, and condition. By assessing this color display in potential mates, females find males who are likely to be good partners in the care of young and perhaps males who carry better-than-average genes for offspring. Comparative data suggest that changes in the cost of producing ornamental plumage coloration have driven the evolution of color display in populations of House Finches.

These facts, communicated so effortlessly in seven sentences, are the product of many thousands of hours of observation and experimentation on both captive and wild House Finches. In writing this book, I have paused to compile what is known about ornamental plumage coloration in the House Finch. Many times I questioned whether this was the appropriate time for such a summary project. After all, my research on House Finches has not ended. However, the direction of my research has changed substantially in recent years, and the end of the century and the millennium seemed like a good time to summarize what has been discovered so far about the function and evolution of colorful plumage. Thus, in no way do I view the work presented in this book as a culmination. The studies conducted to date on plumage coloration in the House Finch, and on birds in general, have merely confirmed the most basic assumptions about the present function and proximate control of carotenoid-based ornamentation. They leave unanswered many more questions than they have answered. We are not yet near a comprehensive understanding of carotenoid signaling in the House Finch or any animal for that matter.

One of the primary reasons that I decided that it was a good time to summarize studies on colorful plumage in the House Finch is that we are entering a new era in the study of carotenoid-based display. New and powerful analytical techniques for identification and quantification of the carotenoid pigments in color displays are becoming widely available to biologists interested in plumage coloration. With such tools now available, I think that it will no longer be acceptable to summarize laboratory experiments on carotenoid expression simply by assessing the color of the integumentary structures that are produced. Such studies based on color assessment, which have been the foundation of my studies of plumage coloration, had their place in advancing our understanding of color plumage. In future studies, however, researchers will be expected to quantify the carotenoid pigments in the feathers of their study organisms. Through such improved assessment of carotenoid displays, we are bound to achieve a better understanding of how parasites, nutritional condition, and dietary access to carotenoids affect expression of plumage coloration.

Another area in which old ideas are giving way to new approaches is in how to quantify complex traits like colorful plumage. Carotenoid-based coloration has traditionally been treated as a single trait that can be quantified as a single number. Although at various times I measured the hue, saturation, patch size, and pigment symmetry of colorful plumage, in most of my studies I summarized variation in plumage coloration as a single number, either hue or overall plumage brightness. Collapsing plumage variation to a single metric was necessary as a first approach to testing basic assumptions about the function of plumage coloration. It had great heuristic value in allowing me to keep my studies simple and comprehensible.

It is becoming increasingly clear, however, that "plumage coloration" refers not to a single trait; it is a term that is used to describe a multi-dimensional visual display. Very different characteristics of birds, which reflect fundamentally different behavioral or physiological properties, might be studied by different researchers studying "plumage coloration." Consider that there is a large literature on melanin-based plumage coloration that is composed almost entirely of studies on the extent of patches of black coloration. In these studies, the blackness of feathers within the patches of coloration is, with few exceptions, not considered. In the smaller literature on carotenoid-based coloration, the focus is almost entirely on the quality of pigmentation—hue or saturation. Few studies of carotenoid pigmentation consider the area of plumage with coloration. Thus, those biologists focused on melanin-based coloration and those focused on carotenoid-based coloration are looking not only at different classes of feather pigments, but they are studying fundamentally different components of plumage coloration. Numerous stand-alone reviews and introductions to research papers, however, have tried to draw generalities from the studies of "plumage coloration," lumping together studies of melanin and carotenoid coloration, patch size and plumage hue. Rarely do authors reflect on the fact that "plumage coloration" means different things to different researchers.

I think that the recognition that plumage coloration is a complex trait, with components under very different genetic and environmental control, will lead to a better understanding of how colorful plumage serves as a signal to females. In a recent paper, Alex Badyaev and I conducted an analysis in which we looked at pairing success, fecundity, and survival in a Montana population of House Finches relative to four components of carotenoid coloration: hue, patch size, pigment symmetry, and patch-size symmetry (Badyaev et al. 2001) (Figure 12.1). Hue, patch size, and pigment symmetry have been the focus of various studies by my students and me on Michigan and Alabama populations of House Finches (studies presented in chapters 5–11), but never before did I attempt to simultaneously look at the fitness consequences of multiple components of plumage coloration in one population. The results were very enlightening.

The four components of plumage coloration were partially independent of each other, and each had distinct fitness consequences. In contrast to what I have observed in eastern populations, in the Montana population, hue, pigment symmetry, and patch size were not associated with pairing success; differences in male pairing success were related to patch-size symmetry. Fecundity selection acted quite differently. Much as I found in studies focused on hue alone in Michigan and Alabama, selection for higher fecundity in the Montana population favored an increase in redness of male plumage. At the same time, fecundity selection favored more perfect symmetry of pigmentation and patch area, but it did not act on patch area. The targets of viability selection were yet again different. Viability selection favored larger and more symmetrical ornamental patches, but did not act on the hue or pigment symmetry of ornamental coloration. Overall, net fitness of males in the Montana population was most strongly correlated with patch size and patch-size symmetry with color hue contributing a significant but smaller effect (Badyaev et al. 2001) (Figure 12.1).

In most of the research that I have conducted on the fitness consequences of ornamental plumage coloration in wild populations of House Finches, I've focused on one component of coloration—plumage hue—and one component of fitness—pairing success (see chapters 6 and 7). If pigment quality and pairing success had been the exclusive focus of the Montana study, however, we would have found no net selection on plumage coloration. By considering multiple components of plumage coloration, we observed a complex pattern in which different components of coloration, including hue, affected different aspects of male fitness (Badyaev et al. 2001). Failure to consider different components of plumage coloration is a likely explanation for the disparate results that have been found by different researchers

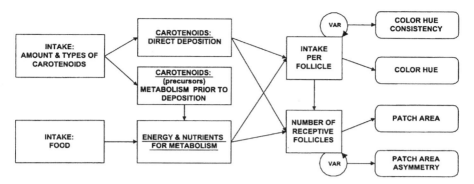

Figure 12.1. Pathways for the development of components of carotenoid-based plumage coloration. Direct use of dietary pigments moves carotenoids along the top of the figure and avoids most dependency of body condition. Use of carotenoids metabolized from dietary precursors leads to greater dependency on the energy and health state of the individual. The elaboration of pigmentation (hue and intensity of coloration) is thus variably dependent on dietary pigments and body condition. Expression of patch area is also dependent on dietary pigments and body condition, but the contribution of these factors to patch size can be quite different than the contribution of these factors to pigment elaboration. Circular arrows indicate the variance in intake per follicle and the number of receptive follicles, which determines pigment consistency and patch area symmetry across the midline of the organism. Reproduced from Badyaev et al. (2001) with permission.

studying plumage coloration. Viewing plumage coloration as a multi-component complex trait should help make sense of much of the confusion that currently exists in the plumage coloration literature when different components of coloration are unwittingly compared and contrasted.

Hue, patch size, pigment symmetry, and patch-size symmetry provide one useful framework from which to look at components of plumage coloration and this approach has provided new insights into the fitness consequences of plumage ornamentation in a House Finch population. There is yet more dimensionality to carotenoid pigmentation, however, that has not been adequately considered in any study. Pigment quality is another multi-dimensional descriptor of plumage coloration (Wedekind et al. 1998). As I outlined in chapter 3, color as perceived by the human eye can be described by three variables—hue, saturation, and tone—called tristimulus descriptors. In my research on plumage coloration in House Finches I have generally either combined the three tristimulus color descriptors into an overall plumage color score, or I have considered only hue, which is the most variable descriptor in House Finches. In the House Finch and other species with carotenoid-based coloration, however, separate consideration of hue, saturation, and tone may provide better insight into the signal content of carotenoid displays (Wedekind et al. 1998). For instance, in most species with red coloration, including the House Finch, red plumage pigments appear to be derived primarily from yellow precursor pigments that are ingested. Plumage redness, then, reflects not just the ability to ingest the yellow precursor but also the ability to metabolize yellow pigments into red. A male with bright yellow coloration may be signaling that it is a good forager, having ingested a large quantity of pigments, but that it lacks the energy or physiological mechanisms to convert the yellow pigments into red pigments. By combining hue and saturation or by considering just one of these components, researchers may be obscuring some of the most important patterns of carotenoid display. The best approach would be to combine a study of different components of plumage coloration with a better understanding of the mechanisms by which birds utilize carotenoids. Such consideration could lead to a better understanding of how females perceive and respond to the complex variations in expression of color display shown by males.

With these looming changes in the techniques for quantifying plumage pigments and in the approaches used to study plumage complexity, I thought that now was the time to summarize the simple set of observations and experiments that I have conducted to date. Rather than a eulogy to the study of colorful plumage, I hope that this book will show how many fundamental questions related to honest signaling in general and carotenoid pigmentation in particular remain to be answered. I hope that this book will serve as a starting point for future research. I do not view any of the new approaches to the study of colorful plumage as threats to the body of research that I have completed. Instead, they represent new challenges, and they hold great promise for pushing forward our understanding of plumage color. The list of questions that lay unanswered relative to carotenoid display is immense, and the potential for studies addressing these questions to expand not only our understanding of carotenoid signaling but also our grasp of the general process of sexual selection is truly exciting.

Let's roll up our sleeves. There is work to be done.

# GLOSSARY

I've worked to make this text as free from jargon as possible. By necessity, however, I have used a few avian terms that non-ornithologists might not know (e.g., banding terminology) and a few technical terms that readers who are not research biologists might find unclear. Below I define these terms and concepts.

## Levels of Investigation

PROXIMATE VERSUS ULTIMATE EXPLANATIONS: A question like "Why do birds have colorful plumage?" can be addressed at two very different levels. First, one can study traits from the standpoint of the mechanisms that create them. For instance, the feathers of House Finches are colorful because they have carotenoid pigments deposited in them. Some males are brighter than others because of the difficulty of acquiring or utilizing dietary carotenoid pigments. This approach to explaining the traits of organsims is a proximate approach. Alternatively, one could focus on the fitness advantages associated with the trait. For instance, males of some species of birds are colorful because bright coloration gives them an advantage in attracting mates. This approach to explaining the traits of organsims is an ultimate approach. Proximate explanations focus on how traits are produced; ultimate explanations focus on why traits exist. Both approaches are valid and provide a level of understanding of natural phenomena. In my research I addressed both ultimate and proximate explanations for colorful plumage in the House Finch, and I used an understanding of the proximate constraints on expression of plumage coloration to better understand the ultimate reason for their existance in their present form.

## Ornithological Terms

HATCH-YEAR (HY) BIRD: A bird in the calendar year of its hatch. In northern temperate regions, where birds hatch only in the spring and summer, all HY birds are within a few months of their hatch and have yet to breed for the first time. In this book I sometimes informally called these HY birds "young" or "juvenal."

AFTER HATCH-YEAR (AHY) BIRD: A bird in a calendar year subsequent to the calendar year of its hatch. AHY birds could be only a few months old, if they hatched the previous summer, to as much as 63 years old—the greatest age ever recorded for a wild bird. To make the terminology more precise, sometimes specific age categories (SY, ASY) are used when they are known (see below).

SECOND YEAR (SY) BIRD: A bird in the calendar year following its hatch year. Note that an SY bird is not in its second breeding season as the name might imply. Note also that an SY bird is also an AHY bird, but SY is a specific age designation while AHY includes all ages except the year of hatching. Most songbirds, including most House Finches, breed as SY birds. In species with delayed plumage maturation, SY males have plumage that is more female-like than older males. In this book, I frequently refer to SY birds as "yearlings" which I find to be a more easily understood age designation.

AFTER SECOND-YEAR (ASY) BIRD: A bird in its third or subsequent calendar year. For many species of birds including populations of House Finches with delayed plumage maturation, males can be grouped into birds that are known to be SY and males that are known to be older than their second year (ASY). In the text I sometimes informally refer to ASY males as "adults" or "older" males, especially in comparisons with SY males.

DELAYED PLUMAGE MATURATION: A developmental pattern whereby males don't achieve fully adult plumage until after their first breeding season, so they look to some degree like a female in their first season. Although these young males don't have fully adult plumage, they typically produce sperm and can father offspring. In a few populations or subspecies of House Finches, males show delayed plumage maturation.

DEFINITIVE PLUMAGE: Plumage with a color and pattern that culminates a series of transitional plumages with different colors or patterns. This sequence is most complex in some gulls and raptors in which definitive plumage is not reached until several years are spent in subadult plumages. In most subspecies and populations of House Finches, the plumage sequence for both males and females is simple: in their first fall birds molt from juvenal plumage into definitive plumage. For those few subspecies and population in which males have delayed plumage maturation, definitive plumage is not achieved by males until their second prebasic molt; they spend their first breeding season in a subadult plumage.

CARDUELINE FINCHES: About 130 species of seed-eating songbirds in the family Fringillidae, subfamily Carduelinae. This group includes goldfinches, canaries, siskins, hawfinches, redpolls, rosy finches, bullfinches, and, of course, House

Finches. Data indicate that this is a natural or monophyletic group of birds. In other words, all cardueline finches are descended from a single common ancestor. Thus, all of the finches in this group are more closely related to each other than they are to any species of bird outside the group.

## Statistical Terminology

STATISTICAL SIGNIFICANCE: If one measures some parameter (for example height) for two groups (for example professors and basketball players), the means for the two groups are bound to be different. The question is whether the differences are likely to be due to sampling error alone or whether the differences are due to fundamental differences between the groups that were sampled. Statistics help one decide. Statisticians have devised statistical tests that can determine the probability of different outcomes of sampling. The results of such a statistical test is a probability or $P$-value that ranges from zero to 1. By convention, if $P < 0.05$—if there is less than 5% chance of getting a difference as big or bigger than the one observed if there really are no differences between the groups—then we call the pattern statistically significant. If $P > 0.05$ we call the pattern non-significant.

It is important to realize that a small proportion of statistically significant patterns will not be real and hence will not be biologically meaningful. Statistics provide a useful guideline for deciding which patterns are relevant and which patterns are not and they are powerful tools in scientific investigation. Statistics are only tools for assessing data, however, and careful interpretation by an experienced scientist is an even more important part of data analysis.

SUMMARY STATISTICS: In my figure legends, after a statement about whether differences between groups are significant or not significant, I typically give a set of numbers and symbols in parentheses, for example ($t = 1.00$, $n = 20$, $P = 0.05$). The first number given in this series is typically the test statistic (see below). This is the outcome of the specific statistical test used in the analysis. The second number(s) typically is (are) the sample size(s) ($n$). The third number is the probability that the observed pattern could have occurred by chance. The test statistic along with the sample size and $P$-value allow one to determine whether a pattern is statistically significant and biologically meaningful.

PARAMETRIC AND NON-PARAMETRIC STATISTICS: Throughout this book I present both parametric and non-parametric statistics (usually reproducing the statistics that were used in the published paper that is referenced). Parametric statistics compare the actual values of the measured variable. This group of statistical tests requires that the variables being tested have distributions that are approximately normal (appear as a bell curve) and usually that the variances of the groups being compared are approximately equal. Many comparisons made by behavioral ecologists, and many of the comparisons that I make in this book, violate one or more of the requirements for using parametric statistics. In these cases, I have used non-parametric statistics, which are a set of statistical tests that convert measured values into ranks and then compare the ranks. Non-parametric statistical tests make no assumptions about the distribution or variance of the original data and are especially useful when small sample sizes prevent the critical testing of assumptions of parametric statistics.

Here are three of the parametric tests and their equivalent non-parametric tests that are used most commonly in comparisons made in this book:

| Parametric test | Test statistic | Non-parametric test | Test statistic |
|---|---|---|---|
| Pearson correlation coefficient | $r$ | Spearman rank correlation coefficient | $r_s$ |
| Student's $t$-test | $t$ | Mann–Whitney $U$-test | $U^*$ |
| Analysis of variance (ANOVA) | $F$ | Kruskal–Wallis test | $K^*$ |

*Sometimes $Z$ is used as a normal approximation of the test statistic when sample sizes are relatively large.

## Evolutionary Terminology

TAXON (PLURAL, TAXA): A taxonomic group. Taxa can be populations, species, or any higher level groups through phyla and kingdoms.

BIOGEOGRAPHY: The field of science that deals with the distributions of taxa, and the historical and environmental factors responsible for differences among the taxa.

PHYLOGENY: Every human has a genealogy—a branching tree showing the relationships of one's ancestors through generations. A phylogeny is the same sort of depiction of relationships except instead of people, the units being connected are populations, species, or higher level taxa and the time frame across which branches occur is not generations but thousands to millions of years. Also, unlike genealogies, which are based on recorded history, phylogenies must be deduced from available evidence, so phylogenies are necessarily hypotheses of relationships.

COMPARATIVE APPROACH: An approach to the study of the evolution of traits in which the researcher compares the expression of traits across taxa. Typically in comparative studies, one looks for associations between the traits of organisms and the environments of those organims. For instance, in looking for an explanation for the pelage color of mammals, one could compare the latitude (within the Northern Hemisphere) at which the animals live to the darkness of the animals' pelage. There would be a conspicuous tendency for animals living in the north to be lighter than those living in the south. Such comparisons among taxa can be used to test ideas about the evolution of traits in much the same way that experiments can be used to test other sorts of hypotheses (see Brooks and McClennan 1991 or Harvey and Pagel 1991 for a detailed discussion of comparative biology).

# REFERENCES

Able, K. P., and J. R. Belthoff. 1998. Rapid "evolution" of migratory behaviour in the introduced house finch of eastern North America. Proceedings of the Royal Society of London B265:2063–2071.

Allen, J. A. 1893. Keeler on the "Evolution of the Color of North American Land Birds." Auk 10:189–199.

Allen, P. C. 1987a. Effect of *Eimeria acervulina* infection on chick (*Gallus domesticus*) high density lipoprotein composition. Comparative Biochemical Physiology 87B:313–319.

Allen, P. C. 1987b. Physiological response of chicken gut tissue to coccidial infection: Comparative effects of *Eimeria acervulina* and *Eimeria mitis* on mucosla mass, carotenoid content, and brush border enzyme activity. Poultry Science 66:1306–1315.

Allen, P. C. 1992. Effect of coccidiosis on the distribution of dietary lutein in the chick. Poultry Science 71:1457–1463.

Amundsen, T. 2000. Why are female birds ornamented? Trends in Ecology and Evolution 15: 149–155.

Andersson, M. 1982a. Female choice selects for extreme tail length in a widowbird. Nature 299:818–820.

Andersson, M. 1982b. Sexual selection, natural selection and quality advertisement. Biological Journal of the Linnean Society 17:375–393.

Andersson, M. 1986a. Evolution of condition-dependent sex ornaments and mating preferences: sexual selection based on viability differences. Evolution 40: 804–816.

Andersson, M. 1986b. Sexual selection and the importance of viability differences: a reply. Journal of Theoretical Biology 120:251–254.

Andersson, M. 1994. Sexual selection. Princeton University Press, Princeton.

A.O.U. (American Ornithologists' Union). 1957. Check-list of North American birds, 5th edn. Allen Press, Lawrence, Kansas.

Arnaiz-Vellena, A., M. Alvarez-Tejado, V. Ruiz-del-Valle, C. Garcia-de-la-Torre, P. Varela, M. J. Recio, S. Ferre, and J. Martinez-Laso. 1998. Phylogeny and rapid

Northern and Southern Hemisphere speciation of goldfinches during the Miocene and Pliocene Epochs. Cellular and Molecular Life Science 54:1031–1041.

Arnaiz-Vellena, A., M. Alvarez-Tejado, V. Ruiz-del-Valle, C. Garcia-de-la-Torre, P. Varela, M. J. Recio, S. Ferre, and J. Martinez-Laso. 1999. Rapid radiation of canaries (Genus *Seriinus*). Molecular Biology and Evolution 16:2–11.

Badyaev, A. V. 1997. Altitudinal variation in sexual dimorphism: a new pattern and alternative hypotheses. Behavioral Ecology 8:675–690.

Badyaev, A. V., and G. E. Hill. 2000a. Evolution of sexual dichromatism: Contribution of carotenoid- versus melanin-based plumage coloration. Biological Journal of the Linnean Society 69:153–172.

Badyaev, A. V., and G. E. Hill. 2000b. The evolution of sexual dimorphism in the House Finch. I. Population divergence in morphological covariance structure. Evolution 54:1784–1794.

Badyaev, A. V., G. E. Hill, P. O. Dunn, and J. C. Glen. 2001. Plumage color as a composite trait: developmental and functional integration of sexual ornamentation. American Naturalist 158:221–235.

Baird, S. F., T. M. Brewer, and R. Ridgway. 1874. A history of North American birds. Land birds. Little, Brown, and Company, Boston.

Balph, M. H., D. F. Balph, and H. C. Romesburg. 1979. Social status signaling in winter flocking birds: an examination of a current hypothesis. Auk 96:78–93.

Bartholomew, G. A., and T. J. Cade. 1956. Water consumption of House Finches. Condor 58:406–412.

Basolo, A. L. 1990. Female preference predates the evolution of the sword in swordtail fish. Science 250:808–810.

Beal, F. E. L. 1907. Birds of California in relation to fruit industry. United States Department of Agriculture Biological Survey Bulletin 30:13–23.

Beddard, F. E. 1892. Animal coloration. Swan Sonnenschein, London.

Bell, G. 1978. The handicap principle in sexual selection. Evolution 32:872–885.

Belthoff, J. R., and S. A. Gauthreaux, Jr. 1991a. Aggression and dominance in house finches. Condor 93:1010–1013.

Belthoff, J. R., and S. A. J. Gauthreaux. 1991b. Partial migration and differential winter distribution of house finches in the eastern USA. Condor 93:374–382.

Belthoff, J. R., and P. A. Gowaty. 1996. Male plumage coloration affects dominance and aggression in female house finches. Bird Behavior 11:1–7.

Belthoff, J. R., A. M. Dufty, Jr., and S. A. Gauthreaux, Jr. 1994. Plumage variation, plasma steroids and social dominance in male House Finches. Condor 96:614–625.

Bendich, A. 1989. Carotenoids and the immune response. Journal of Nutrition 119:112–115.

Bendich, A., and S. S. Shapiro. 1986. Effect of β-carotene and canthaxanthin on the immune response of the rat. Journal of Nutrition 116:2254–2262.

Benner, W. L. 1991. Mitochondrial DNA variation in the House Finch (*Carpodacus mexicanus*). Master's Thesis, Cornell University.

Bennett, A. T. D., I. C. Cuthill, and K. J. Norris. 1994. Sexual selection and the mismeasure of color. American Naturalist 144:848–860.

Bennett, G. F., J. R. Caines, and M. A. Bishop. 1988. Influence of blood parasites on the body mass of passerine birds. Journal of Wildlife Disease 24:339–343.

Bergtold, W. H. 1913. A study of the House Finch (*Carpodacus mexicanus frontalis*). Auk 30:40–73.

Binford, L. C. 2000. Re-evaluation of the House Finch subspecies *Carpodacus mexicanus roseipectus* from Oaxaca, Mexico. Bulletin of the British Ornithologists' Club 120:120–128.

Birkhead, T. R., and A. P. Møller. 1992. Sperm competition in birds: evolutionary causes and consequences. Academic Press, London.

Birkhead, T., H. Schwabl, and T. Burke. 2000. Testosterone and maternal effects: Integrating mechanisms and function. Trends in Ecology and Evolution 15:86–87.

Björklund, M. 1990. Mate choice is not important for female reproductive success in the common rosefinch (*Carpodacus erythrinus*). Auk 107:35–44.

Blaisdell, M. L. 1992. Darwinism and its data: the adaptive coloration of animals. Garland Publishing, Inc., New York.

Blanco, G., J. L. Tella, and J. Potti. 1997. Feather mites on group-living red-billed choughs: a non-parasitic interaction? Journal of Avian Biology 28:197–206.

Blank, J. L., and V. J. Nolan. 1983. Offspring sex ratio in Red-winged Blackbirds is dependent on maternal age. Proceedings of the National Academy of Sciences USA 80:6141–6145.

Bletner, J. K., R. P. Mitchell, and R. L. Tugwell. 1966. The effect of *Eimeria maxima* on broiler pigmentation. Poultry Science 45:689–694.

Blount, J. D., D. C. Houston, and A. P. Møller. 2000. Why egg yolk is yellow. Trends in Ecology and Evolution 15:47–49.

Bortolotti, G., J. J. Negro, J. L. Tella, T. A. Marchant, and D. M. Bird. 1996. Sexual dichromatism in birds independent of diet, parasites and androgens. Proceedings of the Royal Society of London B263:1171–1176.

Bradbury, J. W., and M. B. Andersson. 1987. Sexual selection: testing the alternatives. John Wiley & Sons, New York.

Brawner, W. R., III. 1997. The effects of coccidial and mycoplasmal infection on plumage pigmentation in male House Finches (*Carpodacus mexicanus*): a test of the Hamilton-Zuk hypothesis. Master of Science Thesis, Auburn University.

Brawner, W. R., III., and G. E. Hill. 1999. Temporal variation in shedding of coccidal oocysts: implications for sexual-selection studies. Canadian Journal of Zoology 77:347–350.

Brawner, W. R., III., G. E. Hill, and C. A. Sundermann. 2000. Effects of coccidial and mycoplasmal infections on carotenoid-based plumage pigmentation in male House Finches. Auk 117:952–963.

Brockmann, H., and O. Völker. 1934. Der Gelbe Federfarbstoff des Kanarienvogels (*Serinus canaria canaria* (L.)) und das Vorkommen von Carotinoiden bei Vögeln. Hoppe-Seyler's Zeitschrift für Physiologische Chemie 224:193–215.

Brooks, D. R., and D. A. McLennan. 1991. Phylogeny, ecology, and behavior : a research program in comparative biology. University of Chicago Press, Chicago.

Brotons, L. 1998. Status signalling in the coal tit (*Parus ater*): the role of previous knowledge of individuals. Etologia 6:49–52.

Brown, M. B., and C. R. Brown. 1988. Access to winter food resources by bright- versus dull-colored house finches. Condor 90:729–731.

Brush, A. H. 1967. Pigmentation in the scarlet tanager, *Piranga olivacea*. Condor 69:549–559.

Brush, A. H. 1968. Pigmentation and feather structure in genetic variants of the Gouldian Finch, *Poephila gouldiae*. Auk 85:416–430.

Brush, A. H. 1978. Avian pigmentation. In A. H. Brush (ed.), Chemical zoology, pp. 141–161. Academic Press, New York.

Brush, A. H. 1981. Carotenoids in wild and captive birds. In J. C. Bauernfeind (ed.), Carotenoids as colorants and vitamin A precursors, pp. 539–562. Acdemic Press, London.

Brush, A. H. 1990. Metabolism of carotenoid pigments in birds. Federation of American Societies for Experimental Biology Journal 4:2969–2977.

Brush, A. H., and D. M. Power. 1976. House Finch pigmentation: carotenoid metabolism and the effect of diet. Auk 93:725–739.

Burley, N. 1977. Parental investment, mate choice, and mate quality. Proceedings of the National Academy of Sciences USA 74:3476–3479.

Burley, N. 1981. Mate choice by multiple criteria in a monogamous species. American Naturalist 117:515–528.

Burley, N. 1986. Sexual selection for aesthetic traits in species with biparental care. American Naturalist 127:415–445.

Burley, N. 1988. The differential-allocation hypothesis: an experimental test. American Naturalist 132:611–628.

Burley, N., G. Krantzberg, and P. Radman. 1982. Influence of colour-banding on the conspecific preferences of zebra finches. Animal Behaviour 30:444–455.

Burley, N., S. C. Tidemann, and K. Halupka. 1991. Bill colour and parasite levels of zebra finches. In J. E. Loye and M. Zuk (eds.), Bird–parasite interactions. Oxford University Press, Oxford.

Burley, N., D. A. Enstrom, and L. Chitwood. 1994. Extra-pair relations in zebra finches: differential male success results from female tactics. Animal Behaviour 48:1031–1041.

Burley, N. T., P. G. Parker, and K. Lundy. 1996. Sexual selection and extrapair fertilization in a socially monogamous passerine, the zebra finch (*Taeniopygia guttata*). Behavioral Ecology 7:218–226.

Burton, G. W. 1989. Antioxidant action of carotenoids. Journal of Nutrition 119:109–111

Butcher, G. S., and S. Rohwer. 1989. The evolution of conspicuous and distinctive colouration for communication in birds. Current Ornithology 6:51–108.

Cannings, R. A., R. J. Cannings, and G. S. Cannings. 1987. Birds of the Okanagan Valley. Royal British Columbia Museum, Victoria, B.C.

Cant, G., and H. P. Geis. 1961. The House Finch: a new east coast migrant? Eastern Bird Banding News 102–107.

Cheesman, D. F., W. L. Lee, and P. F. Zagalsky. 1967. Carotenoproteins in invertebrates. Biological Review of the Cambridge Philosophical Society 42:132–160.

Chen, D., and T. H. Goldsmith. 1986. Four spectral classes of cone in the retinas of birds. Journal of Comparative Physiology A 159:473–479.

Chen, D., J. S. Collins, and T. H. Goldsmith. 1984. The ultraviolet receptor in bird retinas. Science 225:337–340.

Chew, B. P. 1993. Role of carotenoids in the immune response. Journal of Dairy Science 76:2804–2811.

Clayton, D. H. 1991. The influence of parasites on host sexual selection. Parasitology Today 7:329–334.

Cleland, T. M. 1921. A grammar of color. The Strathmore Paper Company, Mittineague, Mass.

Clement, P., A. Harris, and J. Davis. 1993. Finches and sparrows. Princeton University Press, Princeton.

Clutton-Brock, T. H., F. E. Guinness, and S. D. Albon. 1982. Red deer: behavior and ecology of two sexes. University of Chicago Press, Chicago.

Cooke, F., and F. G. Cooch. 1968. The genetics of polymorphim in the Snow Goose *Anser vaerulescens*. Evolution 22:289–300.

Cornwall, D. G., F. A. Kruger, and H. B. Robinson. 1962. Studies on the absorption of beta-carotene and the distribution of total carotenoid in human serum lipoproteins after oral administration. Journal of Lipid Research 3:65–70.

Cott, H. B. 1940. Adaptive coloration in animals. Methuen, London.

Cowen, I. M. 1937. The House Finch at Victoria, British Columbia. Condor 39:225.

Crawford, R. D. 1977. Breeding biology of year-old and older female Red-winged and Yellow-headed Blackbirds. Wilson Bulletin 89:73–80.

Cronin, H. 1991. The ant and the peacock. Cambridge University Press, Cambridge, UK.

Curio, E. 1983. Why do young birds reproduce less well? Ibis 125:400–404.

Cuthill, I. C., J. C. Partridge, and A. T. D. Bennet. 2000. Avian UV vision and sexual selection. In Y. Epsmark, T. Amundsen, and G. Rosenqvist (eds.), Animal signals: signalling and signal design in animal communication, pp. 61–82. Tapir Academic Press, Trondheim.

Darwin, C. 1859. On the origin of species by natural selection or the preservation of favoured races in the struggle for life. John Murray, London.

Darwin, C. 1871. The descent of man, and selection in relation to sex. John Murray, London.

Darwin, C. 1882. Autobiography. With original omissions restored. Collins, London.

Darwin, C., and A. R. Wallace. 1858. On the tendency of species to form varieties; and on the perpetuation of varieties and species by means of natural selection. Proceedings of the Zoological Society of London 3:45–62.

Davis, J. W. F., and P. O'Donald. 1976. Sexual selection for a handicap: a critical analysis of Zahavi's model. Journal of Theoretical Biology 57:345–354.

Dawkins, R. 1980. Good strategy or evolutionarily stable strategy? In G. Barlow and J. Silverberg (eds.), Sociobiology: beyond nature/nurture? Westview Press, Boulder, Colorado.

Dawkins, R., and T. R. Carlisle. 1976. Parental investment, mate desertion and a fallacy. Nature 262:131–133.

Dawson, W. L. 1923. The birds of California: a complete, scientific and popular account of the 580 species and sub-species of birds found in the state. Birds of California Publishing Company, San Francisco.

Dijkstra, C., A. Bult, S. Bijlsma, S. Daan, T. Meijer, and M. Zijlstra. 1990. Brood size manipulations in the kestrel (*Falco tinnunculus*): effects on offspring and parent survival. Journal of Animal Ecology 59:269–285.

Dobzhansky, T. 1937. Genetics and the origin of species. Columbia University Press, New York.

Dominey, W. J. 1983. Sexual selection, additive genetic variance and the "Phenotypic Handicap." Journal of Theoretical Biology 101:495–502.

Double, M. C., D. Dawson, T. Burke, and A. Cockburn. 1997. Finding the fathers in the least faithful bird: a microsatellite-based genotyping system for the superb fairy-wren *Malurus syaneus*. Molecular Ecology 6:691–693.

Drachmann, J. 1998. Sexual Selection and Reproductive Success in the Linnet (*Carduelis cannabina*). Ph.D. Dissertation, University of Aarhus, Denmark.

Duckworth, R. A., M. T. Mendonça, and G. E. Hill. 2001. A condition dependent link between testosterone and disease resistance in the House Finch. Proceedings of the Royal Society of London B268:2467–2472.

Duckworth, R. A., M. Mendonça, and G. E. Hill. in prep. Effect of testosterone on male dominance rank in captive male House Finches.

Dufva, R. 1996. Blood parasites, health, reproductive success, and egg volume in female great tits *Parus major*. Journal of Avian Biology 27:83–87.

Dunn, P. O., and A. Cockburn. 1999. Extrapair mate choice and honest signaling in cooperatively breeding superb fairy-wrens. Evolution 53:938–946.

Eberhard, W. G. 1993. Evaluating models of sexual selection: genitalia as a test case. American Naturalist 142:564–571.

Eisner, E. 1960. The relationship of hormones to reproductive behavior of birds, referring especially to parental care: a review. Animal Behavior 8:155–179.

Elliot, J. J., and R. S. J. Arbib. 1953. Origin and status of the house finch in the eastern United States. Auk 70:31–37.

Endler, J. A. 1980. Natural and sexual selection on color patterns in *Poecilia reticulata*. Evolution 34:76–91.

Endler, J. A. 1983. Natural and sexual selection on color patterns in poeciliid fishes. Environmental Biology of Fishes 9:173–190.

Endler J. A. 1990. On the measurement and classification of color in studies of animal color patterns. Biological Journal of the Linnean Society 41: 315–352.

Endler, J. A., and A. L. Basolo. 1998. Sensory ecology, receiver biases and sexual selection. Trends in Ecology and Evolution 13:415–420.

Endler, J. A., and A. M. Lyles. 1989. Bright ideas about parasites. Trends in Ecology and Evolution 4:246–248.

Ens, G. J., U. N. Safriel, and M. P. Harris. 1993. Divorce in the long-lived and monogamous oystercatcher, *Haematophagus ostralegus*. Incompatibility or choosing a better option? Animal Behaviour 45:1199–1217.

Enstrom, D. A., E. D. Ketterson, and J. V. Nolan. 1997. Testosterone and mate-choice in the dark-eyed junco. Animal Behaviour 54:1135–1146.

Erdman, J. W., Jr., T. L. Bierer, and E. T. Gugger. 1993. Absorption and transport of carotenoids. Annals of the New York Academy of Sciences 691:76–85.

Evans, M. R., A. R. Goldsmith, and S. R. A. Norris. 2000. The effects of testosterone on antibody production and plumage coloration in male house sparrows (*Passer domesticus*). Behavioral Ecology and Sociobiology 47:156–163.

Evenden, F. G. 1957. Observations on nesting behavior of the House Finch. Condor 59:112–117.

Farris, J. S. 1970. Methods for computing Wagner Trees. Systematic Zoology 19:83–92.

Finger, E. 1995. Visible and UV coloration in birds: Mie scattering as the basis of color in many bird feathers. Naturwissenschaften 82:570–573.

Finger, E., D. Burkhardt, and J. Dyck. 1992. Avian plumage colors: origin of UV reflection in a black parrot. Naturwissenschaften 79:187–188.

Finn, F. 1907. Ornithological and other oddities. John Lane, London.

Fisher, R. A. 1915. The evolution of sexual preference. Eugenics Review 7:184–192.

Fisher, R. A. 1930. The genetical theory of natural selection. Dover, New York.

Fisher, R. A. 1958. The genetical theory of natural selection, 2nd edn. Dover, New York.

Folstad, I., and A. J. Karter. 1992. Parasites, bright males, and the immunocompetence handicap. American Naturalist 139:603–622.

Fox, D. L. 1962. Metabolic fractionation, storage and display of cartenoid pigments by flamingoes. Comparative Biochemistry and Physiology 6:1–40.

Fox, D. L. 1976. Animal biochromes and structural colours. University of California Press, Berkeley, CA.

Fox, D. L., and T. S. Hopkins. 1966. Comparative metabolic fractionation of carotenoids in three flamingo species. Comparative Biochemistry and Physiology 17:841–856.

Fox, D. L., and T. S. Hopkins. 1967. β-Carotenoid fractionation in the scarlet ibis. Comparative Biochemistry and Physiology 19:267–278.

Fox, D. L., and J. W. McBeth. 1970. Some dietary and blood carotenoid levels in flamingos. Comparative Biochemistry and Physiology 34:707–713.

Fox, D. L., T. S. Hopkins, and D. B. Zilversmit. 1965. Blood carotenoids of the Roseate Spoonbill. Comparative Biochemistry and Physiology 14:641–649.

Fox, D. L., V. E. Smith, and A. A. Wolfson. 1967. Carotenoid selectivity in blood and feathers of Lesser (African), Chilean and Greater (European) Flamingos. Comparative Biochemistry and Physiology 23:225–232.

Fox, D. L., A. A. Wolfson, and J. W. McBeth. 1969. Metabolism of β-carotene in the American Flamingo, *Phoinicopterus ruber*. Comparative Biochemistry and Physiology 29:1223–1229.

Fox, D. L., J. W. McBeth, and G. MacKinney. 1970. Some dietary carotentoids and blood-carotenoid levels in flamingos. II. γ-carotene and α-carotene consumed by the American Flamingo. Comparative Biochemistry and Physiology 36:253–262.

Fox, H. M., and G. Vevers. 1960. The nature of animal colors. Macmillan, New York.

Furr, H. C., and R. M. Clark. 1997. Intestinal absorption and tissue distribution of carotenoids. Nutritional Biochemistry 8:364–377.

Giacomo, R., P. Stefania, T. Ennio, V. C. Giorgina, B. Giovanni, and R. Giacomo. 1997. Mortality in Black Siskins (*Carduelis atrata*) with systemic coccidiosis. Journal of Wildlife Diseases 33:152–157.

Gil, D., J. Graves, N. Hazon, and A. Wells. 1999. Male attractiveness and differential testosterone investment in zebra finch eggs. Science 286:126–128.

Goodwin, T. W. 1950. Carotenoids and reproduction. Biological Review 25:391–413.

Goodwin, T. W. 1980. The biochemistry of carotenoids. Volume 1, Plants, 2nd edn. Chapman & Hall, New York.

Goodwin, T. W. 1984. The biochemistry of carotenoids. Volume 2, Animals, 2nd edn. Chapman & Hall, New York.

Goodwin, T. W., and G. Britton. 1988. Distribution and analysis of carotenoids. In T. W. Goodwin (ed.), Plant pigments. Academic Press, New York.

Grafen, A. 1990a. Biological signals as handicaps. Journal of Theoretical Biology 144:517–546.

Grafen, A. 1990b. Sexual selection unhandicapped by the Fisher process. Journal of Theoretical Biology 144:473–516.

Grant, B. R., and P. R. Grant. 1987. Mate choice in Darwin's finches. Biological Journal of the Linnean Society 32:247–270.

Gray, D. A. 1996. Carotenoids and sexual dichromatism in North American passerine birds. American Naturalist 148:3:453–480.

Grether, G. F., J. Hudon, and D. F. Millie. 1999. Carotenoid limitation of sexual coloration along an environmental gradient in guppies. Proceedings of the Royal Society of London B266:1317–1322.

Grinnell, J. 1911. The Linnet of the Hawaiian Islands: A problem in speciation. University of California Publications in Zoology 7:179–195.

Grinnell, J. 1912. A name for the Hawaiian linnet. Auk 29:24–25.

Gross, J. 1987. Pigments in fruits. Academic Press, New York.

Groth, J. G. 1998. Molecular phyologenetics of finches and sparrows: Consequences of character state removal in cytochrome b sequences. Molecular Phylogenetics and Evolution 10:377–390.

Grubb, T. C. J. 1989. Ptilochronology: feather growth bars as indicators of nutritional status. Auk 106:314–320.

Grubb, T. C. J. 1991. A deficient diet narrows growth bars on induced feathers. Auk 108:725–727.

Grubb, T. C., and D. A. Cimprich. 1990. Supplementary food improves the nutritional condition of wintering woodland birds: Evidence from ptilochronology. Ornis Scandinavica 21:277–281.

Hahn, T. P. 1996. Cassin's Finch (*Carpodacus cassinii*). In A. Poole and F. Gill (eds.), The birds of North America, Vol. 240. The Academy of Natural Sciences, Philadelphia.

Hamilton, T. R. 1992. House Finch winter range expansion as documented by Christmas Bird Counts, 1950–1990. Indiana Audubon Quarterly 70:147–153.

Hamilton, T. R., and C. D. Wise. 1991. The effect of the increasing House Finch population on House Sparrows, American Goldfinches, and Purple Finches in Indiana. Indiana Audubon Quarterly 69:251–254.

Hamilton, W. D., and M. Zuk. 1982. Heritable true fitness and bright birds: a role for parasites? Science 218:384–386.

Hanotte, O., C. Zanon, A. Pugh, C. Greig, A. Dixon, and T. Burke. 1994. Isolation and characterization of microsatellite loci in a passerine bird: the reed bunting *Emberiza schoeniclus*. Molecular Ecology 3:529–530.

Harper, D. G. C. 1999. Feather mites, pectoral muscle condition, wing length and plumage coloration of passerines. Animal Behaviour 58:553–562.

Harvey, P. H., and M. D. Pagel. 1991. The comparative method in evolutionary biology. Oxford University Press, Oxford.

Harvey, P. H., P. J. Greenwood, C. M. Perrins, and A. R. Martin. 1979. Breeding success of great tits *Parus major* in relation to age of male and female parent. Ibis 121:216–219.

Hassan, J. O., and R. I. Curtiss. 1996. Effect of vaccination of hens with an avirulent strain of *Salmonella typhimurium* on immunity of progeny challenged with wild-type *Salmonella* strains. Infection and Immunity 64:938–944.

Hegner, R. E., and J. C. Wingfield. 1987. Effects of experimental manipulation of testosterone levels on parental investment and breeding success in male house sparrows. Auk 104:462–469.

Hill, G. E. 1988a. Age, plumage brightness, territory quality and reproductive success in the Black-headed Grosbeak. Condor 90:379–388.

Hill, G. E. 1988b. The function of delayed plumage maturation in male Black-headed Grosbeaks. Auk 105:1–10.

Hill, G. E. 1989. Late spring arrival and dull nuptial plumage: aggression avoidance by yearling males? Animal Behaviour 37:665–673.

Hill, G. E. 1990. Female house finches prefer colourful males: sexual selection for a condition-dependent trait. Animal Behaviour 40:563–572.

Hill, G. E. 1991. Plumage coloration is a sexually selected indicator of male quality. Nature 350:337–339.

Hill, G. E. 1992. Proximate basis of variation in carotenoid pigmentation in male house finches. Auk 109:1–12.

Hill, G. E. 1993a. Geographic variation in the carotenoid plumage pigmentation of male house finches (*Carpodacus mexicanus*). Biological Journal of the Linnean Society 49:63–86.

Hill, G. E. 1993b. House Finch (*Carpodacus mexicanus*). In A. Poole and F. Gill (eds.), The birds of North America, No. 46. Academy of Natural Sciences, Philadelphia, and American Ornithologists' Union, Washington, D.C.

Hill, G. E. 1993c. Male mate choice and the evolution of female plumage coloration in the house finch. Evolution 47:1515–1525.

Hill, G. E. 1993d. The proximate basis of inter- and intra-population variation in female plumage coloration in the House Finch. Canadian Journal of Zoology 71:619–627.

Hill, G. E. 1994a. Geographic variation in male ornamentation and female mate preference in the house finch: a comparative test of models of sexual selection. Behavioral Ecology 5:64–73.

Hill, G. E. 1994b. House Finches are what they eat: a reply to Hudon. Auk 111:221–225.

Hill, G. E. 1994c. Trait elaboration via adaptive mate choice: sexual conflict in the evolution of signals of male quality. Ethology, Ecology and Evolution 6:351–370.

Hill, G. E. 1995a. Evolutionary inference from patterns of female preference and male display. Behavioral Ecology 6:350–351.

Hill, G. E. 1995b. Seasonal variation in circulating carotenoid pigments in the House Finch. Auk 112:1057–1061.

Hill, G. E. 1996a. Redness as a measure of the production cost of ornamental coloration. Ethology, Ecology and Evolution 8:157–175.

Hill, G. E. 1996b. Subadult plumage in the house finch and tests of models for the evolution of delayed plumage maturation. Auk 113:858–874.

Hill, G. E. 1998a. An easy, inexpensive means to quantify plumage colouration. Journal of Field Ornithology 69:353–363.

Hill, G. E. 1998b. Plumage redness and pigment symmetry in the house finch. Journal of Avian Biology 29:86–92.

Hill, G. E. 1999a. Is there an immunological cost to carotenoid-based ornamental coloration? American Naturalist 154:589–595.

Hill, G. E. 1999b. Mate choice, male quality, and carotenoid-based plumage coloration. In N. Adams and R. Slotow (eds.), Proceedings of the 22nd International Ornithological Congress, pp. 1654–1668. University of Natal, Durban.

Hill, G. E. 2000. Energetic constraints on expression of carotenoid-based plumage coloration. Journal of Avian Biology 31:559–566.

Hill, G. E. 2001. Pox and plumage coloration in the House Finch: a critique of Zahn and Rothstein. Auk 118:256–260.

Hill, G. E., and C. W. Benkman. 1995. Exceptional response by female Red Crossbills to dietary carotenoid supplementation. Wilson Bulletin 107:555–557.

Hill, G. E., and W. R. Brawner, III. 1998. Melanin-based plumage colouration in the house finch is unaffected by coccidial infection. Proceedings of the Royal Society of London B265:1105–1109.

Hill, G. E., and R. Montgomerie. 1994. Plumage colour signals nutritional condition in the house finch. Proceedings of the Royal Society of London B258:47–52.

Hill, G. E., R. Montgomerie, C. Y. Inouye, and J. Dale. 1994a. Influence of dietary carotenoids on plasma and plumage colour in the house finch: intra- and intersexual variation. Functional Ecology 8:343–350.

Hill, G. E., R. Montgomerie, C. Roeder, and P. Boag. 1994b. Sexual selection and cuckoldry in a monogamous songbird: Implications for sexual selection theory. Behavioral Ecology and Sociobiology 35:193–199.

Hill, G. E., P. M. Nolan, and A. M. Stoehr. 1999. Pairing success relative to male plumage redness and pigment symmetry in the house finch: temporal and geographic constancy. Behavioral Ecology 10:48–53.

Hill, G. E., C. Y. Inouye, and R. Montgomerie. 2002. Dietary carotenoids predict plumage coloration in wild house finches. Proceedings of the Royal Society of London B, in press.

Hingston, R. W. G. 1933. The meaning of animal colour and adornment. Edward Arnold, London.

Hochachka, W. M., and A. A. Dhont. 2000. Density-dependent decline of host abundance resulting from a new infectious disease. Proceedings of the National Academy of Sciences USA 97:5303–5306.

Höglund, J., and B. C. Sheldon. 1998. The cost of reproduction and sexual selection. Oikos 83:478–483.

Högstad, O. 1989. Social organization and dominance behavior in some *Parus* species. Wilson Bulletin 101:254–262.

Högstad, O., and R. T. Kroglund. 1993. The throat badge as a status signal in juvenal male Willow Tits *Parus montanus*. Journal für Ornithologie 134:413–423.

Hooge, P. N. 1990. Maintenance of pair-bonds in the House Finch. Condor 92:1066–1067.

Houde, A. E. 1987. Mate choice based upon naturally occurring color-pattern variation in a guppy population. Evolution 41:1–10.

Houde, A. E. 1993. Evolution by sexual selection: what can population comparisons tell us? American Naturalist 141:796–803.

Houde, A. E., and J. A. Endler. 1990. Correlated evolution of female mating preferences and color patterns in the Guppy *Poecilia reticulata*. Science 248:1405–1407.

Houde, A. E., and A. J. Torio. 1992. Effect of parasitic infection on male color pattern and female choice in guppies. Behavioral Ecology 3:346–351.

Howell, T. R., R. A. Paynter, Jr., and A. L. Rand. 1968. Subfamily Carduelinae. In R. A. Paynter, Jr. (ed.), Check-list of birds of the world, Vol. XIV, pp. 207–306. Museum of Comparative Zoology, Cambridge, Massachusetts.

Hudon, J. 1991. Unusual carotenoid use by the Western Tanager (*Piranga ludoviciana*) and its evolutionary implications. Canadian Journal of Zoology 69:2311–2320.

Hudon, J. 1994. Showiness, carotenoids, and captivity: a comment on Hill (1992). Auk 111:218–221.

Hudon, J., and A. H. Brush. 1989. Probable dietary basis of a color variant of the Cedar Waxwing. Journal of Field Ornithology 60:361–368.

Hudon, J., H. Ouellet, E. Benito Espinal, and A. H. Brush. 1996. Characterization of an orange variant of the Bananaquit (*Coereba flaveola*) on La Desirade, Guadeloupe, French West Indies. Auk 113:715–718.

Huxley, J. S. 1938. The present standing of the theory of sexual selection. In G. R. de Beer (ed.), Evolution: essays on aspects of evolutionary biology, presented to Professor E. S. Goodrich on his seventieth birthday. Clarendon Press, Oxford.

Huxley, J. S. 1942. Evolution: the modern synthesis. Harper, New York.

Inouye, C. Y. 1999. The physiological bases for carotenoid color variation in the House Finch, *Carpodacus mexicanus*. Ph.D. Dissertation, University of California at Los Angeles.

Inouye, C. Y., G. E. Hill, R. D. Stradi, and R. Montgomerie. 2001. Carotenoid pigments in male house finch plumage in relation to age, subspecies, and ornamental coloration. Auk 118:900–915.

Iwasa, Y., and A. Pomiankowski. 1994. The evolution of mate preferences for multiple sex ornaments. Evolution 48:853–867.

Jarvi, T., and M. Bakken. 1984. The function of the variation in the breast stripe of the great tit (*Parus major*). Animal Behaviour 32:590–596.

Jarvi, T., O. Walso, and M. Bakken. 1987. Status signalling by *Parus major*: an experiment in deception. Ethology 76:334–342.

Jehl, J. R. J. 1971. The status of *Carpodacus mcgregori*. Condor 73:375–376.

Johnsen, T. S., J. D. Hengeveld, J. L. Blank, K. Yasukawa, and J. Nolan, V. 1996. Epaulet brightness and condition in female Red-winged Blackbirds. Auk 113:356–362.

Johnson, K., R. Dalton, and N. Burley. 1993. Preferences of female American goldfinches (*Carduelis tristis*) for natural and artificial male traits. Behavioral Ecology 4:138–143.

Johnson, K., E. DuVal, M. MKielt, and C. Hughes. 2000. Male mating strategies and the mating system of great-tailed grackles. Behavioral Ecology 11:132–141.

Johnson, S. G. 1991. Effects of predation, parasites, and phylogeny on the evolution of bright coloration in North American male passerines. Evolutionary Ecology 5:52–62.

Kalinoski, R. 1975. Intra- and interspecific aggression in House Finches and House Sparrows. Condor 77:375–384.

Keeler, C. A. 1893. Evolution of the colors of North American land birds. Occasional Papers of the California Academy of Sciences (no. 3). California Academy of Sciences, San Francisco.

Kempenaers, B., G. R. Verheyen, M. Van den Broek, T. Burke, C. Van Broeckhoven, and A. A. Dhondt. 1992. Extrapair paternity results from female preference for high-quality males in the blue tit. Nature 357:494–496.

Kenyon, L. S. 1902. [Untitled.] Birds and Nature 12:24.

Ketterson, E. D., V. Nolan, L. Wolf, and C. Ziegenfus. 1992. Testosterone and avian life histories: effects of experimentally elevated testosterone on behavior and correlates of fitness in the dark-eyed junco (*Junco hyemalis*). American Naturalist 140:980–999.

Keyser, A., and G. E. Hill. 1999. Condition-dependent variation in the blue-ultraviolet coloration of a structurally based plumage ornament. Proceedings of the Royal Society of London B266:771–777.

Keyser, A. J., and G. E. Hill. 2000. Structurally based plumage coloration is an honest signal of quality in male blue grosbeaks. Behavioral Ecology 11:202–209.

Kimball, R. T., and J. T. Ligon. 1999. Evolution of avian plumage dichromatism from a proximate perspective. American Naturalist 142:182–193.

Kirby, W. 1833. On the power, wisdom, and goodness of God as manifested in the creation of animals and in their history, habits and instincts. William Pickering, London.

Kirkpatrick, M. 1982. Sexual selection and the evolution of female choice. Evolution 36:1–12.

Kirkpatrick, M. 1985. Evolution of female choice and male parental investment in polygynous species: the demise of the "sexy son." American Naturalist 125:788–810.

Kirkpatrick, M., and M. J. Ryan. 1991. The evolution of mating preferences and the paradox of the lek. Nature 350:33–38.

Kitching, I. J., P. L. Forey, C. J. Humphries, and D. Williams. 1988. Cladistics: theory and practice of parsimony analysis. Oxford University Press, New York.

Kodric-Brown, A. 1985. Female preference and sexual selection for male coloration in the guppy (*Poecilia reticulata*). Behavioral Ecology and Sociobiology 17:199–205.

Kodric-Brown, A., and J. H. Brown. 1984. Truth in advertising: the kinds of traits favored by sexual selection. American Naturalist 124:309–323.

Kokko, H. 1998. Should advertising parental care be honest? Proceeding of the Royal Society of London B265:1871–1878.

Kornerup, A., and J. H. Wanscher. 1983. Methuen handbook of colour. Methuen, London.

Krebs, J. R. 1979. Bird colours. Nature 282:14–16.

Kricher, J. C. 1983. Correlations between House Finch increase and House Sparrow decline. American Birds 37:358–360.

Krinsky, N. 1989. Carotenoids and cancer in animal models. Journal of Nutrition 119:123–126.

Krinsky, N. I., D. G. Cornwall, and J. L. Oncley. 1958. The transport of vitamin A and carotenoids in human plasma. Archives of Biochemistry and Biophysics 78:233–246.

Kritzler, H. 1943. Carotenoids in the display and eclipse plumage of bishop birds. Physiological Zoology 16:241–245.

Lack, D. M. 1968. Ecological adaptations for breeding in birds. Methuen, London.

Lande, R. 1980. Sexual dimorphism, sexual selection, and adaptation in polygenic characters. Evolution 34:292–305.

Lande, R. 1981. Models of speciation by sexual selection on polygenic traits. Proceedings of the National Academy of Sciences USA 78:6:3721–3725.

Lande, R., and S. Arnold. 1985. Evolution of mating preference and sexual dimorphism. Journal of Theoretical Biology 117:651–664.

Lank, D. B., C. M. Smith, O. Hanotte, T. Burke, and F. Cooke. 1995. Genetic polymorphism for alternative mating behaviour in lekking male ruff *Philomachus pugnax*. Nature 378:59–62.

Lee, W. L. 1966. Pigmentation of the marine isopod *Idothea montereyensis*. Comparative Biochemistry and Physiology 18:17–36.

Lemel, J., and K. Wallin. 1993. Status signaling, motivational condition and dominance: an experimental study in the great tit, *Parus major*. Animal Behaviour 45:549–558.

Levine, N. D. 1982. Taxonomy and life cycles of coccidia. In P. L. Long (ed.), The biology of coccidia. Edward Arnold, London.

Ley, D. H., J. E. Berkhoff, and J. M. McLaren. 1996. *Mycoplasma gallisepticum* isolated from house finches (*Carpodacus mexicanus*) with conjunctivitis. Avian Diseases 40:480–483.

Ligon, J. D. 1999. The evolution of avian breeding systems. Oxford University Press, Oxford.

Lindstrom, A., G. H. Visser, and S. Daan. 1993. The energetic cost of feather synthesis is proportional to basal metabolic rate. Physiological Zoology 66:490–510.

Linville, S. U., and R. Breitwisch. 1997. Carotenoid availability and plumage coloration in a wild population of Northern Cardinals. Auk 114:796–800.

Linville, S. U., R. Breitwisch, and A. J. Schilling. 1998. Plumage brightness as an indicator of parental care in northern cardinals. Animal Behaviour 55:119–127.

Lipar, J., L., E. Ketterson, D., and V. Nolan, Jr. 1999. Intraclutch variation in testosterone content of Red-winged Blackbirds eggs. Auk 116:231–235.

Lombardo, M. P. 1998. On the evolution of sexually transmitted diseases in birds. Journal of Avian Biology 29:314–321.

Lönnberg, E. 1938. The occurrence and importance of carotenoid substances in birds. Proceedings of the 8th International Ornithological Congress, pp. 410–424. Oxford University Press, Oxford.

Lozano, G. A. 1994. Carotenoids, parasites, and sexual selection. Oikos 70:309–311.

Lucas, A. M., and P. R. Stettenheim. 1972. Avian anatomy, integument. U.S. Government Printing Office, Washington D.C.

Lundberg, A., and R. Alatalo. 1992. The Pied Flycatcher. T. & A. E. Poyser, London.

Luttrell, M. P., J. R. Fischer, D. E. Stallknecht, and S. H. Kleven. 1996. Field investigation of *Mycoplasma gallisepticum* infections in house finches (*Carpodacus mexicanus*) from Maryland and Georgia. Avian Diseases 40:335–341.

Lyon, B. E., and R. D. Montgomerie. 1986. Delayed plumage maturation in passerine birds: reliable signaling by subordinate males? Evolution 40:605–615.

Machlin, L. J., and A. Bendich. 1987. Free radical tissue damage: protective role of antioxidant nutrients. Federation of American Societies for Experimental Biology Journal 1:441–445.

Maddison, W. P., and D. R. Maddison. 1993. MacClade. Sinauer, Massachusetts.

Mangels, A. R., J. M. Holden, G. R. Beecher, M. R. Forman, and E. Lanza. 1993. Carotenoid content of fruits and vegetables: an evaluation of analytic data. Journal of the American Dietetic Association 93:284–296.

Marten, J. A., and N. K. Johnson. 1986. Genetic relationships of North American Cardueline finches. Condor 88:409–420.

Martin, T. E., and A. V. Badyaev. 1996. Sexual dichromatism in birds: importance of nest predation and nest location for females versus males. Evolution 50:2454–2460.

Marusich, W. L., E. Schildknecht, E. F. Ogrinz, P. R. Brown, and M. Mitrovic. 1972. Effect of coccidiosis on pigmentation in broilers. British Poultry Science 13:577–585.

Maynard-Smith, J. 1976. Sexual selection and the handicap principle. Journal of Theoretical Biology 57:239–242.

Maynard-Smith, J. 1978a. The evolution of sex. Cambridge University Press, Cambridge.

Maynard-Smith, J. 1978b. The Handicap Principle—A comment. Journal of Theoretical Biology 70:251–252.

Maynard-Smith, J. 1985. Sexual selection, handicaps, and true fitness. Journal of Theoretical Biology 115:1–8.

Maynard-Smith, J. 1991. Theories of sexual selection. Trends in Ecology and Evolution 6:146–151.

Maynard-Smith, J., and D. G. C. Harper. 1988. The evolution of aggression: can selection generate variability? Philosophical Transactions of the Royal Society of London B319:557–570.

Mayr, E. 1942. Systematics and the origin of species. Columbia University Press, New York.

McGraw, K. J., and G. E. Hill. 2000a. Carotenoid-based ornamentation and status signalling in the house finch. Behavioral Ecology 11:520–527.

McGraw, K. J., and G. E. Hill. 2000b. Differential effects of endoparasitism on the expression of carotenoid- and melanin-based ornamental coloration. Proceedings of the Royal Society of London B267:1525–1531.

McGraw, K. J., and G. E. Hill. 2000c. Plumage brightness and breeding-season dominance in the House Finch: a negatively correlated handicap? Condor 102:456–461.

McGraw, K. J., and G. E. Hill. 2001. Carotenoid access and intraspecific variation in plumage pigmentation in male American goldfinches (*Carduelis tristis*) and northern cardinals (*Cardinalis cardinalis*). Functional Ecology 15:732–739.

McGraw, K. J., and G. E. Hill. 2002. Testing reverses sexual dominance from an ontogenetic perspective: juvenal female House Finches *Carpodacus mexicanus* are dominant to juvenal males. Ibis 144:139–142.

McGraw, K. J., A. M. Stoehr, P. M. Nolan, and G. E. Hill. 2001. Plumage redness predicts breeding onset and reproductive success in the House Finch: a validation of Darwin's theory. Journal of Avian Biology 32:90–94.

McGraw, K. J., G. E. Hill, R. Stradi, and R. S. Parker. 2002. The effect of dietary carotenoid access on sexual dichromatism and plumage pigment composition in the American goldfinch. Comparative Biochemistry and Physiology, B131:261–269.

McPherson, J. M. 1988. Preferences of cedar waxwings in the laboratory for fruit species, colour and size: a comparison with field observations. Animal Behaviour 36:961–969.

Mendonça, M. T., S. D. Chernetsky, K. E. Nester, and G. L. Gardner. 1996. Effects of gonadal sex steroids on sexual behavior in the big brown bat, *Eptesicus fuscus*, upon arousal from hibernation. Hormones and Behavior 30:153–161.

Merila, J., B. C. Sheldon, and K. Lindstrom. 1999. Plumage brightness in relation to haematozoan infections in the greenfinch *Carduelis chloris*: bright males are a good bet. Ecoscience 6:12–18.

Michener, H., and J. R. Michener. 1931. Variation in color of male House Finches. Condor 33:12–19.

Michener, H., and J. R. Michener. 1938. Bars in flight feathers. Condor 40:149–160.

Michener, H., and J. R. Michener. 1940. The molt of house finches in the Pasadena Region, California. Condor 42:140–153.

Milinski, M., and T. C. M. Bakker. 1990. Female sticklebacks use male coloration in mate choice and hence avoid parasitized males. Nature 344:330–333.

Miskimen, M. 1980. Red-winged Blackbirds: I. Age-related epaulet color changes in captive females. Ohio Journal of Science 80:232–235.

Møller, A. P. 1988. Female choice selects for male sexual tail ornaments in the monogamous swallow. Nature 332:640–642.

Møller, A. P. 1990a. Effects of a haematophagous mite on the barn swallow (*Hirundo rustica*): a test of the Hamilton and Zuk hypothesis. Evolution 44:771–784.

Møller, A. P. 1990b. Fluctuating asymmetry in male sexual ornaments may reliably reveal male quality. Animal Behaviour 40:1185–1187.

Møller, A. P. 1990c. Male tail length and female mate choice in the monogamous swallow *Hirindo rustica*. Animal Behaviour 39:458–465.

Møller, A. P. 1991. Sexual selection in the monogamous barn swallow (*Hirundo rustica*). I. Determinants of tail ornament size. Evolution 45:1823–1836.

Møller, A. P. 1994. Sexual selection and the Barn Swallow. Oxford University Press, Oxford.

Møller, A. P., and P. Ninni. 1998. Sperm competition and sexual selection: A meta-analysis of paternity studies of birds. Behavioral Ecology and Sociobiology 43:345–358.

Møller, A. P., and A. Pomiankowski. 1993. Why have birds got multiple sex ornaments? Behavioral Ecology and Sociobiology 32:167–176.

Møller, A. P., C. Biard, J. D. Blount, D. C. Houston, P. Ninni, N. Saino, and P. F. Surai. 2000. Carotenoid-dependent signals: Indicators of foraging efficiency, immunocompetence or detoxification ability? Avian and Poultry Biology Reviews 11:137–159.

Montgomerie, R. D., and R. Thornhill. 1989. Fertility advertisement in birds: a means of inciting male-male competition. Ethology 81:209–220.

Moore, R. T. 1939. A review of the house finches of the subgenus *Burrica*. Condor 41:177–205.

Moreno, J., J. P. Veiga, P. J. Cordero, and E. Minguez. 1999. Effects of paternal care on reproductive success in the polygynous spotless starling *Sturnus unicolor*. Behavioral Ecology and Sociobiology 47:47–53.

Morgan, C. L. 1900. Animal behaviour. Edward Arnold, London.

Muller, P. L. S. 1776. Natursystematic Supplement 165.

Mulvihill, R. S., K. C. Parkes, R. C. Leberman, and D. S. Wood. 1992. Evidence supporting a dietary basis for orange-tipped rectrices in the cedar waxwing. Journal of Field Ornithology 63:212–216.

Muma, K. E., and P. J. Weatherhead. 1991. Plumage variation and dominance in captive female Red-winged Blackbirds. Canadian Journal of Zoology 69:49–54.

Murphy, M. E., and T. Taruscio. 1995. Sparrows increase their rate of tissue and whole-body protein synthesis during the annual molt. Comparative Biochemistry and Physiology 111A:385–396.

Murton, R. K., and N. J. Westwood. 1977. Avian breeding cycles. Clarendon Press, Oxford.

National Geographic Society. 1999. Field guide to the birds of North America. National Geographic Society, Washington, D. C.

Navara, K. J., and G. E. Hill. in prep. Dietary carotenoid pigments and immune function in a songbird with extensive carotenoid-based plumage coloration.

Newton, I. 1972. Finches. Collins, London.

Nice, M. M. 1939. The watcher at the nest. Macmillan Co., New York.

Nisbet, I. C. T. 1973. Courtship-feeding, egg-size and breeding success in Common Terns. Nature 241:141–142.

Noble, G. K. 1934. Experimenting with the courtship of lizards. Natural History 34:3–15.

Noble, G. K., and H. T. Bradley. 1933. The mating behavior of lizards; its bearing on the theory of sexual selection. Annals of the New York Academy of Sciences 35:25–106.

Noble, G. K., and B. Curtis. 1939. The social behavior of the jewel fish, *Hemichromis bimaculatus* Gill. Bulletin of the American Museum of Natural History 76:1–46.

Nolan, P. M., G. E. Hill, and A. M. Stoehr. 1998. Sex, size, and plumage redness predict house finch survival in an epidemic. Proceedings of the Royal Society of London B265:961–965.

Nolan, P. M., A. M. Stoehr, G. E. Hill, and K. J. McGraw. 2001. The number of provisioning visits to the nest predicts the mass of food delivery. Condor 103:851–855.

Nolan, V. J., E. D. Ketterson, C. Ziegenfus, D. P. Cullen, and C. R. Chandler. 1992. Testosterone and avian life histories: effects of experimentally elevated testosterone on prebasic molt and survival in male dark-eyed juncos. Condor 94:364–370.

Nowicki, S., D. Hasselquist, S. Bensch, and S. Peters. 2000. Nestling growth and song repertoire size in great reed warblers: evidence for song learning as an indicator mechanism in mate choice. Proceedings of the Royal Society of London B267:2419–2424.

Nunez-De La Mora, A., H. Drummond, and J. C. Wingfield. 1996. Hormonal correlates of dominance and starvation-induced aggression in chicks of the blue-footed booby. Ethology 102:748–761.

Nur, N. 1984. The consequences of brood size for breeding Blue Tits I. Adult survival, weight change and the cost of reproduction. Journal of Animal Ecology 53:479–496.

Nur, N., and O. Hasson. 1984. Phenotypic plasticity and the handicap principle. Journal of Theoretical Biology 110:275–297.

Oberholser, H. C. 1974. The bird life of Texas. University of Texas Press, Austin.

OConner, B. M. 1992. Evolutionary ecology of astigmatid mites. Annual Review of Entomology 27:385–409.

O'Donald, P. 1962. The theory of sexual selection. Heredity 17:541–552.

O'Donald, P. 1967. A general model of natural and sexual selection. Heredity 22:499–518.

O'Donald, P. 1980. Genetic models of sexual selection. Cambridge University Press, Cambridge.

O'Donald, P. 1983. The Artic Skua: A study of the ecology and evolutoin of a seabird. Cambridge University Press, Cambridge.

Olson, V. A., and I. P. F. Owens. 1998. Costly sexual signals: are carotenoids rare, risky or required? Trends in Ecology and Evolution 13:510–514.

Ots, I., and P. Horak. 1996. Great tits Parus major trade health for reproduction. Proceedings of the Royal Society of London B263:1443–1447.

Otter, K., and L. Ratcliffe. 1996. Female initiated divorce in a monogamous songbird: abandoning mates for males of higher quality. Proceedings of the Royal Society of London B263:351–354.

Owens, I. P. F., and I. R. Hartley. 1991. "Trojan Sparrows": evolutionary consequences of dishonest invasion for the badges-of-status model. American Naturalist 138:1187–1205.

Owens, I. P. F., and R. V. Short. 1995. Hormonal basis of sexual dimorphism in birds: implications for new theories of sexual selection. Trends in Ecology and Evolution 10:44–47.

Paley, W. 1802. Natural Theology: or, evidences of the existence and attributes of the Deity, collected from the appearances of nature. Faulder, London.

Palokangas, P., E. Korpimaki, H. Hakkarainen, E. Huhta, P. Tolonen, and R. V. Alatalo. 1994. Female kestrels gain reproductive success by choosing brightly ornamented males. Animal Behaviour 47:443–448.

Parker, G. A. 1983. Mate quality and mating decisions. In P. Bateson (ed.), Mate choice, pp. 141–166. Cambridge University Press, Cambridge.

Parker, R. S. 1996. Absorption, metabolism, and transport of carotenoids. Federation of American Societies for Experimental Biology Journal 10:542–551.

Parker, R. S. 1997. Bioavailability of carotenoids. European Journal of Clinical Nutrition 51:S86-S90.

Part, T. 1995. Does breeding experience explain increased reproductive success with age? An experiment. Proceedings of the Royal Society of London B260:113–117.

Partali, V., S. Liaaen-Jensen, T. Slagsvold, and J. T. Lifjeld. 1987. Carotentoids in food chain studies—II. The food chain of *Parus* spp. monitored by carotenoid analysis. Comparative Biochemical Physiology 87B:885–888.

Payne, R. B. 1972. Mechanisms and control of molt. In D. S. Farner and J. R. King (eds.), Avian biology, Vol. 2, pp. 103–155. Academic Press, New York.

Perrins, C. M., and R. H. McCleery. 1985. The effect of age and pair bond on the breeding success of Great Tits *Parus major*. Ibis 127:306–315.

Peterson, R. T. 1980. A field guide to the birds : a completely new guide to all the birds of eastern and central North America. Houghton Mifflin Co., Boston.

Peto, R., R. Doll, J. D. Buckley, and M. D. Sporn. 1981. Can dietary beta-carotene materially reduce human cancer rates? Nature 290:201–208.

Piersma, T., and J. Jukema. 1993. Red breasts as honest signals of migratory quality in a long-distance migrant, the bar-tailed godwit. Condor 95:163–177.

Pomiankowski, A. 1987. The costs of choice in sexual selection. Journal of Theoretical Biology 128:195–218.

Potti, J., and S. Merino. 1996. Decreased levels of blood trypanosome infection correlate with female expression of a male secondary sexual trait: implications for sexual selection. Proceedings of the Royal Society of London B263:1199–1204.

Poulton, E. B. 1890. The colours of animals: their meaning and use, especially considered in the case of insects. Kegan Paul, Trench, Trübner, London.

Power, D. M. 1979. Evolution in peripheral isolated populations: *Carpodacus* finches on the California Islands. Evolution 33:834–847.

Primmer, C. R., A. P. Moller, and H. Ellegren. 1995. Resolving genetic relationships with microsatellite markers: a parentage testing system for the swallow *Hirundo rustica*. Molecular Ecology 4:493–498.

Pyle, P., S. N. G. Howell, R. P. Yunick, and D. F. DeSante. 1987. Identification Guide to North American Passerines. Slate Creek Press, Bolinas, California.

Qvarnström, A. 2001. Context-dependent genetic benefits from mate choice. Trends in Ecology and Evolution 16:5–7.

Qvarnström, A., and E. Forsgren. 1998. Should females prefer dominant males? Trends in Ecology and Evolution 13:498–501.

Ramenofsky, M., J. M. Gray, and R. B. Johnson. 1992. Behavioral and physiological adjustments of birds living in winter flocks. Ornis Scandinavica 23:371–380.

Ratti, O., R. Dufva, and R. V. Alatalo. 1993. Blood parasites and male fitness in the pied flycatcher. Oecologia 96:410–414.

Rawles, M. E. 1960. The integumentary system. In A. J. Marshall (ed.), Biology and comparative physiology of birds, Vol. 1. Academic Press, New York.

Read, H. F. 1988. Sexual selection and the role of parasites. Trends in Ecology and Evolution 3:97–102.

Real, L. 1990. Search theory and mate choice. I. Models of single-sex discrimination. American Naturalist 136:376–405.

Riddle, O. 1908. The genesis of fault bars in feathers and the cause of alternation of light and dark fundamental bars. Biological Bulletin 14:328–370.

Ridgway, R. 1901. Birds of North and Middle America, Part 1. United States Government Printing Office, Washington, D.C.

Robertson, G. J., F. Cooke, R. I. Goudie, and W. S. Boyd. 1998. Moult speed predicts pairing success in male harlequin ducks. Animal Behaviour 55:1677–1684.

Rohwer, S. 1975. The social significance of avian winter plumage variability. Evolution 29:593–610.

Rohwer, S. 1977. Status signaling in Harris' sparrows: some experiments in deception. Behaviour 61:107–128.

Rohwer, S. 1982. The evolution of reliable and unreliable badges of fighting ability. American Zoologist 22:531–546.

Rohwer, S. 1985. Dyed birds achieve higher social status than controls in Harris' sparrows. Animal Behaviour 33:1325–1331.

Rohwer, S., and G. S. Butcher. 1988. Winter versus summer explanations of delayed plumage maturation in temperate passerine birds. American Naturalist 131:556–572.

Rohwer, S., and G. S. Butcher, 1989. The evolution of conspicuous and distinctive coloration for communication in birds. Current Ornithology 6:51–108.

Rohwer, S., and P. W. Ewald. 1981. The cost of dominance and advantage of subordination in a badge signaling system. Evolution 35:441–454.

Rohwer, S., and F. C. Rohwer. 1978. Status signaling in Harris' sparrows: experimental deceptions achieved. Animal Behaviour 26:1012–1022.

Rohwer, S., S. D. Fretwell, and D. M. Niles. 1980. Delayed maturation in passerine plumages and the deceptive acquisition of resources. American Naturalist 131:556–572.

Rohwer, S., P. W. Ewald, and F. C. Rohwer. 1981. Variation in size, appearance and dominance within and among the sex and age classes of Harris' Sparrows. Journal of Field Ornithology 52:291–303.

Romanes, G. J. 1892. Darwin, and after Darwin: an exposition of the Darwinian theory and a discussion of post-Darwinian questions. The Open Court Publishing Company, Chicago.

Romanes, G. J. 1895. Darwin, and after Darwin: an exposition of the Darwinian theory and a discussion of post-Darwinian questions. The Open Court Publishing Company, Chicago.

Ruff, M. D., W. M. Reid, and J. K. Johnson. 1974. Lowered blood carotenoid levels in chickens infected with coccidia. Poultry Science 53:1801–1809.

Ryan, M. J., and A. Keddy-Hector. 1992. Directional patterns of female mate choice and the role of sensory biases. American Naturalist 139:S4-S35.

Ryan, M. J., J. H. Fox, W. Wilczynski, and A. S. Rand. 1990. Sexual selection for sensory exploitation in the frog *Physalaemus pustulosus*. Nature 343:66–67.

Saetre, G. P., T. Fossnes, and T. Slagsvold. 1995. Food provisioning in the pied flycatcher: do females gain direct benefits from choosing bright-coloured males? Journal of Animal Ecology 64:21–30.

Saino, N., and A. P. Moller. 1995. Testosterone correlates of mate guarding, singing and aggressive behaviour in male barn swallows, *Hirundo rustica*. Animal Behaviour 49:465–472.

Samson, F. B. 1977. Social dominance in winter flocks of Cassin's Finch. Wilson Bulletin 89:57–66.

Sauer, J. R., J. E. Hines, and J. Fallon. 2001. The North American breeding bird survey, results and analysis 1966–2000. Version 2001.2. USGS Patuxent Wildlife Research Center, Laurel, Maryland.

Schereschewsky, H. 1929. Einige Beiträge zum Problem der Verfärbung des Gefieders beim Gempel. Wihelm Roux' Archiv für Entwicklungsmechanik der Organismen 115:110–153.

Schiedt, K., F. J. Leuenberger, M. Vecchi, and E. Glinz. 1985. Absorption, retention, and metabolic transformations of carotenoids in rainbow trout, salmon and chicken. Pure and Applied Chemistry 57:685–692.

Schlinger, B. A., A. J. Fivizzani, and G. V. Callard. 1989. Aromatase, 5α- and 5β-reductase in brain, pituitary and skin of the sex-role reversed Wilson's phalarope. Journal of Endocrinology 122:573–581.

Schwabl, H. 1993. Yolk is a source of maternal testosterone for developing birds. Proceedings of the National Academy of Sciences USA 90:11446–11450.

Schwabl, H. 1996a. Environment modifies the testosterone levels of a female bird and its eggs. Journal of Experimental Zoology 276:157–163.

Schwabl, H. 1996b. Maternal testosterone in the avian egg enhances postnatal growth. Comparative Biochemistry and Physiology 114:271–276.

Searcy, W. A. 1979. Male characteristics and pairing success in red-winged blackbirds. Auk 96:353–363.

Searcy, W. A., and K. Yasukawa. 1995. Polygyny and sexual selection in Red-winged Blackbirds. Princeton University Press, Princeton, New Jersey.

Selander, R. K. 1972. Sexual selection and dimorphism in birds. In B. Campbell (ed.), Sexual selection and the descent of man, 1871–1971. Aldine, Chicago.

Senar, J. C. 1999. Plumage coloration as a signal of social status. In N. Adams and R. Slotow (eds.), Proceedings of the 22nd International Ornithological Congress, pp. 1644–1654. University of Natal, Durban.

Senar, J. C., and M. Camerino. 1998. Status signalling and the ability to recognize dominants: an experiment with siskins (*Carduelis spinus*). Proceedings of the Royal Society of London B265:1515–1520.

Senar, J. C., J. L. Copete, and N. B. Metcalfe. 1990. Dominance relationships between resident and transient wintering Siskins. Ornis Scandinavica 21:129–132.

Senar, J. C., J. L. Camerino, and N. B. Metcalfe. 1993. Variation in black bib of the Eurasian Siskin(*Carduelis spinus*) and its role as a reliable badge of dominance. Auk 110:924–927

Seroni, G. 1994. Concurrent calicivirus and *Isospora lacazei* infections in goldfinches (*Carduelis carduelis*). Veterinary Record 134:196.

Seutin, G. 1994. Plumage redness in redpoll finches does not reflect hemoparasitic infection. Oikos 70:280–286.

Sharp, R. B. 1888. Catalogue of the Passeriformes or perching birds. Catalogue of the Birds of the British Musuem 12:1–871.

Shedd, D. H. 1990. Agressive interactions in wintering House Finches and Purple Finches. Wilson Bulletin 102:174–178.

Sheldon, B. C. 2000. Differential allocation: tests, mechanisms, and implications. Trends in Ecology and Evolution 15:397–402.

Siegel, H. S. 1980. Physiological stress in birds. Bioscience 30:529–533.

Silverin, B. 1980. Effects of long-acting testosterone treatment on free-living pied flycatchers, *Ficedula hypoleuca*, during the breeding period. Animal Behaviour 28:906–912.

Slagsvold, T., and J. T. Lifjeld. 1985. Variation in plumage coloration of the great tit *Parus major* in relation to habitat, season, and food. Journal of Zoology 206A:321–328.

Slagsvold, T., and J. T. Lifjeld. 1988. Plumage colour and sexual selection in the pied flycatcher *Ficedula hypoleuca*. Animal Behaviour 36:395–407.

Slotow, R., J. Alcock, and S. I. Rothstein. 1993. Social status signalling in white-crowned sparrows: an experimental test of the social control hypothesis. Animal Behaviour 46:977–989.

Smith, N. C., M. Wallach, C. M. D. Miller, R. Morgenstern, R. Braun, and J. Eckert. 1994. Maternal transmission of immunity to *Eimeria maxima*: enzyme-linked immunosorbent assay analysis of protective antibodies induced by infection. Infection and Immunity 62:1348–1357.

Smith, P. W. 1987. The Eurasian Collared-Dove arrives in the Americas. American Birds 5:1371–1379.

Stangel, P. W. 1985. Incomplete first prebasic molt of Massachusetts House Finches. Journal of Field Ornithology 56:1–8.

Stockton-Shields, C. 1997. Sexual selection and the dietary color preferences of House Finches. Master's Thesis, Auburn University.

Stoehr, A. M., and G. E. Hill. 2000. Testosterone and the allocation of reproductive effort in male house finches (*Carpodacus mexicanus*). Behavioral Ecology and Sociobiology 48:407–411.

Stoehr, A. M., and G. E. Hill. 2001. The effects of elevated testosterone on plumage hue in male house finches. Journal of Avian Biology 32:153–158.

Stoehr, A. M., K. J. McGraw, P. M. Nolan, and G. E. Hill. 2001. Parental care in relation to brood size in the House Finch. Journal of Field Ornithology 72:412–418.

Stradi, R. 1998. The colour of flight: carotenoids in bird plumage. University of Milan Press, Milan, Italy.

Stradi, R., G. Celentano, E. Rossi, G. Rovati, and M. Pastore. 1995. Carotenoids in bird plumage I. The carotenoid pattern in a series of Palearctic Carduelinae. Comparative Biochemisty and Physiology 110B:131–143.

Stradi, R., E. Rossi, G. Celentano, and B. Bellardi. 1996. Carotenoids in bird plumage: the pattern in three *Loxia* species and in *Pinicola enucleator*. Comparative Biochemistry and Physiology 113B:427–432.

Stradi, R., G. Celetano, M. Boles, and F. Mercato. 1997. Carotenoids in bird plumage: the pattern in a series of red-pigmented Carduelinae. Comparative Biochemistry and Physiology 117B:85–91.

Studd, M. V., and R. J. Robertson. 1985a. Life span, competition, and delayed maturation in male passerines: the breeding threshold hypothesis. American Naturalist 126:101–115.

Studd, M. V., and R. J. Robertson. 1985b. Sexual selection and variation in reproductive strategy in male yellow warblers (*Dendroica petechia*). Behavioral Ecology and Sociobiology 17:101–109.

Sundberg, J. 1995a. Female yellowhammers (*Emberiza citrinella*) prefer yellower males: a laboratory experiment. Behavioral Ecology and Sociobiology 37:275–282.

Sundberg, J. 1995b. Parasites, plumage coloration and reproductive success in the yellowhammer, *Emberiza citrinella*. Oikos 74:331–339.

Sundberg, J., and A. Dixon. 1996. Old, colourful male yellowhammers, *Emberiza citrinella*, benefit from extra-pair copulations. Animal Behaviour 52:113–122.

Sundberg, J., and C. Larsson. 1994. Male coloration as an indicator of parental quality in the yellowhammer, *Emberiza citrinella*. Animal Behaviour 48:885–892.

Svensson, E., and J. Nilsson. 1996. Mate quality affects offspring sex ratio in blue tits. Proceedings of the Royal Society of London B263:357–361.

Tate, R. C. 1925. The House Finch in the Oklahoma panhandle. Condor 27:176.

Test, F. H. 1969. Relation of wing and tail color of the woodpeckers *Colaptes auratus* and *C. cafer* to their food. Condor 71:206–211.

Thommen, H. 1971. Metabolism. Birkhauser Verlag, Basel, Switzerland.

Thompson, C. W. 1991. The sequence of molts and plumages in Painted Buntings and implications for theories of delayed plumage maturation. Condor 93:209–235.

Thompson, C. W., N. Hillgarth, M. Leu, and H. E. McClure. 1997. High parasite load in house finches (*Carpodacus mexicanus*) is correlated with reduced expression of a sexually selected trait. American Naturalist 149:270–294.

Thompson, W. L. 1960a. Agonistic behavior in the House Finch. Part I: Annual cycle and display patterns. Condor 62:245–271.

Thompson, W. L. 1960b. Agonistic behavior in the House Finch. Part II: Factors in aggressiveness and sociality. Condor 62:378–402.

Thorneycroft, H. B. 1966. Chromosomal polymorphsim in the White-throated Sparrow, *Zonotrichia albicollis*. Science 154:1571–1572.

Thornhill, R. 1992. Fluctuating asymmetry and the mating system of the Japanese scorpionfly *Panorpa japonica*. Animal Behaviour 44:867–879.

Tovee, M. J. 1995. Ultra-violet photoreceptors in the animal kingdom: their distribution and function. Trends in Ecology and Evolution 10:455–460.

Trams, E. G. 1969. Carotenoid transport in the plasma of the scarlet ibis (*Eudocimus ruber*). Comparative Biochemistry and Physiology 28:117–118.

Trivers, R. L. 1972. Parental investment and sexual selection. In B. Campbell (ed.), Sexual selection and the descent of man, 1871–1971, pp. 136–179. Aldine, Chicago.

Tyczkowski, J. K., and P. B. Hamilton. 1986. Evidence for differential absorption of zeacarotene, cryptoxanthin, and lutein in young broiler chickens. Poultry Science 65:1137–1140.

Tyczkowski, J. K., P. B. P.B. Hamilton, and M. D. Ruff. 1991. Altered metabolism of carotenoids during pale-bird syndrome in chicks infected with *Eimeria acervulina*. Poultry Science 70:2074–2081.

Tyler, J. D. 1992. History of the House Finch in Oklahoma, 1919–1991. Proceedings of the Oklahoma Academy of Science 72:33–35.

Van Rossem, A. J. 1936. Birds of the Charleston Mountains, Nevada. Pacific Coast Avifauna 24:52–53.

Van Valen, L. M. 1973. A new evolutionary law. Evolutionary Theory 1:1–30.

Vazquez-Phillips, M. A. 1992. Population differentiation of the House Finch (*Carpodacus mexicanus*) in North America and the Hawaiian Islands. Master of Science Thesis, University of Toronto.

Visser, M. E., and N. Verboven. 1999. Long-term fitness effects of fledging date in great tits. Oikos 85:445–450.

Völker, O. 1934. Die Abhangigkeit der lipochrombildung bei vögeln von pflanzlichen carotinoiden. Journal für Ornithologie 82:439.

Völker, O. 1938. The dependence of lipochrome-formation in birds on plant carotenoids. Proceedings of the 8th International Orthinologists Congress, pp. 425–426.

von Schantz, T., T. Bensch, M. Grahn, D. Hasselquist, and H. Wittzell. 1999. Good genes, oxidative stress and condition-dependent sexual signals. Proceedings of the Royal Society of London B266:1–12.

Wallace, A. R. 1864. The origin of human races and the antiquity of man deduced from the theory of "natural selection." Anthropological Review and Journal of the Anthropological Society of London 2:158–187.

Wallace, A. R. 1870. Contributions to the theory of natural selection: a series of essays. Macmillan, London.

Wallace, A. R. 1878. Tropical nature and other essays. Macmillan, London.

Wallace, A. R. 1889. Darwinism. Macmillan, London.

Wallace, A. R. 1905. My life: a record of events and opinioins. Chapman & Hall, London.

Wallace, G. J. 1955. An introduction to ornithology. Macmillan Co., New York.

Walls, G. L. 1942. The vertebrate eye and its adaptive radiation. Crambrook Institute of Science, Bloomfield Hills, Michigan.

Wang, O. T. 1992. Experimental assessment of male attractiveness: female choice or Hobson's choice? American Naturalist 139:433–441.

Watson, P. J., and R. Thornhill. 1994. Fluctuating asymmetry and sexual selection. Trends in Ecology and Evolution 9:21–25.

Watt, D. J. 1986. Relationship of plumage variability, size and sex to social dominance in Harris' sparrows. Animal Behaviour 34:16–27.

Weatherhead, P. J. 1990. Secondary sexual traits, parasites, and polygyny in Red-winged Blackbirds. Behavioral Ecology 1:125–130.

Weatherhead, P. J., and G. F. Bennett. 1992. Ecology of parasitism of Brown-headed Cowbirds by haematozoa. Canadian Journal of Zoology 70:1–7.

Weatherhead, P. J., K. J. Metz, G. F. Bennett, and R. E. Irwin. 1993. Parasite faunas, testosterone and secondary sexual traits in male red-winged blackbirds. Behavioral Ecology and Sociobiology 33:13–23.

Weber, H. 1961. Über die Ursache des Verlustes der roten Federfarbe bei gekäfigten Birkenzeisigen. Journal für Ornithologie 102:158–163.

Wedekind, C., and I. Folstad. 1994. Adaptive or nonadaptive immunosuppression by sex hormones? American Naturalist 143:936–938.

Wedekind, C., P. Meyer, M. Frischknect, U. A. Niggli, and H. Pfander. 1998. Different carotenoids and potential information content of red coloration of male three-spined sticklebacks. Journal of Chemical Ecology 24:787–801.

Weis, A. E., and B. Bisbey. 1947. The relation of the carotenoid pigments of the diet to growth of young chicks and tostorage in their tissues. University of Missouri Research Bulletin 405.

Welty, J. C. 1982. The life of birds, 3rd edn. Saunders College Publishing, New York.

Westneat, D. F., and T. R. Birkhead. 1998. Alternative hypotheses linking the immune system and mate choice for good genes. Proceedings of the Royal Society of London B265:1065–1073.

Westneat, D. F., P. W. Sherman, and M. L. Morton. 1990. The ecology and evolution of extra-pair copulations in birds. Current Ornithology 7:331–369.

Whitfield, D. P. 1987. Plumage variability, status signaling and individual recognition in avian flocks. Trends in Ecology and Evolution 2:13–18.

Wiens, J. J. 1998. The accuracy of methods for coding and sampling higher-level taxa for phylogenetic analysis: a simulation study. Systematic Biology 47:397–413.

Williams, G. C. 1966. Adaptation and natural selection: a critique of some current evolutionary thought. Princeton University Press, Princeton, New Jersey.

Williams, G. C. 1978. Mysteries of sex and recombination. Quarterly Review of Biology 53:287–289.

Wingfield, J. C. 1979. Avian endocrinology. In S. Ishi, T. Hirano and M. Wada (eds.), Hormones, adaptation, and evolution. Springer, Berlin.

Wingfield, J. C. 1984. Androgens and mating systems: testosterone-induced polygyny in normally monogamous birds. Auk 101:665–671.

Wingfield, J. C. 1985. Short term changes in plasma levels of hormones during establishment and defense of a breeding territory in male Song Sparrows, *Melospiza melodia*. Hormones and Behavior 19:174–187.

Wingfield, J. C., and M. C. Moore. 1987. Hormonal, social, and environmental factors in the reproductive biology of free-living male birds. In D. Crews (ed.), Psychobiology of reproductive behavior: an evolutionary perspective, pp. 149–175. Prentice-Hall, Princeton, New Jersey.

Wingfield, J. C., J. P. Smith, and D. S. Farner. 1982. Endocrine responses of White-crowned Sparrows to environmental stress. Condor 84:399–409.

Wingfield, J. C., G. F. Ball, J. Dufty, A M, R. E. Hegner, and M. Ramenofsky. 1987. Testosterone and aggression in birds. American Scientist 75:602–608.

Wingfield, J. C., R. F. Hegner, A. M. J. Dufty, and G. F. Ball. 1990. The "challenge hypothesis": theoretical implications for patterns of testosterone secretion, mating systems, and breeding strategies. American Naturalist 136:829–846.

Winterbottom, J. M. 1929. Studies in sexual phenomena. VII. The transference of male secondary sexual characters to the female. Journal of Genetics 21:367–387.

Witmer, M. 1996. Consequences of an alien shrub on the plumage coloration and ecology of Cedar Waxwings. Auk 113:735–743.

Witschi, E. 1961. Sex and secondary sexual characters. In A. J. Marshall (ed.), Biology and comparative physiology of birds, Vol. 2, pp. 115–168. Academic Press, New York.

Wolfenbarger, L. L. 1999a. Female mate choice in Northern Cardinals: is there a preference for redder males? Wilson Bulletin 111:76–83.

Wolfenbarger, L. L. 1999b. Is red coloration of male northern cardinals beneficial during the nonbreeding season: a test of status signaling. Condor 101:655–663.

Wolfenbarger, L. L. 1999c. Red coloration of male northern cardinals correlates with mate quality and territory quality. Behavioral Ecology 10:80–90.

Wootton, J. T. 1996. Purple Finch (Carpodacus purpureus). In A. Poole and F. Gill (eds.), The birds of North America, Vol. 208. The Academy of Natural Sciences, Philadelphia.

Yunick, R. P. 1987. Age determination of male House Finches. North American Bird Bander 12:8–11.

Zahavi, A. 1975. Mate selection—a selection for a handicap. Journal of Theoretical Biology 67:603–605.

Zahavi, A. 1977. The cost of honesty (further remarks on the handicap principle). Journal of Theoretical Biology 67:603–605.

Zahn, S. N., and S. I. Rothstein. 1999. Recent increase in male House Finch plumage variation and its possible relationship to avian pox disease. Auk 116:35–44.

Ziegler, R. G. 1989. A review of epidemiologic evidence that carotenoids reduce the risk of cancer. Journal of Nutrition 199:116–122.

# INDEX

## DATE DUE

| | | | |
|---|---|---|---|
| | | | |
| | | | |
| | | | |
| | | | |
| | | | |
| | | | |
| | | | |
| | | | |
| | | | |
| | | | |
| | | | |
| | | | |
| | | | |
| | | | |
| | | | |
| | | | |
| | | | |
| | | | |
| | | | |
| GAYLORD | | | PRINTED IN U.S.A. |